Probability and its Applications

A Series of the Applied Probability Trust

Editors: J. Gani, C.C. Heyde, T.G. Kurtz

Springer Science+Business Media, LLC

Probability and its Applications

Richard F. Bass

Diffusions and Elliptic Operators

 Springer

Richard F. Bass
Department of Mathematics
University of Washington
Seattle, WA 98195
USA

Series Editors

J. Gani	C.C. Heyde	T.G. Kurtz
Stochastic Analysis Group	Stochastic Analysis Group	Department of Mathematics
Australian National University	Australian National University	University of Wisconsin 480 Lincoln Drive
Canberra ACT 0200	Canberra ACT 0200	Madison, WI 53706
Australia	Australia	USA

Library of Congress Cataloging-in-Publication Data
Bass, Richard F.
    Diffusions and elliptic operators / Richard F. Bass.
      p.   cm. — (Probability and its applications)
    Includes bibliographical references and index.

    DOI 10.1007/978-0-387-22604-0
    1. Differential equations, Stochastic — Numerical solutions.
    2. Diffusion processes.   3. Elliptic operators. I. Title.
    II. Series.
    QA377.B293   1997
    519.2 — dc21   97-26428

Printed on acid-free paper.

Production managed by Anthony K. Guardiola; manufacturing supervised by Thomas King.
Photocomposed pages prepared from the author's LaTeX files.

9 8 7 6 5 4 3 2 1
www.springer.com/mycopy

To the memory of my father, Jay Bass

(1911–1997)

PREFACE

The interplay of probability theory and partial differential equations forms a fascinating part of mathematics. Among the subjects it has inspired are the martingale problems of Stroock and Varadhan, the Harnack inequality of Krylov and Safonov, the theory of symmetric diffusion processes, and the Malliavin calculus. When I first made an outline for my previous book *Probabilistic Techniques in Analysis*, I planned to devote a chapter to these topics. I soon realized that a single chapter would not do the subject justice, and the current book is the result.

The first chapter provides the probabilistic machine needed to drive the subject, namely, stochastic differential equations. We consider existence, uniqueness, and smoothness of solutions and stochastic differential equations with reflection.

The second chapter is the heart of the subject. We show how many partial differential equations can be solved by simple probabilistic expressions. The Dirichlet problem, the Cauchy problem, the Neumann problem, the oblique derivative problem, Poisson's equation, and Schrödinger's equation all have solutions that are given by appropriate probabilistic expressions. Green functions and fundamental solutions also have simple probabilistic representations.

If an operator has smooth coefficients, then equations with these operators will have smooth solutions. This theory is discussed in Chapter III. The chapter is largely analytic, but probability allows some simplification in the arguments.

Chapter IV considers one-dimensional diffusions and the corresponding second-order ordinary differential equations. Every one-dimensional diffusion can be derived from Brownian motion by changes of time and scale.

What is covered in the first four chapters is mostly classical and well known. The next four chapters discuss material that has appeared only in much more specialized places.

Chapter V concerns operators in nondivergence form. After some preliminaries, the discussion turns to the Harnack inequality of Krylov and Safonov and then to approximating operators with nonsmooth coefficients by those with smooth coefficients. Even in the nonsmooth case, solutions to these equations will have at least some regularity.

Chapter VI concerns the existence and uniqueness of the martingale problem for operators in nondivergence form. If the coefficients are continuous, there exists only one process corresponding to a given operator. A similar assertion can be made in certain other cases.

In Chapter VII we turn to divergence form operators. Our main goals are to derive Moser's Harnack inequality, upper and lower bounds for the heat kernel, and path properties of the associated processes.

Finally, in Chapter VIII we consider two different approaches to the Malliavin calculus. We show how each one can be used to prove a version of Hörmander's theorem.

In this book we consider only linear second-order elliptic and parabolic operators. This is not to imply that probability has nothing to say about nonlinear or higher-order equations, but these topics are not discussed in this book.

It is assumed that the reader knows some probability theory; the first chapter of Bass [1] (referenced in this book by "PTA") is more than sufficient. References are given for the theorems from probability and analysis that are required.

Each chapter ends with some notes that describe where I obtained the material and suggestions for further reading. These are not meant to be a history of the subject and are totally inadequate for that purpose.

Most of the material covered has previously been the subject of courses I have given at the University of Washington, and I would like to thank the students who attended and pointed out errors. In addition, I would like to give special thanks to Heber Farnsworth and Davar Khoshnevisan, who read through the text and made valuable suggestions. Partial support for this project has been provided by the National Science Foundation.

Some notation

We will let $B(x,r)$ denote the open ball in $\mathbb{R}^d$ with center x and radius r. We use $|\cdot|$ for the Euclidean norm of points of $\mathbb{R}^d$, for the norm of vectors, and for the norm of matrices. To be more precise, let e_i denote the unit vector in the x_i direction. If $v = \sum_{i=1}^d b_i e_i$ and A is a matrix, then

$$|v| = \Big(\sum_{i=1}^{d} b_i^2 \Big)^{1/2}, \qquad |A| = \sup_{|v|=1} |Av|.$$

The inner product in $\mathbb{R}^d$ of x and y will be written $x \cdot y$. If A is a matrix, then A^T denotes the transpose of a. Kronecker's delta δ_{ij} is 1 if $i = j$ and 0 otherwise. The complement of a set B is denoted B^c.

∂t is an abbreviation for $\partial/\partial t$ and ∂_i an abbreviation for $\partial/\partial x_i$. The L^p norm of a function f will be denoted $\|f\|_p$. We define the Fourier transform of a function f by

$$\widehat{f}(\xi) = \int_{\mathbb{R}^d} e^{i\xi \cdot x} f(x) \, dx.$$

A smooth function is one such that the function and its partial derivatives of all orders are continuous and bounded. The notation 1_A represents the function or random variable that takes the value 1 on the set A and 0 on the complement of A.

If X_t is a stochastic process and A a Borel subset of $\mathbb{R}^d$, we write

$$T_A = T(A) = \inf\{t > 0 : X_t \in A\}$$

and

$$\tau_A = \tau(A) = \inf\{t > 0 : X_t \notin A\}$$

for the first hitting time and first exit time of A, respectively.

The letter c with a subscript indicates a constant whose exact value is unimportant. We renumber in each theorem, lemma, proposition, and corollary.

The reference PTA refers to Bass [1].

Seattle, Washington   Richard F. Bass

CONTENTS

I
STOCHASTIC DIFFERENTIAL EQUATIONS

The partial differential equations we will consider have solutions that can be represented as functionals of certain stochastic processes. In this chapter we will construct these processes and examine some of their properties.

The processes we study are the solutions to stochastic differential equations (SDEs). After a section of notation and definitions, in Section 2 we discuss what a SDE is and what it means to be a pathwise solution to a SDE. Proposition 2.1 provides the first link between SDEs and PDE.

In Section 3 we prove that pathwise solutions exist and are unique if the coefficients of the SDE are Lipschitz. Under less regularity of the coefficients there are other notions of existence and uniqueness that are appropriate; see Section 4.

For each point x there will be a different solution to the SDE for the process starting at x. Taken together these solutions will form a strong Markov process; this is discussed in Section 5.

Stronger conclusions can be reached when the dimension is one. These are demonstrated in Section 6, whereas in Section 7 a number of examples are given.

The core material of this chapter is completed in Section 8, where some basic estimates of solutions to SDEs are given, e.g., the support theorem (Theorem 8.5). The reader who is eager to get to applications to partial differential equations could read Sections 2 through 5 and Section 8 and then proceed to Chapter II.

Section 9 is concerned with the Stratonovich integral and Section 10 with flows. Although flows make a brief reappearance in Chapter III, these two sections are needed only for the Malliavin calculus in Chapter VIII.

Sections 11 and 12 are about SDEs where there is reflection on the boundary of some domain. Section 11 describes the framework and Section 12 gives a proof of pathwise uniqueness. The reader who is not interested in the Neumann problem or the oblique derivative problem may safely skip Sections 11 and 12.

1. Preliminaries

We start by introducing some notation and recalling a few basic definitions. A *filtration* is an increasing collection of σ-fields $\mathcal{F}_t$, $0 \leq t \leq \infty$, that are right continuous and complete: $\cap_{\varepsilon > 0}\mathcal{F}_{t+\varepsilon} = \mathcal{F}_t$ for all t and $N \in \mathcal{F}_t$ for all t whenever $\mathbb{P}(N) = 0$. A process X_t is a *martingale* if for each t and $s < t$ the random variable X_t is integrable and adapted to $\mathcal{F}_t$ and $\mathbb{E}\left[X_t \mid \mathcal{F}_s\right] = X_s$ a.s. The process X_t is a *local martingale* if there exist stopping times $T_n \uparrow \infty$ such that $X_{T_n \wedge t}$ is a martingale for each n. A process is a *semimartingale* if it is the sum of a local martingale and a process that is locally of finite bounded variation (i.e., finite bounded variation on every interval $[0, t]$). We will be dealing almost exclusively with continuous processes, so unless stated otherwise, all of our processes will have continuous paths. If X_t is a local martingale, the *quadratic variation* of X is the unique increasing continuous process $\langle X \rangle_t$ such that $X_t^2 - \langle X \rangle_t$ is a local martingale. If $X_t = M_t + A_t$, where M_t is a local martingale and A_t has paths of locally finite bounded variation, then $\langle X \rangle_t$ is defined to be $\langle M \rangle_t$. If X and Y are two semimartingales, we define

$$\langle X, Y \rangle_t = \frac{1}{2}\big(\langle X + Y \rangle_t - \langle X \rangle_t - \langle Y \rangle_t\big).$$

A one-dimensional *Brownian motion* adapted to $\mathcal{F}_t$ is a process W_t with continuous paths such that W_t is adapted to $\{\mathcal{F}_t\}$ and if $s < t$, then $W_t - W_s$ is independent of $\mathcal{F}_s$ and has the law of a mean zero normal random variable with variance $t - s$. Recall $\langle W \rangle_t = t$. A d-dimensional Brownian motion is a d-dimensional process whose components are independent one-dimensional Brownian motions.

If M_t is a local martingale, H_t is adapted to the filtration $\{\mathcal{F}_t\}$, and $\int_0^t H_s^2 \, d\langle M \rangle_s < \infty$ for all t, we define the *stochastic integral* $N_t = \int_0^t H_s \, dM_s$ to be the local martingale such that $\langle N, L \rangle_t = \int_0^t H_s \, d\langle M, L \rangle_s$ for all martingales L_t adapted to $\{\mathcal{F}_t\}$. If $N_t = \int_0^t H_s \, dM_s$, then $\langle N \rangle_t = \int_0^t H_s^2 \, d\langle M \rangle_s$. Recall that if $H_s(\omega) = F(\omega)1_{[a,b]}(s)$ where F is $\mathcal{F}_a$ measurable, then

$$\int_0^t H_s \, dM_s = F(\omega)[M_{t \wedge b}(\omega) - M_{t \wedge a}(\omega)];$$

we extend this construction to more general H_s by linearity and taking limits in L^2.

For $X_t = M_t + A_t$ a semimartingale, $\int_0^t H_s\, dX_s$ is defined by

$$\int_0^t H_s\, dX_s = \int_0^t H_s\, dM_s + \int_0^t H_s\, dA_s,$$

where the first integral on the right is a stochastic integral and the second integral on the right is a Lebesgue-Stieltjes integral.

If X_t is a semimartingale and $f \in C^2(\mathbb{R})$, *Itô's formula* is the equation

$$f(X_t) - f(X_0) = \int_0^t f'(X_s)\, dX_s + \frac{1}{2}\int_0^t f''(X_s)\, d\langle X\rangle_s. \tag{1.1}$$

If $X_t = (X_t^1, \ldots, X_t^d)$ is a d-dimensional semimartingale, that is, a process in $\mathbb{R}^d$, each of whose components is a semimartingale, the higher-dimensional analogue of Itô's formula says that if $f \in C^2(\mathbb{R}^d)$,

$$f(X_t) - f(X_0) \tag{1.2}$$
$$= \int_0^t \sum_{i=1}^d \partial_i f(X_s)\, dX_s^i + \frac{1}{2}\int_0^t \sum_{i,j=1}^d \partial_{ij} f(X_s)\, d\langle X^i, X^j\rangle_s.$$

Throughout this book we write ∂_i for $\partial/\partial x_i$ and ∂_{ij} for $\partial^2/\partial x_i \partial x_j$. If X and Y are real-valued semimartingales, the *product formula* is

$$X_t Y_t = X_0 Y_0 + \int_0^t X_s\, dY_s + \int_0^t Y_s\, dX_s + \langle X, Y\rangle_t. \tag{1.3}$$

Doob's inequality ([PTA, Theorem I.4.7]) says that if M_t is a right-continuous martingale and $p > 1$, there exists a constant $c_1 = c_1(p)$ such that

$$\mathbb{E}\sup_{s \leq t} |M_s|^p \leq c_1(p)\mathbb{E}\,|M_t|^p. \tag{1.4}$$

We will use the *Burkholder-Davis-Gundy inequalities* ([PTA, Theorem I.6.8]), which say that for $p > 0$ there exist constants $c_1(p)$ such that if M_t is a continuous martingale and T a stopping time, then

$$\mathbb{E}\sup_{s \leq T} |M_s|^p \leq c_1(p)\mathbb{E}\,\langle M\rangle_T^{p/2}. \tag{1.5}$$

Lévy's theorem ([PTA, Theorem I.5.9]) is the following: suppose M_t is a continuous local martingale with $\langle M\rangle_t = t$ for all t; then M_t is a Brownian motion.

There is a higher-dimensional analogue of this ([PTA, Corollary I.5.10]). If X_t is a d-dimensional process, each of whose coordinates is a continuous local martingale, and $\langle X^i, X^j\rangle_t = \delta_{ij}t$, then X_t is a d-dimensional Brownian motion.

A consequence of Lévy's theorem is the following. If M_t is a continuous local martingale with $\langle M\rangle_\infty = \infty$, then M_t is a time change of a Brownian motion. See [PTA, Theorem I.5.11] for a proof.

The *Girsanov theorem* ([PTA, Theorem I.6.4]) is the result that if X_t and M_t are continuous martingales under $\mathbb{P}$ with $M_0 = 0$ $\mathbb{P}$-a.s. and we define a new probability measure $\mathbb{Q}$ by setting the restriction of $d\mathbb{Q}/d\mathbb{P}$ to $\mathcal{F}_t$ to be $\exp(M_t - \langle M \rangle_t/2)$, then $X_t - \langle X, M \rangle_t$ is a martingale under $\mathbb{Q}$ and the quadratic variation of X_t is the same under $\mathbb{P}$ and $\mathbb{Q}$.

2. Pathwise solutions

Let W_t be a one-dimensional Brownian motion. We will be concerned with the *stochastic differential equation* (SDE)

$$dX_t = \sigma(X_t)\,dW_t + b(X_t)\,dt, \qquad X_0 = x. \tag{2.1}$$

This is a shorthand way of writing

$$X_t = x + \int_0^t \sigma(X_s)\,dW_s + \int_0^t b(X_s)\,ds. \tag{2.2}$$

Here σ and b are measurable real-valued functions. We will say (2.1) or (2.2) has a solution if there exists a continuous adapted process X_t satisfying (2.2). X_t is necessarily a semimartingale. Later on we will talk about various types of solutions, so to be more precise, we say that X_t is a *pathwise solution*. We say that we have *pathwise uniqueness* for (2.1) or (2.2) if whenever X_t and X_t' are two solutions, then there exists a set N such that $\mathbb{P}(N) = 0$ and for all $\omega \notin N$, we have $X_t = X_t'$ for all t.

The definitions for the higher-dimensional analogues of (2.1) and (2.2) are the same. Let σ_{ij} be measurable functions for $i, j = 1, \ldots, d$ and b_i measurable functions for $i = 1, \ldots, d$. Let W_t be a d-dimensional Brownian motion. We consider the equation

$$dX_t = \sigma(X_t)\,dW_t + b(X_t)\,dt, \qquad X_0 = x, \tag{2.3}$$

or equivalently, for $i = 1, \ldots, d$,

$$X_t^i = x_i + \int_0^t \sum_{j=1}^d \sigma_{ij}(X_s)\,dW_s^j + \int_0^t b_i(X_s)\,ds. \tag{2.4}$$

Here $X_t = (X_t^1, \ldots, X_t^d)$ is a semimartingale on $\mathbb{R}^d$.

The connection between stochastic differential equations and partial differential equations comes about through the following theorem, which is simply an application of Itô's formula. Let σ^T denote the transpose of the matrix σ and let a be the matrix $\sigma\sigma^T$. Let $C^2(\mathbb{R}^d)$ be the functions on $\mathbb{R}^d$ whose first and second partial derivatives are continuous and let $\mathcal{L}$ be the operator on $C^2(\mathbb{R}^d)$ defined by

$$\mathcal{L}f(x) = \frac{1}{2} \sum_{i,j=1}^d a_{ij}(x)\partial_{ij}f(x) + \sum_{i=1}^d b_i(x)\partial_i f(x). \tag{2.5}$$

(2.1) Proposition. *Suppose X_t is a solution to (2.3) with σ and b bounded and measurable and let $f \in C^2(\mathbb{R}^d)$. Then*

$$f(X_t) = f(X_0) + M_t + \int_0^t \mathcal{L}f(X_s)\,ds, \tag{2.6}$$

where

$$M_t = \int_0^t \sum_{i,j=1}^d \partial_i f(X_s)\sigma_{ij}(X_s)\,dW_s^j \tag{2.7}$$

is a martingale.

Proof. Since the components of the Brownian motion W_t are independent, we have $d\langle W^k, W^\ell\rangle_t = 0$ if $k \neq \ell$. Therefore

$$d\langle X^i, X^j\rangle_t = \sum_k \sum_\ell \sigma_{ik}(X_t)\sigma_{jl}(X_t)\,d\langle W^k, W^\ell\rangle_t$$

$$= \sum_k \sigma_{ik}(X_t)\sigma_{kj}^T(X_t)\,dt = a_{ij}(X_t)\,dt.$$

We now apply Itô's formula:

$$f(X_t) = f(X_0) + \sum_i \int_0^t \partial_i f(X_s)\,dX_s^i + \frac{1}{2}\int_0^t \sum_{i,j} \partial_{ij} f(X_s)\,d\langle X^i, X^j\rangle_s$$

$$= f(X_0) + M_t + \sum_i \int_0^t \partial_i f(X_s)b_i(X_s)\,ds$$

$$+ \frac{1}{2}\int_0^t \sum_{i,j} \partial_{ij} f(X_s)a_{ij}(X_s)\,ds$$

$$= f(X_0) + M_t + \int_0^t \mathcal{L}f(X_s)\,ds. \qquad \square$$

We will say that a process X_t and an operator $\mathcal{L}$ are *associated* if X_t satisfies (2.3) for $\mathcal{L}$ given by (2.5) and $a = \sigma\sigma^T$. We call the functions b the *drift coefficients* of X_t and of $\mathcal{L}$, and we call σ and a the *diffusion coefficients* of X_t and $\mathcal{L}$, respectively.

3. Lipschitz coefficients

We now proceed to show existence and uniqueness for the SDE (2.1) when the coefficients σ and b are Lipschitz continuous. For notational simplicity, we first consider the case where the dimension is one. Recall that a function f is *Lipschitz* if there exists a constant c_1 such that $|f(x) - f(y)| \leq c_1|x - y|$ for all x, y.

(3.1) Theorem. *Suppose σ and b are Lipschitz and bounded. Then there exists a pathwise solution to the SDE (2.1).*

Proof. We use Picard iteration. Define $X^0(t) \equiv x$ and define inductively

$$X^{i+1}(t) = x + \int_0^t \sigma(X^i(s))\, dW_s + \int_0^t b(X^i(s))\, ds \tag{3.1}$$

for $i = 0, 1, \ldots$ Note

$$X^{i+1}(t) - X^i(t) = \int_0^t [\sigma(X^i(s)) - \sigma(X^{i-1}(s))]\, dW_s \tag{3.2}$$

$$+ \int_0^t [b(X^i(s)) - b(X^{i-1}(s))]\, ds.$$

Let $g_i(t) = \mathbb{E}\left[\sup_{s \leq t} |X^{i+1}(s) - X^i(s)|^2\right].$

If F_t denotes the first term on the right-hand side of (3.2), then by Doob's inequality (1.4),

$$\mathbb{E}\sup_{s \leq t} F_s^2 \leq c_1 \mathbb{E} \int_0^t |\sigma(X^i(s)) - \sigma(X^{i-1}(s))|^2\, ds \tag{3.3}$$

$$\leq c_2 \int_0^t \mathbb{E}\,|X^i(s) - X^{i-1}(s)|^2\, ds$$

$$\leq c_2 \int_0^t g_{i-1}(s)\, ds.$$

If G_t denotes the second term on the right-hand side of (3.2), then by the Cauchy-Schwarz inequality,

$$\mathbb{E}\sup_{s \leq t} G_s^2 \leq \mathbb{E}\left(\int_0^t |b(X^i(s)) - b(X^{i-1}(s))|\, ds\right)^2 \tag{3.4}$$

$$\leq \mathbb{E}\, t \int_0^t |b(X^i(s)) - b(X^{i-1}(s))|^2\, ds$$

$$\leq c_3 t \int_0^t \mathbb{E}\,|X^i(s) - X^{i-1}(s)|^2\, ds$$

$$\leq c_3 t \int_0^t g_{i-1}(s)\, ds.$$

So (3.2), (3.3), (3.4), and the inequality $(x+y)^2 \leq 2x^2 + 2y^2$ tell us that there exists A such that

$$g_i(t) \leq 2\mathbb{E}\sup_{s \leq t} F_s^2 + 2\mathbb{E}\sup_{s \leq t} G_s^2 \leq A(1+t)\int_0^t g_{i-1}(s)\, ds. \tag{3.5}$$

Since σ and b are bounded, arguments to those in the derivation of (3.3) and (3.4) show that $g_0(t)$ is bounded by $B(1+t)$ for some constant B. Iterating (3.5),

$$g_1(t) \leq A \int_0^t B(1+s)\, ds \leq AB(1+t)^2/2$$

for all t, so

$$g_2(t) \leq A \int_0^t (AB(1+s)^2)/2\, ds \leq A^2 B(1+t)^3/3!$$

for all t. By induction,

$$g_i(t) \leq A^i B(1+t)^{i+1}/(i+1)!$$

Hence $\sum_{i=0}^{\infty} g_i(t)^{1/2} < \infty$. Fix t and define the norm

$$\|Y\| = (\mathbb{E} \sup_{s \leq t} |Y_s|^2)^{1/2}. \tag{3.6}$$

We then have that

$$\|X^n - X^m\| \leq \sum_{i=n}^{m-1} g_i(t)^{1/2} \to 0$$

if $m > n$ and $m, n \to \infty$. Therefore X^n is a Cauchy sequence with respect to this norm. It is clear that there is a process X such that $\|X^n - X\| \to 0$ as $n \to \infty$. For each t, we can look at a subsequence so that $\sup_{s \leq t} |X(s) - X^{n_j}(s)| \to 0$ a.s., which implies that $X(s)$ has continuous paths. Letting $i \to \infty$ in (3.1), we see that $X(s)$ satisfies (2.2). $\qquad\square$

For use in Chapter VIII we need the following corollary.

(3.2) Corollary. *If $p \geq 2$, then for all $t \geq 0$, $\mathbb{E} \sup_{s \leq t} |X^n(s) - X_s| \to 0$ as $n \to \infty$.*

Proof. Let

$$g_i(t) = \mathbb{E} \sup_{s \leq t} |X^{i+1}(s) - X^i(s)|^p.$$

In place of (3.3) we use the Burkholder-Davis-Gundy inequalities (1.5). Let F_t and G_t be as in the proof of Theorem 3.1. Suppose $t_0 > 0$ and we consider $t \leq t_0$. We then write

$$\mathbb{E} \sup_{s \leq t} |F_s|^p \leq c_2 \mathbb{E} \left[\int_0^t |\sigma(X^i(s)) - \sigma(X^{i-1}(s))|^2\, ds \right]^{p/2}$$

$$\leq c_3(t_0) \mathbb{E} \int_0^t |\sigma(X^i(s)) - \sigma(X^{i-1}(s))|^p\, ds$$

$$\leq c_4 \mathbb{E} \int_0^t |X^i(s) - X^{i-1}(s)|^p\, ds$$

$$\leq c_4 \int_0^t g_{i-1}(s)\, ds,$$

and in place of (3.4)

$$\mathbb{E}\sup_{s\leq t}|G_s|^p \leq \mathbb{E}\left(\int_0^t |b(X^i(s)) - b(X^{i-1}(s))|\,ds\right)^p$$

$$\leq c_5(t_0)\mathbb{E}\int_0^t |b(X^i(s)) - b(X^{i-1}(s))|^p\,ds$$

$$\leq c_6\mathbb{E}\int_0^t |X^i(s) - X^{i-1}(s)|^p\,ds$$

$$\leq c_6\int_0^t g_{i-1}(s)\,ds.$$

With these changes and the inequality $(x+y)^p \leq 2^{p-1}x^p + 2^{p-1}y^p$ when $x, y > 0$, we have as before that $g_i(t) \leq A^i B(1+t)^{i+1}/(i+1)!$ We conclude that $\sum_{i=n}^{m-1} g_i(t)^{1/p} \to 0$ as $m, n \to \infty$. This shows that X^n converges to X in L^p. $\qquad\square$

Uniqueness will be shown next. We first examine a portion of the proof that is known as Gronwall's lemma, since this elementary lemma will be used repeatedly in what follows.

(3.3) Lemma. (Gronwall's lemma) *Suppose $g : [0, \infty) \to \mathbb{R}$ is bounded on each finite interval, is measurable, and there exist A and B such that for all t*

$$g(t) \leq A + B\int_0^t g(s)\,ds. \tag{3.7}$$

Then $g(t) \leq Ae^{Bt}$ for all t.

Proof. Iterating the inequality for g,

$$g(t) \leq A + B\int_0^t \left[A + B\int_0^s g(r)\,dr\right]ds$$

$$\leq A + ABt + B^2\int_0^t\int_0^s \left[A + B\int_0^r g(q)\,dq\right]ds\,dt$$

$$= A + ABt + AB^2t^2/2 + B^3\int_0^t\int_0^s\int_0^r g(q)\,dq\,dr\,ds$$

$$\leq \cdots.$$

Since g is bounded on $[0, t]$, say by C, then

$$\int_0^t g(s)\,ds \leq Ct, \qquad \int_0^t\int_0^s g(r)\,dr\,ds \leq \int_0^t Cs\,ds \leq Ct^2/2!,$$

and so on. Hence

$$g(t) \leq Ae^{Bt} + B^n Ct^n/n!$$

for each n. Letting $n \to \infty$ completes the proof. $\qquad\square$

(3.4) Theorem. *Suppose σ and b are Lipschitz and bounded. Then the solution to the SDE (2.1) is pathwise unique.*

Proof. Suppose X_t and X'_t are two pathwise solutions to (2.1). Let

$$g(t) = \mathbb{E} \sup_{s \leq t} |X_s - X'_s|^2.$$

Since X_t and X'_t both satisfy (2.1), their difference satisfies

$$X_t - X'_t = H_t + I_t,$$

where

$$H_t = \int_0^t [\sigma(X_s) - \sigma(X'_s)]\, dW_s, \qquad I_t = \int_0^t [b(X_s) - b(X'_s)]\, ds.$$

As in the proof of Theorem 3.1, there exist c_1 and c_2 such that

$$\mathbb{E} \sup_{s \leq t} H_s^2 \leq c_1 \int_0^t g(s)\, ds, \qquad \mathbb{E} \sup_{s \leq t} I_s^2 \leq c_2 t \int_0^t g(s)\, ds.$$

Hence, if t_0 is a positive real and $t \leq t_0$, there exists a constant c_3 depending on t_0 such that

$$g(t) \leq 2\mathbb{E} \sup_{s \leq t} H_s^2 + 2\mathbb{E} \sup_{s \leq t} I_s^2 \leq c_3 \int_0^t g(s)\, ds.$$

By Lemma 3.3, $g(t) = 0$ for $t \leq t_0$. Since t_0 is arbitrary, uniqueness is proved.

$\square$

It is often useful to be able to remove the boundedness assumption on σ and b. We still want σ and b to be Lipschitz, so this can be phrased as follows.

(3.5) Theorem. *Suppose σ and b are Lipschitz and there exists a constant c_1 such that*

$$|\sigma(x)| + |b(x)| \leq c_1(1 + |x|).$$

Then there exists a pathwise solution to (2.1) and the solution is pathwise unique.

Proof. Let σ_n and b_n be bounded Lipschitz functions that agree on $[-n, n]$ with σ and b, respectively, and let X_t^n be the solution to (2.1) with σ and b replaced by σ_n and b_n, respectively. Let $T_n = \inf\{t : |X_t^n| \geq n\}$.

Note

$$X_t^n = X_t^m \quad \text{if} \quad t \leq T_m \wedge T_n. \tag{3.8}$$

To see this, set $g(t) = \mathbb{E} \sup_{s \leq t_0 \wedge T_m \wedge T_n} |X_s^n - X_s^m|^2$ and proceed as in the proof of Theorem 3.4.

Set $X_t = X_t^n$ for $t \leq T_n$. We will show $T_n \to \infty$ a.s. Once we have this, the existence and uniqueness of the solution to (2.1) follow easily from (3.8) and Theorems 3.1 and 3.4.

Let $g_n(t) = \mathbb{E} \sup_{s \leq t \wedge T_n} |X_s^n|^2$. Fix $t_0 > 0$. Then as in the proof of Theorem 3.1, for $t \leq t_0$

$$
\begin{aligned}
\mathbb{E} \sup_{s \leq t \wedge T_n} (X_s^n)^2 &\leq c_2 |x|^2 + c_2 \mathbb{E} \int_0^t \sigma_n(X_s^n)^2 \, ds \\
&\quad + c_3 t_0 \mathbb{E} \int_0^t b_n(X_s^n)^2 \, ds \\
&\leq c_2 |x|^2 + c_4 + c_5 \mathbb{E} \int_0^t |X_s^n|^2 \, ds,
\end{aligned}
$$

or

$$
g_n(t) \leq c_2 |x|^2 + c_4 + c_5 \int_0^t g_n(s) \, ds,
$$

where c_2, c_4, and c_5 do not depend on n. By Gronwall's lemma, $g_n(t) \leq (c_2 |x|^2 + c_4) e^{c_5 t}$ for $t \leq t_0$. Using Chebyshev's inequality,

$$
\begin{aligned}
\mathbb{P}(T_n < t_0) = \mathbb{P}(\sup_{s \leq t_0 \wedge T_n} |X_s^n| \geq n) \\
\leq \mathbb{E} \sup_{s \leq t_0 \wedge T_n} |X_s^n|^2 / n^2 \leq g_n(t_0)/n^2 \to 0
\end{aligned}
$$

as n tends to infinity. Since $T_n \uparrow$ by (3.8) and t_0 is arbitrary, the result follows. $\qquad\square$

We remark that as a consequence of the above proof, $X_t = X_t^n$ if $T_n > t$, so X_t does not *explode*, that is, X_t does not tend to infinity in finite time.

We have considered the case of $\mathbb{R}$-valued processes for simplicity, but with only trivial changes the proofs work when the state space is $\mathbb{R}^d$ (and even infinite dimensions if properly formulated), so we can state

(3.6) Theorem. *Suppose $d \geq 1$ and suppose σ and b are Lipschitz. Then the SDE (2.3) has a pathwise solution and this solution is pathwise unique.*

In the above, we required σ and b to be functions of X_t only. Only cosmetic changes are required if we allow σ and b to be functions of t and X_t and consider

$$
dX_t = \sigma(t, X_t) \, dW_t + b(t, X_t) \, dt. \tag{3.9}
$$

4. Types of uniqueness

When the coefficients σ and b fail to be Lipschitz, it is sometimes the case that (2.1) may not have a pathwise solution at all, or it may not be unique. We define some other notions of existence and uniqueness that are useful. We now assume that the dimension of the state space may be larger than one.

We say a *strong solution exists* to the SDE (2.3) if given the Brownian motion W_t there exists a process X_t satisfying (2.3) such that X_t is adapted to the filtration generated by W_t. A *weak solution exists* if there exists a pair of processes (X_t, W_t) such that W_t is a Brownian motion and the equation (2.3) holds. There is *weak uniqueness* holding if whenever (X_t, W_t) and (X'_t, W'_t) are two weak solutions, then the joint laws of the processes (X, W) and (X', W') are equal. When this happens, we also say that the solution to (2.3) is *unique in law*.

Let us explore some of the relationships between the various definitions just given. Pathwise existence and the existence of a strong solution are very close, differing only in unimportant measurability concerns. If the solution to (2.3) is pathwise unique, then weak uniqueness holds. For the proof, see Yamada and Watanabe [1]. In the case that σ and b are Lipschitz, the proof is much simpler.

(4.1) Proposition. *Suppose σ and b are Lipschitz. Then the solution to (2.3) is a strong solution. Weak uniqueness holds for (2.3).*

Proof. For notational simplicity we consider the case of dimension one. The Picard iteration in Theorem 3.1 and the definition via stopping times in Theorem 3.4 preserve measurability, so the solution constructed in these two theorems is adapted to the filtration generated by W_t. Thus the solution is a strong solution.

Suppose (X_t, W_t) and (X'_t, W'_t) are two solutions to (2.1). Let X''_t be the process that is constructed from W'_t analogously to how X_t was constructed from W_t, namely, by Picard iteration and stopping times. It follows that (X, W) and (X'', W') have the same law. By the pathwise uniqueness, $X'' = X'$, so the result follows. $\qquad\square$

There is a generalization of the argument of Proposition 4.1 that is not as well known as it deserves to be. Here we are not assuming that σ and b are Lipschitz.

(4.2) Theorem. *Suppose the matrix σ has an inverse that is bounded. Suppose σ and b are bounded and measurable. If (2.3) has a strong solution and the solution to (2.3) is weakly unique, then pathwise uniqueness holds for (2.3).*

Proof. Let a Brownian motion W_t be given, and let X be a strong solution to (2.3). Suppose X' is another solution to (2.3). We must show $X_t = X'_t$ for all t a.s.

X is a strong solution, which implies that there exists a function F on the space of continuous functions so that $X = F(W)$ a.s. Let $M_t = X_t - \int_0^t b(X_s)\, ds$, so that $W_t = \int_0^t \sigma^{-1}(X_s)\, dM_s$. Hence W_t is measurable with respect to the filtration generated by the semimartingale X. Similarly, W is adapted to the filtration generated by X'. Since weak uniqueness holds, X is equal to X' in law. It follows that the pair (X, W) is equal in law to the pair (X', W). Thus $X' = F(W)$ a.s. as well. The conclusion follows from combining: $X' = F(W) = X$ a.s. $\square$

We now give an example to show that weak uniqueness might hold even if pathwise uniqueness does not. Let $\sigma(x)$ be equal to 1 if $x \geq 0$ and -1 otherwise. We take b to be identically 0. We consider solutions to

$$X_t = \int_0^t \sigma(X_s)\, dW_s. \tag{4.1}$$

Weak uniqueness holds since X_t must be a martingale, and the quadratic variation of X is $d\langle X \rangle_t = \sigma(X_t)^2\, dt = dt$; by Lévy's theorem (Section 1), X_t is a Brownian motion. Given a Brownian motion X_t and letting $W_t = \int_0^t \sigma^{-1}(X_s)\, dX_s$ where $\sigma^{-1} = 1/\sigma$, then again by Lévy's theorem, W_t is a Brownian motion; thus weak solutions exist.

On the other hand, pathwise uniqueness does not hold (so no strong solution exists). To see this, let $Y_t = -X_t$. We have

$$Y_t = \int_0^t \sigma(Y_s)\, dW_s - 2 \int_0^t 1_{\{0\}}(X_s)\, dW_s. \tag{4.2}$$

The second term on the right has quadratic variation $4 \int_0^t 1_{\{0\}}(X_s)\, ds$, which is equal to 0 almost surely because X is a Brownian motion. Therefore the second term on the right of (4.2) equals 0 almost surely, or Y is another pathwise solution to (4.1).

This example is not satisfying because one would like σ to be positive and even continuous if possible. Such examples exist, however; see Barlow [1]. See also the earlier example of Tsirelson [1].

5. Markov properties

One of the more important applications of SDEs is to Markov processes. A Markov process is one where the probability of future events depends on the past history only through the present position. In order to be more precise, we need to introduce some notation. Rather than having one probability measure and a collection of processes, it is more convenient to have one process and a collection of measures.

Define Ω' to be the set of all continuous functions from $[0, \infty)$ to $\mathbb{R}^d$. We define $Z_t(\omega) = \omega(t)$ for $\omega \in \Omega'$. We call Z_t the *canonical process*. Suppose that for each starting point x the SDE (2.3) has a solution that is unique in law. Let us denote the solution by $X(x, t, \omega)$. For each x define a probability measure $\mathbb{P}^x$ on Ω' so that

$$\mathbb{P}^x(Z_{t_1} \in A_1, \ldots, Z_{t_n} \in A_n)$$
$$= \mathbb{P}(X(x, t_1, \omega) \in A_1, \ldots, X(x, t_n, \omega) \in A_n)$$

whenever $t_1, \ldots, t_n \in [0, \infty)$ and $A_1, \ldots, A_n$ are Borel sets in $\mathbb{R}^d$. The measure $\mathbb{P}^x$ is determined on the smallest σ-field containing these cylindrical sets. Let $\mathcal{G}_t^{00}$ be the σ-algebra generated by Z_s, $s \leq t$. We complete these σ-fields by considering all sets that are in the $\mathbb{P}^x$ completion of $\mathcal{G}_t^{00}$ for all x. (This is not quite the same as the completion with respect to $\mathbb{P}^x$, but it will be good enough for our purposes.) Finally, we obtain a right continuous filtration by letting $\mathcal{F}_t' = \cap_{\varepsilon>0}\mathcal{G}_{t+\varepsilon}^{00}$. We then extend $\mathbb{P}^x$ to $\mathcal{F}_\infty'$.

One advantage of Ω' is that it is equipped with *shift operators* $\theta_t : \Omega' \to \Omega'$ defined by $\theta_t(\omega)(s) = \omega(t + s)$. Another way of writing this is $Z_t \circ \theta_s = Z_{t+s}$. For stopping times T we set $\theta_T(\omega) = \theta_{T(\omega)}(\omega)$.

The *strong Markov property* is the assertion that

$$\mathbb{E}^x[Y \circ \theta_T \mid \mathcal{F}_T'] = \mathbb{E}^{Z_T}[Y], \qquad \text{a.s. } (\mathbb{P}^x) \tag{5.1}$$

whenever $x \in \mathbb{R}^d$, Y is bounded and $\mathcal{F}_\infty'$ measurable, and T is a finite stopping time. The *Markov property* holds if the above equality holds whenever T is a fixed (i.e., nonrandom) time. If the strong Markov property holds, we say $(\mathbb{P}^x, Z_t)$ is a *strong Markov process*.

As in the discussion in [PTA, Section I.3], to prove the strong Markov property it suffices to show

$$\mathbb{E}^x[f(Z_{T+t}) \mid \mathcal{F}_T'] = \mathbb{E}^{Z_T} f(Z_t), \qquad \text{a.s. } (\mathbb{P}^x) \tag{5.2}$$

for all $x \in \mathbb{R}^d$, f a bounded and continuous function on $\mathbb{R}^d$, and T a bounded stopping time. See also [PTA, Section I.3] for some examples of how to interpret (5.1).

It turns out that if pathwise uniqueness holds for (2.3) for every x, then $(\mathbb{P}^x, Z_t)$ form a strong Markov process. However, aside from technicalities involving regular conditional probabilities, it is no more difficult to prove that weak uniqueness also implies the strong Markov property. We will have need of this stronger result and we now proceed to its proof.

Let T be a bounded stopping time. A *regular conditional probability* for $\mathbb{E}[\cdot \mid \mathcal{F}_T]$ is a kernel $Q_T(\omega, d\omega')$ such that

(i) $Q_T(\omega, \cdot)$ is a probability measure on Ω' for each ω;

(ii) for each $\mathcal{F}_\infty'$ measurable set A, $Q_T(\cdot, A)$ is a $\mathcal{F}_\infty'$ measurable random variable;

(iii) for each $\mathcal{F}_\infty'$ measurable set A and each $\mathcal{F}_T'$ measurable set B,

$$\int_B \mathbb{Q}_T(\omega, A)\, \mathbb{P}(d\omega) = \mathbb{P}(A \cap B).$$

Regular conditional probabilities need not always exist, but if the probability space is regular enough, as Ω' is, then they do. We defer the proof until Theorem 5.2.

We have the equation

$$Z_t = Z_0 + \int_0^t \sigma(Z_r)\, dW_r + \int_0^t b(Z_r)\, dr, \qquad (5.3)$$

where W_r is a Brownian motion, not necessarily the same as the one in (2.3). If we let $\widetilde{Z}_t = Z_{T+t}$ and $\widetilde{W}_t = W_{T+t} - W_T$, it is plausible that $\widetilde{W}$ is a Brownian motion with respect to the measure $Q_T(\omega, \cdot)$ for almost every ω. We show this below in Proposition 5.3. We write (5.3) with t replaced by $T + t$ and then write (5.3) with t replaced by T. Taking the difference and using a change of variables, we obtain

$$\widetilde{Z}_t = \widetilde{Z}_0 + \int_0^t \sigma(\widetilde{Z}_r)\, \widetilde{W}_r + \int_0^t b(\widetilde{Z}_r)\, dr. \qquad (5.4)$$

(5.1) Theorem. *Suppose the solution to (2.3) is weakly unique for each x. Then $(\mathbb{P}^x, Z_t)$ is a strong Markov process.*

Proof. Fix x and let Q_T denote the regular conditional probability for $\mathbb{E}^x[\cdot \mid \mathcal{F}'_T]$. Except for ω in a null set, under $Q_T(\omega, \cdot)$ we have from (5.4) and Proposition 5.3 that $\widetilde{Z}$ is a solution to (2.3) with starting point $\widetilde{Z}_0 = Z_T$. So if $\mathbb{E}_{Q_T}$ denotes the expectation with respect to Q_T, the uniqueness in law tells us that

$$\mathbb{E}_{Q_T} f(\widetilde{Z}_t) = \mathbb{E}^{Z_T} f(Z_t), \qquad \text{a.s. } (\mathbb{P}^x).$$

On the other hand,

$$\mathbb{E}_{Q_T} f(\widetilde{Z}_t) = \mathbb{E}_{Q_T} f(Z_{T+t}) = \mathbb{E}^x[f(Z_{T+t}) \mid \mathcal{F}'_T], \qquad \text{a.s. } (\mathbb{P}^x),$$

which proves (5.2). $\qquad\qquad\qquad\qquad\qquad\qquad\qquad\qquad\qquad\qquad \square$

It remains to prove that regular conditional probabilities exist and that under Q_T the process $\widetilde{W}$ is a Brownian motion.

(5.2) Theorem. *Suppose $(\Omega, \mathcal{F}, \mathbb{P})$ is a probability space, $\mathcal{G} \subset \mathcal{F}$, and Ω is a complete and separable metric space. Then a regular conditional probability for $\mathbb{P}(\cdot \mid \mathcal{G})$ exists.*

Proof. We can embed Ω in $[0,1]^\infty$, a compact set (see, for example, [PTA, the proof of Theorem I.7.4]). Let $\{f_j\}$ be a countable collection of linearly independent uniformly continuous functions on Ω whose closure is dense in the class of uniformly continuous functions on Ω; let us assume $f_1 \equiv 1$.

Let $g_j = \mathbb{E}[f_j \mid \mathcal{G}]$. If $r_1, \ldots, r_n$ are rationals with $r_1 f_1 + \cdots r_n f_n \geq 0$, let

$$N(r_1, \ldots, r_n) = \{\omega : r_1 g_1(\omega) + \cdots r_n g_n(\omega) < 0\}.$$

Clearly, $\mathbb{P}(N(r_1, \ldots, r_n)) = 0$. Let N_1 be the union of all such $N(r_1, \ldots, r_n)$ with $n \geq 1$, the r_j rational. Then $N_1 \in \mathcal{G}$ and $\mathbb{P}(N_1) = 0$.

Fix $\omega \in \Omega - N_1$. Define a functional L on the finite linear combinations of the f_j by $L(f) = t_1 g_1(\omega) + \cdots + t_n g_n(\omega)$ if $f = t_1 f_1 + \cdots + t_n f_n$. We claim L is a positive linear functional. If $f = t_1 f_1 + \cdots + t_n f_n \geq 0$ and $\varepsilon > 0$ is rational, then there exists rational $r_1, \ldots, r_n$ such that $r_1 f_1 + \ldots + r_n f_n \geq -\varepsilon$, or $(r_1 + \varepsilon) f_1 + r_2 f_2 + \cdots + r_n f_n \geq 0$. Since $\omega \notin N_1$, then $(r_1 + \varepsilon) g_1 + r_2 g_2 + \cdots + r_n g_n \geq 0$. Letting $\varepsilon \to 0$, it follows that $t_1 g_1 + \cdots + t_n g_n \geq 0$. Since $L(f_1) = 1$, L can be extended to a positive linear functional on the closure of the collection of finite linear combinations of the f_j. Any uniformly continuous function on Ω can be extended uniquely to $\overline{\Omega}$, so L can be considered as a positive linear functional on $C(\overline{\Omega})$. By the Riesz representation theorem, there exists a probability measure $Q(\omega, \cdot)$ such that $L(f) = \int f(\omega') Q(\omega, d\omega')$.

The mapping $\omega \to L(f)$ is measurable with respect to $\mathcal{F}$ for each finite linear combination of the f_j, hence for all uniformly continuous functions on Ω by a limit argument. If $B \in \mathcal{G}$,

$$\int_B \left[\int (t_1 f_1 + \cdots + t_n f_n)(\omega') Q(\omega, d\omega') \right] \mathbb{P}(d\omega)$$

$$= \int_B (t_1 g_1 + \cdots + t_n g_n)(\omega) \mathbb{P}(d\omega)$$

$$= \int_B (t_1 f_1 + \cdots + t_n f_n)(\omega) \mathbb{P}(d\omega)$$

or $\int f(\omega') Q(\omega, d\omega')$ is a version of $\mathbb{E}[f | \mathcal{G}]$ if f is a finite linear combination of the f_j. By a limit argument, the same is true for all f that are of the form $f = 1_A$ with $A \in \mathcal{F}$.

Let G_{ni} be a sequence of balls of radius $1/n$ (with respect to the metric of Ω) contained in Ω and covering Ω. Choose i_n such that $\mathbb{P}(\cup_{i \leq i_n} G_{ni}) > 1 - 1/n2^n$. The set $H_n = \cap_{n \geq 1} \cup_{i \leq i_n} G_{ni}$ is totally bounded; let K_n be the closure of H_n in Ω. Since Ω is complete, K_n is complete and totally bounded, and hence compact, and $\mathbb{P}(K_n) \geq 1 - 1/n$. So

$$\mathbb{E}[Q(\cdot, \cup_n K_n); \Omega - N_1] \geq \mathbb{E}[Q(\cdot, K_n); \Omega - N_1] = \mathbb{P}(K_n) \to 1,$$

or $Q(\omega, \cup_n K_n) = 1$ a.s. Let N_2 be the null set for which this fails. Thus for $\omega \in \Omega - (N_1 \cup N_2)$, we see that $Q(\omega, d\omega')$ is a probability measure on Ω. For $\omega \in N_1 \cup N_2$, let $Q(\omega, \cdot) = \mathbb{P}(\cdot)$. This Q is the desired regular conditional probability. $\qquad \square$

Finally, we have

(5.3) Proposition. *Let Z_t be a weak solution to (5.3), T a bounded stopping time. Let Q_T be the regular conditional probability for $\mathbb{E}^x[\cdot \mid \mathcal{F}_T']$. Then, except for a $\mathbb{P}$-null set of ω, under $Q_T(\omega, \cdot)$, Z_{T+t} is a weak solution to (5.3) starting from Z_T for almost every ω.*

Proof. The principal step in the proof is to show that if $\widetilde{W}_t = W_{T+t} - W_T$, then under $Q_T(\omega, \cdot)$ the process $\widetilde{W}$ is a Brownian motion, except for a $\mathbb{P}$-null set of ω. Q_T is a probability measure on Ω', so $\widetilde{W}$ is continuous. Let $t_1 < \cdots < t_n$ and

$$N(u_2, \ldots, u_n, t_1, \ldots, t_n)$$

$$= \left\{ \omega : \mathbb{E}_{Q_T} \exp\left(i \sum_{j=2}^{n} u_j \cdot (W_{T+t_j} - W_{T+t_{j-1}}) \right) \right.$$

$$\left. \neq \exp\left(\sum_{j=2}^{n} |u_j|^2 (t_j - t_{j-1})/2 \right) \right\}.$$

Here $\cdot$ denotes the dot product in $\mathbb{R}^d$. By the strong Markov property for W_t, this is a null set with respect to $\mathbb{P}$. Let N be the union of all such $N(u_1, \ldots, u_n, t_1, \ldots, t_n)$ for $n \geq 1$, $u_1, \ldots, u_n$ rational, and $t_1 < \ldots < t_n$ rational. By continuity, if $\omega \notin N$, then the finite dimensional distributions of $\widetilde{W}$ under $Q_T(\omega, \cdot)$ are those of a Brownian motion. By the continuity of $\widetilde{W}$, under Q_T, $\widetilde{W}$ is a Brownian motion, except for a null set of ωs.

That the process Z_{T+t} starts from Z_T under $Q_T(\omega, \cdot)$ for almost every ω is left to the reader. $\qquad\square$

By a slight abuse of notation, we will say $(\mathbb{P}^x, X_t)$ is a strong Markov family when $(\mathbb{P}^x, Z_t)$ is a strong Markov family.

6. One-dimensional case

Although we have often looked at the case where the state space is $\mathbb{R}$ instead of $\mathbb{R}^d$ for the sake of simplicity of notation, everything we have done so far has been valid in $\mathbb{R}^d$ for any d. We now look at some stronger results that hold only in the one-dimensional case.

(6.1) Theorem. *Suppose b is bounded and Lipschitz. Suppose there exists a continuous function $\rho : [0, \infty) \to [0, \infty)$ such that $\rho(0) = 0$, $\int_{0+} \rho^{-2}(u)\, du = \infty$, and σ is bounded and satisfies*

$$|\sigma(x) - \sigma(y)| \leq \rho(|x - y|)$$

for all x and y. Then the solution to (2.1) is pathwise unique.

Proof. Let $a_n \downarrow 0$ be selected so that $\int_{a_n}^{a_{n-1}} du/\rho^2(u) = n$. Let h_n be continuous, supported in (a_n, a_{n-1}), $0 \leq h_n(u) \leq 2/n\rho^2(u)$, and $\int_{a_n}^{a_{n-1}} h_n(u)\, du = 1$ for each n. Let g_n be such that $g_n(0) = g_n'(0) = 0$ and $g_n'' = h_n$. Note $|g_n'(u)| \leq 1$ and $g_n'(u) = 1$ if $u \geq a_{n-1}$, hence $g_n(u) \uparrow u$ for $u \geq 0$.

Let X_t and X_t' be two solutions to (2.1). The function g_n is in C^2 and is 0 in a neighborhood of 0. We apply Itô's formula to $g_n((\varepsilon^2 + |X_t - X_t'|^2)^{1/2})$ and let $\varepsilon \to 0$ to obtain

$$g_n(|X_t - X_t'|) = \text{ martingale } + \int_0^t g_n'(|X_s - X_s'|)[b(X_s) - b(X_s')]\,ds$$

$$+ \frac{1}{2}\int_0^t g_n''(|X_s - X_s'|)[\sigma(X_s) - \sigma(X_s')]^2\,ds.$$

We take the expectation of the right-hand side. The martingale term has 0 expectation. The next term has expectation bounded by

$$c_1 \int_0^t \mathbb{E}\,|X_s - X_s'|\,ds.$$

The final term on the right-hand side is bounded in expectation by

$$\frac{1}{2}\mathbb{E}\int_0^t \frac{2}{n(\rho|X_s - X_s'|)^2}(\rho|X_s - X_s'|)^2\,ds \leq \frac{t}{n}.$$

Letting $n \to \infty$,

$$\mathbb{E}\,|X_t - X_t'| \leq c_1 \int_0^t \mathbb{E}\,|X_s - X_s'|\,ds.$$

By Gronwall's lemma, $\mathbb{E}\,|X_t - X_t'| = 0$ for each t. By the continuity of X_t and X_t', we deduce the uniqueness. $\square$

We shall see in Chapter IV that the integral condition on ρ cannot be weakened. There are, however, other related theorems. If σ is bounded below and is of finite quadratic variation, then pathwise uniqueness holds. See Rogers and Williams [1] for a presentation of this result.

A proof similar to that of Theorem 6.1 gives a useful comparison theorem. Suppose σ satisfies the conditions in Theorem 6.1. Suppose X_t satisfies (2.1) with b Lipschitz. Suppose Y_t is a continuous semimartingale satisfying $dY_t \geq \sigma(Y_t)\,dW_t + B(Y_t)\,dt$, $Y_0 = y$. This means

$$Y_t \geq Y_0 + \int_0^t \sigma(Y_s)\,dW_s + \int_0^t B(Y_s)\,ds.$$

(6.2) Theorem. *Suppose X and Y are as described. If $b(z) \leq B(z)$ for all z and $x \leq y$, then $X_t \leq Y_t$ almost surely for all t.*

Proof. Let h_n and g_n be as in the proof of Theorem 6.1. Since $x \leq y$, then $g_n(x - y) = 0$, and we have

$$g_n(X_t - Y_t) \leq \text{ martingale } + \int_0^t g_n'(X_s - Y_s)[b(X_s) - B(Y_s)]\,ds$$

$$+ \frac{1}{2}\int_0^t g_n''(X_s - Y_s)[\sigma(X_s) - \sigma(Y_s)]^2\,ds.$$

As before, the expectation of the third term is less than t/n, which tends to 0 an $n \to \infty$. The expectation of the second term on the right is bounded by

$$\mathbb{E} \int_0^t g_n'(X_s - Y_s)[b(X_s) - b(Y_s)] \, ds$$

$$+ \mathbb{E} \int_0^t g_n'(X_s - Y_s)[b(Y_s) - B(Y_s)] \, ds$$

$$\leq c_1 \mathbb{E} \int_0^t 1_{[0,\infty)}(X_s - Y_s)|X_s - Y_s| \, ds$$

$$\leq c_1 \mathbb{E} \int_0^t (X_s - Y_s)^+ \, ds.$$

Letting $n \to \infty$,

$$\mathbb{E} \, (X_t - Y_t)^+ \leq c_1 \int_0^t \mathbb{E} \, (X_s - Y_s)^+ \, ds.$$

Gronwall's lemma implies $\mathbb{E} \, (X_t - Y_t)^+ = 0$ for all t. Using the continuity of the paths of X_t and Y_t completes the proof. $\qquad\qquad\square$

Regarding weak uniqueness of one-dimensional SDEs, we will see later (Chapter IV) that weak uniqueness holds if σ and b are bounded and measurable and σ is bounded below.

7. Examples

Ornstein-Uhlenbeck process. The Ornstein-Uhlenbeck process is the solution to the SDE

$$dX_t = dW_t - \frac{X_t}{2} \, dt, \qquad X_0 = x. \tag{7.1}$$

The existence and uniqueness follow from Theorem 3.5, so $(\mathbb{P}^x, X_t)$ is a strong Markov process.

The equation (7.1) can be solved explicitly. Rewriting it and using the product rule,

$$e^{t/2} \, dW_t = e^{t/2} \, dX_t + e^{t/2} \frac{X_t}{2} dt = d[e^{t/2} X_t],$$

or

$$X_t = e^{-t/2} x + e^{-t/2} \int_0^t e^{s/2} \, dW_s. \tag{7.2}$$

Since the integrand of the stochastic integral is deterministic, it follows that X_t is a Gaussian process and the distribution of X_t is that of a normal random variable with mean $e^{-t/2} x$ and variance equal to $e^{-t} \int_0^t e^s \, ds = 1 - e^{-t}$.

If we let $Y_t = \int_0^t e^{s/2} \, dW_s$ and $V_t = Y(\log(t+1))$, then Y_t is a mean 0 continuous Gaussian process with independent increments, and hence so is V_t. Since the variance of $V_u - V_t$ is $\int_{\log(t+1)}^{\log(u+1)} e^s \, ds = u - t$, then V_t is a Brownian motion. Hence $X_t = e^{-t/2} x + e^{-t/2} V(e^t - 1)$. This representation of an Ornstein-Uhlenbeck process in terms of a Brownian motion is useful for, among other things, calculating the exit probabilities of a square root boundary; see Bass and Burdzy [1].

Bessel processes. A Bessel process of order $\nu \geq 0$ will be defined to be a nonnegative solution of the SDE

$$dX_t = dW_t + \frac{\nu - 1}{2X_t} dt, \qquad X_0 = x. \tag{7.3}$$

Let us first prove the existence of a finite solution. Define

$$dX_t^n = dW_t + \left[\left(n \wedge \frac{\nu - 1}{2X_t^n} \right) 1_{(X_t^n > 0)} + n1_{(X_t^n \leq 0)} \right] dt, \qquad X_0^n = x.$$

By Theorem 6.2, $X_t^n \geq X_t^m$ for all t if $n \geq m$. Let $X_t = \sup_n X_t^n$. To see that X_t is finite, use Itô's formula for $f(z) = z^2$ to obtain

$$\begin{aligned}
\mathbb{E} \left(X_{t \wedge T_N}^n \right)^2 &= x^2 + \mathbb{E}\left(t \wedge T_N \right) \\
&\quad + 2\mathbb{E} \int_0^{t \wedge T_N} \left[\left(nX_s^n \wedge \frac{\nu - 1}{2} \right) 1_{(X_s^n > 0)} + nX_s^n 1_{(X_s^n \leq 0)} \right] dt \\
&\leq x^2 + t + 2\mathbb{E} \int_0^{t \wedge T_N} \frac{\nu - 1}{2} dt \\
&\leq x^2 + \nu t,
\end{aligned}$$

where $T_N = \inf\{t : |X_t| \geq N\}$. Letting $N \to \infty$ and then $n \to \infty$, Fatou's lemma tells us that $\mathbb{E} X_t^2 < \infty$. In particular, $\int_0^t (\nu - 1)/(2X_s)\, ds < \infty$ a.s., or X_t is a semimartingale.

Next we show $X_t \geq 0$ a.s. Suppose $\varepsilon > 0$. If b_n is a nonnegative continuous function that is equal to n for $x \leq 0$, equal to 0 for $x > 1/n$, bounded above by n, and $b_n(x) \leq (\nu - 1)/2x$ for all positive x, then

$$f(x) = \int_0^t \exp \left(-2 \int_0^x b_n(y)\, dy \right) dx$$

solves

$$\frac{1}{2} f''(x) + b_n(x) f'(x) = 0$$

(cf. the scale function of Chapter IV). If

$$dY_t^n = dW_t + b_n(Y_t)\, dt, \qquad Y_0 = x,$$

Theorem 6.2 tells us that $X_t \geq X_t^n \geq Y_t^n$ for all t a.s. By Itô's formula, $f(Y_t^n)$ is a martingale. Hence

$$\mathbb{P}(X_t \text{ hits } -\varepsilon \text{ before } N)$$
$$\leq \mathbb{P}(f(Y_t^n) \text{ hits } -(e^{2n\varepsilon} - 1)/2n \text{ before } 2N),$$

which tends to 0 as $n \to \infty$ by [PTA, Corollary I.4.10]. Since N is arbitrary, then $\mathbb{P}(X_t \text{ hits } -\varepsilon) = 0$.

It is easy to see that the X_t we constructed satisfies (7.3).

We now have existence; let us turn to uniqueness. If $Y_t = X_t^2$, then

$$dY_t = 2X_t \, dW_t + (\nu - 1) \, dt + dt,$$

or

$$dY_t = 2(Y_t^+)^{1/2} \, dW_t + \nu \, dt. \tag{7.4}$$

The solution to (7.4) is unique by Theorem 6.1 because $|x^{1/2} - y^{1/2}| \leq \rho(|x - y|)$ with $\rho(z) = z^{1/2}$. Thus there is only one nonnegative solution to (7.3). We also have the existence of a solution to (7.4) by using the existence of a solution to (7.3). The process Y_t that solves (7.4) is also of considerable importance and is known as the square of a Bessel process.

The squares of Bessel processes possess a useful additive property.

(7.1) Proposition. *Suppose X_t^i is the square of a Bessel process of order ν_i starting at x_i, $i = 1, 2$, and X_t^1 and X_t^2 are independent. Then $X_t^1 + X_t^2$ is the square of a Bessel process of order $\nu_1 + \nu_2$ starting at $x_1 + x_2$.*

Proof. If $dX_t^i = (X_t^i)^{1/2} \, dW_t^i + \nu_i \, dt$, $i = 1, 2$, let W_t be defined by

$$dW_t = \left(\frac{X_t^1}{X_t^1 + X_t^2}\right)^{1/2} dW_t^1 + \left(\frac{X_t^2}{X_t^1 + X_t^2}\right)^{1/2} dW_t^2.$$

By Lévy's theorem (Section 1), W_t is a Brownian motion. Noting

$$(X_t^1 + X_t^2)^{1/2} \, dW_t = (X_t^1)^{1/2} \, dW_t^1 + (X_t^2)^{1/2} \, dW_t^{1/2},$$

the result follows by summing the SDEs for X_t^1 and X_t^2. $\qquad\square$

From (7.3), a Bessel process of order 1 is the same as the absolute value of a Brownian motion. By the use of Proposition 7.1, then, the modulus of a d-dimensional Brownian motion is the same as a Bessel process of order d.

Bessel processes have the same scaling properties as Brownian motion. That is, if X_t is a Bessel process of order ν started at x, then $aX_{a^{-2}t}$ is a Bessel process of order ν started at ax. In fact, from (7.3),

$$d(aX_{a^{-2}t}) = a \, dW_{a^{-2}t} + a^2 \frac{\nu - 1}{2aX_{a^{-2}t}} \, d(a^{-2}t),$$

and the assertion follows from the uniqueness and the fact that $aW(a^{-2}t)$ is again a Brownian motion.

Bessel processes are useful for comparison purposes, and so the following is worth recording.

(7.2) Proposition. *Suppose X_t is a Bessel process of order ν.*
(i) If $\nu > 2$, X_t never hits 0 and $|X_t| \to \infty$ a.s.
(ii) If $\nu = 2$, X_t hits every neighborhood of 0 infinitely often, but never hits 0.
(iii) If $0 < \nu < 2$, X_t hits 0 infinitely often.
(iv) If $\nu = 0$, then X_t hits 0 and then remains at 0 thereafter.

When we say that X_t hits 0, we consider only times $t > 0$.

Proof. When $\nu = 2$, X_t has the same law as a 2-dimensional Brownian motion, and (ii) follows from the corresponding facts about 2-dimensional Brownian motion. Suppose $\nu \neq 2$; by Itô's formula, $(X_t)^{2-\nu}$ is a martingale. Assertions (i) and (iii) now follow from the same proof as [PTA, Theorem I.5.8]. Similarly, a Bessel process of order 0 hits 0. If X_t is such a process and $Y_t = X_t^2$, then $dY_t = Y_t^{1/2}\, dW_t$. Starting from 0, $Y_t \equiv 0$ is evidently a solution, so by the uniqueness any solution starting at 0 must remain at 0 forever; (iv) now follows by the strong Markov property. □

Brownian bridge. Brownian motion conditioned to be at 0 at time 1 is called Brownian bridge. Brownian bridge has the same law as $W_t - tW_1$. To see this, the covariance of $W_t - tW_1$ and W_1 is 0; hence they are independent. Therefore the law of W_t conditional on W_1 being 0 is the same as the law of $W_t - tW_1 + tW_1$ conditional on W_1 being 0, which is $W_t - tW_1$ by independence.

We will see shortly that Brownian bridge can be represented as the solution of a SDE

$$dX_t = dW_t - \frac{X_s}{1-s}\, dt, \qquad X_0 = 0. \tag{7.5}$$

Although Theorem 3.5 does not apply because the drift term depends on s as well as the position X_s, the same proof as that of Theorem 3.5 guarantees uniqueness and existence for the solution of (7.5) for $s \leq t$ for any $t < 1$ (cf. remark following Theorem 3.6). As with the Ornstein-Uhlenbeck process, (7.5) may be solved explicitly. We have

$$dW_t = dX_t + \frac{X_t}{1-t}\, dt = (1-t)\, d\!\left[\frac{X_t}{1-t}\right],$$

or

$$X_t = (1-t) \int_0^t \frac{dW_s}{1-s}.$$

Thus X_t is a continuous Gaussian process with mean 0. The variance of X_t is

$$(1-t)^2 \int_0^t (1-s)^{-2}\, ds = t - t^2,$$

the same as the variance of Brownian bridge. A similar calculation shows that the covariance of X_t and X_s is the same as the covariance of $W_t - tW_1$ and $W_s - sW_1$. Hence the law of X_t and Brownian bridge are the same.

Linear equations. The equation $dX_t = AX_t\,dW_t + BX_t\,dt$ may be written $dX_t = X_t\,dY_t$, where $Y_t = AW_t + Bt$, and then it is well known that the solution is $X_t = X_0\exp(Y_t - \langle Y\rangle_t/2)$. Let us here consider the more general equation

$$dX_t = dH_t + X_t\,dY_t, \tag{7.6}$$

where Y_t is a semimartingale $M_t + A_t$ with $d\langle M\rangle_t/dt$ and dA_t/dt bounded. Again we can write an explicit solution, and existence and uniqueness are easy under our assumptions on A_t and M_t. We have the following.

(7.3) Proposition. *Suppose Y_t is as above and X_t is the solution to (7.6) with $X_0 \equiv 1$. Suppose H_t is bounded, of bounded variation, and adapted. Then the solution to*

$$dZ_t = dH_t + Z_t\,dY_t$$

is given by

$$Z_t = X_t\int_0^t X_s^{-1}dH_s.$$

Proof. Since $X_t = \exp(Y_t - \langle Y\rangle_t/2)$, then $X_t > 0$ for all t. By the product formula and the fact that $\int_0^t X_s^{-1}\,dH_s$ is of bounded variation,

$$dZ_t = X_t(X_t^{-1}dH_t) + \left(\int_0^t X_s^{-1}\,dH_s\right)(dX_t).$$

Since $dX_t = X_t\,dY_t$, the right-hand side is $dH_t + Z_t\,dY_t$, as desired. $\square$

The solutions to linear SDEs have moments of all orders. We prove a slightly stronger statement. The proof is very similar to that of Theorem 3.5.

(7.4) Proposition. *Suppose*

$$dX_t = A_t\,dW_t + B_t\,dt, \qquad X_0 = x_0,$$

where $|A_t|, |B_t| \le c_1(1+|X_t|)$. Then for all $p \ge 2$ and $t_0 > 0$ there exists $c_2(p, t_0)$ such that

$$\mathbb{E}\sup_{s\le t_0}|X_s|^p \le c_2(p, t_0).$$

Proof. Let $T_n = \inf\{t : |X_t| \ge n\}$ and let $g_n(t) = \mathbb{E}\sup_{s\le t\wedge T_n}|X_s|^p$. By the Burkholder-Davis-Gundy inequalities (1.5), Doob's inequality, and the triangle inequality,

$$\mathbb{E}\sup_{s\le t\wedge T_n}|X_s|^p \le c_3\mathbb{E}\left(\int_0^{t\wedge T_n}A_s^2\,ds\right)^{p/2} + c_4\mathbb{E}\left(\int_0^{t\wedge T_n}|B_s|\,ds\right)^p$$

$$\le c_5\mathbb{E}\int_0^{t\wedge T_n}|A_s|^p\,ds + c_5\mathbb{E}\int_0^{t\wedge T_n}|B_s|^p\,ds$$

$$\le c_6 + c_6\mathbb{E}\int_0^{t\wedge T_n}|X_s|^p\,ds,$$

or

$$g_n(t) \le c_6 + c_6 \int_0^t g_n(s)\, ds.$$

By Gronwall's lemma, $g_n(t) \le c_6 e^{c_6 t} \le c_6 e^{c_6 t_0}$, where c_6 is independent of n. Letting $n \to \infty$ and using Fatou's lemma completes the proof. $\square$

8. Some estimates

We collect a few facts and estimates about solutions to SDEs that we will need later.

(8.1) Proposition. *Suppose X_t solves (2.3) with σ and b bounded. There exist c_1 and c_2 depending only on $|\sigma|$ such that*

$$\mathbb{P}(\sup_{s \le t} |X_s - X_0| > \lambda + \|b\|_\infty t) \le c_1 \exp(-c_2 \lambda^2 / t).$$

Proof. Since $|\int_0^t b(X_s)\, ds| \le \|b\|_\infty t$, our result will follow if we show that for each i,

$$\mathbb{P}\left(\sup_{s \le t} \left| \int_0^s \sum_j \sigma_{ij}(X_r)\, dW_r^j \right| > \lambda / \sqrt{d} \right) \le c_3 \exp(-c_4 \lambda^2 / dt). \tag{8.1}$$

The martingale $\int_0^t \sum_j \sigma_{ij}(X_r)\, dW_r^j$ has quadratic variation bounded by $\int_0^t |\sigma\sigma^T(X_r)|\, dr$, and (8.1) follows by [PTA, Exercise I.8.13]. $\square$

We will need to know that X_t exits from bounded domains with probability one. Suppose X_t solves (2.3).

(8.2) Proposition. *Suppose X_t solves (2.3) and $\sigma, b,$ and σ^{-1} are bounded. If $N > 0$, then $\mathbb{P}(|X_t| \text{ exits } B(0, N)) = 1$.*

Proof. Without loss of generality, we may assume the process starts at 0. Let $a(x) = (\sigma\sigma^T)_{11}(x)$. We look at the first component of X_t:

$$dX_t^1 = \sum_{j=1}^d \sigma_{1j}(X_t)\, dW_t^j + b_1(X_t)\, dt.$$

Let M_t be the martingale term; M_t has quadratic variation

$$d\langle M \rangle_t = \sum_{j,k} \sigma_{1j}(X_t)\sigma_{1k}(X_t)\, d\langle W^j, W^k \rangle_t$$

$$= \sum_j (\sigma_{1j}\sigma_{j1}^T)(X_t)\, dt = a(X_t)\, dt$$

since $d\langle W^j, W^k\rangle_t = \delta_{jk}\,dt$, where δ_{jk} is 1 if $j = k$ and 0 otherwise. Since σ^{-1} is bounded, $a \geq c_1$ for some constant c_1. If we let $B_t = \inf\{u : \langle M\rangle_u \geq t\}$, then $\widetilde{W}_t = M_{B_t}$ is a continuous martingale with quadratic variation equal to t; hence by Lévy's theorem (Section 1), $\widetilde{W}_t$ is a Brownian motion. So $X_{B_t}^1$ is a semimartingale of the form $\widetilde{W}_t + \int_0^t e_s\,ds$, where e_s is bounded (cf. [PTA, Theorem I.5.11]). Let us define a new probability measure $\mathbb{Q}$ by

$$d\mathbb{Q}/d\mathbb{P} = \exp\left(-\int_0^t e_s\,d\widetilde{W}_s - \frac{1}{2}\int_0^t e_s^2\,ds\right)$$

on $\mathcal{F}_t$. By the Girsanov transformation (see Section 1),

$$X_{B_t}^1 = \widetilde{W}_t + \int_0^t e_s\,ds$$

$$= \widetilde{W}_t - \left\langle \int_0^{\cdot} (-e_s)\,d\widetilde{W}_s, \widetilde{W}\right\rangle_t$$

is a martingale under $\mathbb{Q}$. Moreover, its quadratic variation under $\mathbb{Q}$ is the same as its quadratic variation under $\mathbb{P}$, namely, t. By Lévy's theorem (Section 1), $X_{B_t}^1$ is a Brownian motion under $\mathbb{Q}$. Therefore $X_{B_t}^1$ exits $[-N, N]$, $\mathbb{Q}$-a.s. Since $\mathbb{P}$ and $\mathbb{Q}$ are equivalent measures, under $\mathbb{P}$ the process $X_{B_t}^1$ also exits $[-N, N]$ a.s. Since $d\langle M\rangle_t/dt$ is bounded above and below, it follows that X_t^1 exits $[-N, N]$ a.s.　$\square$

An important property of X_t is that it satisfies a support theorem. Suppose X_t satisfies (2.3). We suppose that σ, σ^{-1}, and b are bounded, but we impose no other smoothness conditions. Let $a = \sigma\sigma^T$.

(8.3) Lemma. *Suppose $Y_t = M_t + A_t$ is a continuous semimartingale with dA_t/dt and $d\langle M\rangle_t/dt$ bounded above by c_1 and $d\langle M\rangle_t/dt$ bounded below by $c_2 > 0$. If $\varepsilon > 0$ and $t_0 > 0$, then*

$$\mathbb{P}(\sup_{s \leq t_0} |Y_s| < \varepsilon) \geq c_3,$$

where $c_3 > 0$ depends only on c_1, c_2, ε, and t_0.

Proof. Let $B_t = \inf\{u : \langle M\rangle_u > t\}$. Then $W_t = M_{B_t}$ is a continuous martingale with quadratic variation equal to t; hence by Lévy's theorem (Section 1), W_t is a Brownian motion. If $Z_t = Y_{B_t} = W_t + E_t$, then $E_t = \int_0^t e_s\,ds$ for some e_s bounded by c_4, where c_4 depends only on c_1 and c_2. Our assertion will follow if we can show

$$\mathbb{P}(\sup_{s \leq c_1 t_0} |Z_s| < \varepsilon) \geq c_3.$$

We now use Girsanov's theorem. Define a probability measure $\mathbb{Q}$ by

$$d\mathbb{Q}/d\mathbb{P} = \exp\left(-\int_0^{t_0} e_s\,dW_s - \frac{1}{2}\int_0^{t_0} e_s^2\,ds\right)$$

on $\mathcal{F}_{t_0}$. Under $\mathbb{P}$, W_t is a martingale, so under $\mathbb{Q}$ we have that

$$W_t - \left\langle \int_0^{\cdot} (-e_s)\, dW, W \right\rangle_t = W_t + \int_0^t e_s\, ds$$

is a martingale with the same quadratic variation as W has under $\mathbb{P}$, namely, t. Then under $\mathbb{Q}$, Z_t is a Brownian motion. By a well known property of Brownian motion (see [PTA, Proposition I.6.5]),

$$\mathbb{Q}(\sup_{s \leq c_1 t_0} |Z_s| < \varepsilon) \geq c_5,$$

for c_5 depending only on ε and $c_1 t_0$. So if C is the event $\{\sup_{s \leq c_1 t_0} |Z_s| < \varepsilon\}$,

$$c_5 \leq \mathbb{Q}(C) = \int_C (d\mathbb{Q}/d\mathbb{P})\, d\mathbb{P} \leq \left(\mathbb{E}\, (d\mathbb{Q}/d\mathbb{P})^2 \right)^{1/2} \left(\mathbb{P}(C) \right)^{1/2}$$

by the Cauchy-Schwarz inequality. The proof is concluded by noting that $d\mathbb{Q}/d\mathbb{P}$ has a second moment depending only on c_4 and t_0. $\qquad\square$

We use this lemma to obtain an analogous result for X_t.

(8.4) Theorem. *Let $\varepsilon \in (0,1)$, $t_0 > 0$. There exists c_1 depending only on the upper bounds of σ, b, and σ^{-1} such that*

$$\mathbb{P}(\sup_{s \leq t_0} |X_s - X_0| < \varepsilon) \geq c_1.$$

Proof. For notational simplicity assume $X_0 = 0$. Let $y = (\varepsilon/4, 0, \ldots, 0)$. Applying Itô's formula with $f(z) = |z - y|^2$ and setting $V_t = |X_t - y|^2$, then $V_0 = (\varepsilon/4)^2$ and

$$dV_t = 2 \sum_i (X_t^i - y_i)\, dX_t^i + \sum_i d\langle X^i \rangle_t.$$

If we set Y_t equal to V_t for $t \leq \inf\{u : |V_u| > (\varepsilon/2)^2\}$ and equal to some Brownian motion for t larger than this stopping time, then Lemma 8.3 applies and

$$\mathbb{P}(\sup_{s \leq t_0} |V_s - V_0| \leq (\varepsilon/8)^2) = \mathbb{P}(\sup_{s \leq t_0} |Y_s - Y_0| \leq (\varepsilon/8)^2) \geq c_2.$$

By the definition of y and V_t, this implies with probability at least c_2 that X_t stays inside $B(0, \varepsilon)$. $\qquad\square$

We can now prove the *support theorem* for X_t.

(8.5) Theorem. *Suppose σ and b are bounded, σ^{-1} is bounded, $x \in \mathbb{R}^d$, and X_t satisfies (2.3) with $X_0 = x$. Suppose $\psi : [0,t] \to \mathbb{R}^d$ is continuous with $\psi(0) = x$ and $\varepsilon > 0$. There exists c_1, depending only on ε, t, the modulus of continuity of ψ, and the bounds on b and σ such that*

$$\mathbb{P}(\sup_{s\leq t}|X_s - \psi(s)| < \varepsilon) \geq c_1.$$

This can be phrased as saying the graph of X_s stays inside an ε-tube about ψ. By this we mean, if $G_\psi^\varepsilon = \{(s,y) : |y - \psi(s)| < \varepsilon, s \leq t\}$, then $\{(s, X_s) : s \leq t\}$ is contained in G_ψ^ε with positive probability.

Proof. We can find a differentiable function $\widehat{\psi}$ such that $\widehat{\psi}(0) = x$ and the $\varepsilon/2$ tube about $\widehat{\psi}$ (which is $G_{\widehat{\psi}}^{\varepsilon/2}$ in the above notation) is contained in G_ψ^ε, the ε-tube about ψ. So without loss of generality, we may assume ψ is differentiable with a derivative bounded by a constant, say c_2.

Define a new probability measure $\mathbb{Q}$ by

$$d\mathbb{Q}/d\mathbb{P} = \exp\left(-\int_0^t \psi'(s)\sigma^{-1}(X_s)\,dW_s - \frac{1}{2}\int_0^t |\psi'(s)\sigma^{-1}(X_s)|^2\,ds\right)$$

on $\mathcal{F}_t$. We see that

$$\left\langle -\int_0^\cdot \psi'(s)\sigma^{-1}(X_s)\,dW_s, X\right\rangle$$
$$= \left\langle -\int_0^\cdot \psi'(s)\sigma^{-1}(X_s)\,dW_s, \int^\cdot \sigma(X_s)\,dW_s\right\rangle$$
$$= -\int_0^t \psi'(s)\,ds = -\psi(t) + \psi(0).$$

So by the Girsanov theorem (Section 1), under $\mathbb{Q}$ each component of X_t is a semimartingale and $X_t^i - \int_0^t b_i(X_s)\,ds - \psi_i(t)$ is a martingale for each i. Furthermore, if

$$\widehat{W}_t = \int_0^t \sigma^{-1}(X_t)\,[dX_t - b(X_t)\,dt - \psi'(t)\,dt],$$

each component of $\widehat{W}$ is a continuous martingale, and a calculation shows that $d\langle \widehat{W}^i, \widehat{W}^j\rangle_t = \delta_{ij}\,dt$ under $\mathbb{Q}$. Therefore $\widehat{W}$ is a d-dimensional Brownian motion under $\mathbb{Q}$. Since

$$d(X_t - \psi(t)) = \sigma(X_t)\,d\widehat{W}_t + b(X_t)\,dt,$$

then by Theorem 8.4,

$$\mathbb{Q}(\sup_{s\leq t}|X_s - \psi(s)| < \varepsilon) \geq c_3.$$

Very similarly to the last paragraph of the proof of Lemma 8.3, we conclude

$$\mathbb{P}(\sup_{s\leq t}|X_s - \psi(s)| < \varepsilon) \geq c_4. \qquad \square$$

We show how X_t scales. By the scaling property of Brownian motion, if W_t is a Brownian motion and $a > 0$, then $\widetilde{W}_t = aW_{t/a^2}$ is again a Brownian motion.

(8.6) Proposition. *Suppose X_t solves (2.3). If $a > 0$, $\widetilde{W}_t = aW_{t/a^2}$, $\sigma_a(x) = \sigma(a^{-1}x)$, and $b_a(x) = a^{-1}b(a^{-1}x)$, then $Y_t = aX_{t/a^2}$ solves*

$$dY_t = \sigma_a(Y_t)\, d\widetilde{W}_t + b_a(Y_t)\, dt, \qquad Y_0 = aX_0.$$

Proof. We write

$$Y_t = aX_{t/a^2} = aX_0 + \int_0^{t/a^2} a\sigma(X_s)\, dW_s + \int_0^{t/a^2} ab(X_s)\, ds$$

$$= aX_0 + \int_0^t a\sigma(X_{r/a^2})\, dW_{r/a^2} + \int_0^t a^{-1}b(X_{r/a^2})\, dr$$

$$= aX_0 + \int_0^t \sigma_a(Y_r)\, d\widetilde{W}_r + \int_0^t b_a(Y_r)\, dr. \qquad \square$$

9. Stratonovich integrals

For stochastic differential geometry and for the Malliavin calculus, the Stratonovich integral is more convenient than the Itô integral. If X and Y are continuous semimartingales, the *Stratonovich integral*, denoted $\int_0^t X_s \circ dY_s$, is defined by

$$\int_0^t X_s \circ dY_s = \int_0^t X_s\, dY_s + \frac{1}{2}\langle X, Y\rangle_t.$$

Both the beauty and the difficulty of Itô's formula are due to the quadratic term. The change of variables for the Stratonovich integral avoids this.

(9.1) Theorem. *Suppose $f \in C^3$ and X is a continuous semimartingale. Then*

$$f(X_t) = f(X_0) + \int_0^t f'(X_s) \circ dX_s.$$

Proof. By Itô's formula applied to the function f and the definition of the Stratonovich integral, it suffices to show that

$$\langle f'(X), X\rangle_t = \int_0^t f''(X_s)\, d\langle X\rangle_s. \tag{9.1}$$

Applying Itô's formula to the function f', which is in C^2,

$$f'(X_t) = f'(X_0) + \int_0^t f''(X_s)\, dX_s + \frac{1}{2} \int_0^t f'''(X_s)\, d\langle X \rangle_s,$$

from which (9.1) follows immediately.   □

If X and Y are continuous semimartingales and we apply the change of variables formula with $f(x) = x^2$ to $X + Y$ and $X - Y$, we obtain

$$d(X_t + Y_t)^2 = 2(X_t + Y_t) \circ d(X_t + Y_t)$$

and

$$d(X_t - Y_t)^2 = 2(X_t - Y_t) \circ d(X_t - Y_t).$$

Summing and then dividing by 4, we have the *product formula for Stratonovich integrals*

$$X_t Y_t = X_0 Y_0 + \int_0^t X_s \circ dY_s + \int_0^t Y_s \circ dX_s. \tag{9.2}$$

The Stratonovich integral $\int H_s \circ dX_s$ can be represented as a limit of Riemann sums.

(9.2) Proposition. *Suppose $\{s_0, \ldots, s_n\}$ are partitions of $[0, t]$ whose mesh size tends to 0 and H_s is a continuous semimartingale. Then $\int_0^t H_s \circ dX_s$ is the limit in probability of*

$$\sum_{i=0}^{n-1} \frac{H_{s_i} + H_{s_{i+1}}}{2} (X_{s_{i+1}} - X_{s_i}).$$

Proof. We write the sum as

$$\sum H_{s_i}(X_{s_{i+1}} - X_{s_i}) + \frac{1}{2}(H_{s_{i+1}} - H_{s_i})(X_{s_{i+1}} - X_{s_i}).$$

The first sum tends to $\int_0^t H_s\, dX_s$ while by [PTA, Theorem I.4.2] the second sum tends to $(1/2)\langle H, X \rangle_t$.   □

10. Flows

Let $X(x, t, \omega)$ denote the solution to

$$dX_t = \sigma(X_t)\, dW_t + b(X_t)\, dt, \qquad X_0 = x.$$

If σ and b are Lipschitz, then $X(x, t)$ will be continuous in x.

(10.1) Theorem. *If σ and b are Lipschitz, then there exist versions of $X(x, t)$ that are jointly continuous in x and t a.s.*

Two processes $X(x,t)$ and $X'(x,t)$ are versions of each other if for each x and t we have $\mathbb{P}(X(x,t) \neq X'(x,t)) = 0$. The null set may depend on x and t.

Proof. We have

$$X(x,t) - X(y,t) = x - y + \int_0^t [\sigma(X(x,s)) - \sigma(X(y,s))]\, dW_s \qquad (10.1)$$

$$+ \int_0^t [b(X(x,s)) - b(X(y,s))]\, ds.$$

Let $t_0 > 0$. Suppose M is the Lipschitz constant of σ, that is, $|\sigma(x) - \sigma(y)| \leq M|x-y|$ for all x and y. If F_t denotes the stochastic integral, p is a positive integer, and $t \leq t_0$, by the Burkholder-Davis-Gundy inequalities,

$$\mathbb{E} \sup_{s \leq t} |F_s|^{2p} \leq c_1 \mathbb{E}\left[\int_0^t \left(\sigma(X(x,s)) - \sigma(X(y,s))\right)^2 ds \right]^p$$

$$\leq c_1 M^{2p} \mathbb{E} \left(\int_0^t |X(x,s) - X(y,s)|^2\, ds \right)^p$$

$$\leq c_2 \mathbb{E} \int_0^t |X(x,s) - X(y,s)|^{2p}\, ds.$$

The expression involving the b terms is handled similarly (cf. proof of Theorem 3.1). So we have, with $g(t) = \mathbb{E} \sup_{s \leq t} |X(x,s) - X(y,s)|^{2p}$, that

$$g(t) \leq c_3 |x - y|^{2p} + c_4 \int_0^t g(s)\, ds, \qquad t \leq t_0.$$

By Gronwall's lemma,

$$\mathbb{E} \sup_{s \leq t_0} |X(x,s) - X(y,s)|^{2p} \leq c_5 |x - y|^{2p}. \qquad (10.2)$$

Recall Kolmogorov's theorem: if $\mathbb{E}|Y_t - Y_s|^p \leq c_6 |t - s|^{1+\varepsilon}$ for some $\varepsilon > 0$ and all $s, t \geq 0$, then $\{Y_t, t \in D\}$ is uniformly continuous a.s. Here D is the dyadic rationals; see, e.g., [PTA, Theorem I.3.11]. The same proof shows that one is not required to have the index set be $[0, \infty)$. If $\mathbb{E}|Y_x - Y_y|^p \leq c_7 |x - y|^{d+\varepsilon}$ for $x, y \in \mathbb{R}^d$, then $\{Y_x, x \in D\}$ is uniformly continuous, where here D is the collection of points in $\mathbb{R}^d$ all of whose coordinates are dyadic rationals. The proof also shows that we may replace $|\cdot|$ by any metric or norm. If we use the norm $\|Y\| = \sup_{s \leq t_0} |Y_s|$, then (10.2) says that $\mathbb{E}\|X(x,\cdot) - X(y,\cdot)\|^{2p} \leq c_5 |x - y|^{2p}$. So taking p large enough, the extension to $\mathbb{R}^d$ of Kolmogorov's theorem implies that $\{X(x,\cdot), x \in D\}$ is uniformly continuous a.s. Define $\widetilde{X}(x,t) = \lim_{s \to t} X(x,s)$. Then $\widetilde{X}$ is jointly continuous in x and t a.s. In view of (10.2), $\widetilde{X}(x,\cdot) = X(x,\cdot)$ a.s., and in particular $\widetilde{X}$ is a solution to the same SDE that X is. $\widetilde{X}$ is the desired version. $\qquad \square$

The collection of processes $X(x,t)$ is called a *flow*.

If σ and b are smoother functions, then $X(x, t)$ will be smoother in x. Let us suppose for now that we are in dimension one. If in (10.1) we divide both sides by $x - y$, let $y \to x$, and use the chain rule, formally we obtain

$$
\begin{aligned}
dX(x, t)/dx \\
= 1 + \int_0^t \sigma'(X(x, s))(dX(x, s)/dx)\, dW_s \\
+ \int_0^t b'(X(x, s))(dX(x, s)/dx)\, ds.
\end{aligned}
$$

To make this more precise, suppose σ and b are in C^2 and are bounded with bounded first and second derivatives and consider the SDE

$$
dY_t = \sigma'(X(x, t))Y_t\, dW_t + b'(X(x, t))Y_t\, dt, \qquad Y_0 = 1. \tag{10.3}
$$

(10.2) Theorem. *A pathwise solution to (10.3) exists and is unique. The solution has moments of all orders. If $(DX)(x, t)$ denotes the solution, versions of $(DX)(x, t)$ exist that are jointly continuous in x and t.*

Proof. Let us prove uniqueness of (10.3). Let $t_0 > 0$ and $N > 0$. If Y_t and Y_t' are two solutions and $T_N = \inf\{t : |Y_t| \text{ or } |Y_t'| \geq N\}$, let $g(t) = \mathbb{E}\sup_{s \leq t \wedge T_N} |Y_s - Y_s'|^2$. Observe

$$
d(Y_t - Y_t') = \sigma'(X(x, t))(Y_t - Y_t')\, dW_t + b'(X(x, t))(Y_t - Y_t')\, dt,
$$

so as in Section 3, $g(t) \leq c_1 \int_0^t g(s)\, ds$, if $t \leq t_0$, hence $g(t) = 0$. Since t_0 and N are arbitrary, this proves uniqueness.

Existence can also be proved by similar modifications to the proofs in Section 3.

Let $t \leq t_0$. If p is a positive integer and $T_N = \inf\{t : |Y_t| \geq N\}$, by the inequalities of Burkholder-Davis-Gundy,

$$
\begin{aligned}
\mathbb{E}\sup_{s \leq t \wedge T_N} |Y_s|^p &\leq c_2 + c_3 \mathbb{E}\left(\int_0^{t \wedge T_N} (\sigma'(X(x, s)))^2 Y_s^2\, ds \right)^{p/2} \\
&\quad + c_4 \mathbb{E}\left(\int_0^{t \wedge T_N} |b'(X(x, s))|\, |Y_s|\, ds \right)^p \\
&\leq c_2 + c_5 \mathbb{E}\int_0^{t \wedge T_N} |Y_s|^p\, ds.
\end{aligned}
$$

By Gronwall's lemma, $\mathbb{E}\sup_{s \leq t \wedge T_N} |Y_s|^p \leq c_6$, where c_6 depends on t_0 but not N. Letting $N \to \infty$ proves the moment assertion.

We now turn to the proof of the existence of jointly continuous versions. We obtain an estimate on $\mathbb{E}\,|(DX)(x, t) - (DX)(y, t)|^p$. Writing the SDE that $(DX)(x, t) - (DX)(y, t)$ satisfies, the stochastic integral term can be written

$$\int_0^t [\sigma'(X(x,s)) - \sigma'(X(y,s))](DX)(x,s)\, dW_s$$

$$+ \int_0^t [(DX)(x,s) - DX(y,s)]\sigma'(X(y,s))\, dW_s.$$

The pth moment of the second integral can be bounded by

$$c_7 \|\sigma'\|_\infty^p \int_0^t |(DX)(x,s) - DX(y,s)|^p\, ds$$

if $t \le t_0$. The pth moment of the first integral can be bounded by

$$c_8 \int_0^t \mathbb{E}\,|\sigma'(X(x,s)) - \sigma'(X(y,s))|^p |(DX)(x,s)|^p\, ds$$

$$\le c_8 \int_0^t (\mathbb{E}\,|\sigma'(X(x,s)) - \sigma'(X(y,s))|^{2p})^{1/2} (\mathbb{E}\,|(DX)(x,s)|^{2p})^{1/2}\, ds$$

$$\le c_9 \int_0^t \|\sigma'\|_\infty^p (\mathbb{E}\,|X(x,s) - X(y,s)|^{2p})^{1/2}\, ds$$

$$\le c_{10}|x - y|^p$$

if $t \le t_0$, using (10.2). The terms involving b' can be handled similarly. Using Gronwall's lemma,

$$\mathbb{E}\,|(DX)(x,t) - (DX)(y,t)|^p \le c_{11}|x - y|^p, \qquad t \le t_0.$$

We then follow the proof of Theorem 10.1 to obtain the joint continuity. $\square$

We now prove the differentiability of $X(x,t)$.

(10.3) Theorem. *Suppose the dimension of the state space is one and σ and b are in C^2 and are bounded with bounded first and second derivatives. For each x,*

$$X(x,t) - X(y,t) = \int_y^x (DX)(z,t)\, dz, \qquad \text{a.s.}$$

Proof. For simplicity we take $b \equiv 0$. Let

$$Z(x,t) = X(x,t) - \int_0^x (DX)(z,t)\, dz.$$

Our goal is to show that Z is constant in x.

Write $Z(x,t) - Z(y,t) = F_t + G_t + H_t$, where

$$F_t = \int_0^t \big[\sigma(X(x,s)) - \sigma(X(y,s)) - \sigma'(X(x,s))(X(x,s) - X(y,s))\big]\, dW_s,$$

$$G_t = \int_0^t \sigma'(X(x,s))(Z(x,s) - Z(y,s))\, dW_s,$$

$$H_t = \int_0^t \Big[\int_y^x (DX)(z,s)(\sigma'(X(z,s)) - \sigma'(X(x,s)))\, dz\Big]\, dW_s.$$

Let $t_0 > 0$. The integrand in F_t is bounded by

$$\|\sigma''\|_\infty (X(x,s) - X(y,s))^2,$$

so by Doob's inequality and Hölder's inequality, for $t \leq t_0$,

$$\mathbb{E} \sup_{s \leq t} F_s^2 \leq c_1 \mathbb{E} \int_0^t |X(x,s) - X(y,s)|^4 \, ds,$$

which is less than $c_2 |x - y|^4$ by (10.2). We have

$$\mathbb{E} \sup_{s \leq t} G_s^2 \leq c_3 \|\sigma'\|_\infty^2 \int_0^t \mathbb{E} |Z(x,s) - Z(y,s)|^2 \, ds.$$

The integrand in H_t is bounded by

$$\|\sigma''\|_\infty \int_y^x |(DX)(z,s)| \, |X(z,s) - X(x,s)| \, dz,$$

so for $t \leq t_0$,

$$
\begin{aligned}
\mathbb{E} \sup_{s \leq t} H_s^2 \\
&\leq c_4 \int_0^t \mathbb{E} \left| \int_y^x (DX)(z,s)(X(z,s) - X(x,s)) \, dz \right|^2 ds \\
&\leq c_4 \int_0^t |x - y| \mathbb{E} \int_y^x |(DX)(z,s)|^2 |X(z,s) - X(x,s)|^2 \, dz \, ds \\
&\leq c_4 |x - y| \int_0^t \int_y^x [(\mathbb{E}((DX)(z,s))^4)^{1/2} (\mathbb{E}|X(z,s) - X(x,s)|^4)^{1/2}] \, dz \, ds \\
&\leq c_5 |x - y| \int_y^x |z - x|^2 \, dz \\
&\leq c_6 |x - y|^4,
\end{aligned}
$$

using (10.2) and Hölder's inequality. Therefore by Gronwall's lemma,

$$\mathbb{E} \sup_{s \leq t} |Z(x,t) - Z(y,t)|^2 \leq c_7 |x - y|^4.$$

We now show that Z is constant in x. Let $\lambda > 0$. For $n > 0$, let $x_i = x + i(y - x)/n$.

$$
\begin{aligned}
\mathbb{P}(|Z(x,t) - Z(y,t)| > \lambda) \\
&\leq \mathbb{P}(\exists i \leq n : |Z(x_{i+1},t) - Z(x_i,t)| > \lambda/n) \\
&\leq n \sup_{i \leq n} \mathbb{P}(|Z(x_{i+1},t) - Z(x_i,t)| > \lambda/n) \\
&\leq n \sup_{i \leq n} \frac{\mathbb{E} |Z(x_{i+1},t) - Z(x_i,t)|^2}{(\lambda/n)^2} \\
&\leq c_8 n \frac{(|y - x|/n)^4}{(\lambda/n)^2}.
\end{aligned}
$$

Since n is arbitrary, the left-hand side must be 0. Since λ is arbitrary, $Z(x,t) = Z(y,t)$ a.s. Thus Z is constant in $x \in D$, the dyadic rationals. By the continuity of Z in x, Z is constant. This implies the result.   □

As has often before been the case, we have taken the case of dimension one for simplicity of notation only. The above proofs were constructed so that they work for any dimension; the principal difference in higher dimensions is describing the derivative. If X_t is d-dimensional, we must consider the d partial derivatives of each of the d components. Thus (DX) becomes a $d \times d$ matrix, and it is the solution to

$$(DX)(x,t) = I + \int_0^t \sum_{k=1}^d (DX)(x,s)\sigma'_k(X(x,s))\, dW_s^k \tag{10.4}$$
$$+ \int_0^t (DX)(x,s)b'(X(x,s))\, ds,$$

where I is the identity matrix, b' is the matrix whose m,j entry is $\partial_m b_j$, and σ'_k is the matrix whose m,j entry is $\partial_m \sigma_{jk}$. $(DX)_{lj}$ represents the partial derivative in the lth direction of $X^j(x,t)$.

As in the case of dimension one, the same proof for higher dimensions shows

(10.4) Proposition. *Suppose σ and b are in C^2 and are bounded with bounded first and second derivatives. For all $t_0 > 0$ and $p \geq 2$, there exists $c_1(p, t_0)$ independent of x such that*

$$\mathbb{E} \sup_{s \leq t_0} |DX(x,s)|^p \leq c_1(p, t_0).$$

Not surprisingly, if σ and b have further smoothness, then $X(x,t)$ will have higher derivatives. If σ and b are C^∞, then $X(x,t)$ will be C^∞ in x also.

One can also show (see Ikeda and Watanabe [1]) that the map $x \to X(x,t)$ is one-to-one and onto $\mathbb{R}^d$.

11. SDEs with reflection

If Y_t is a Brownian motion on the line, then a consequence of Itô's formula ([PTA, (I.6.30)]) says that

$$|Y_t| = W_t + L_t, \tag{11.1}$$

where W_t is another Brownian motion and L_t is a continuous nondecreasing process that increases only when $|Y_t|$ is at 0; this is known as Tanaka's formula. Equation (11.1) can be viewed as an SDE for which existence and uniqueness can be proved.

(11.1) Theorem. *Let W_t be a Brownian motion. There exists a nonnegative continuous process X_t and a continuous nondecreasing process L_t that increases only when X_t equals 0 such that*

$$X_t = W_t + L_t. \tag{11.2}$$

If X_t' is another nonnegative continuous process satisfying $X_t' = W_t + L_t'$, where L_t' increases only when $X_t' = 0$, then $X_t = X_t'$ and $L_t = L_t'$ a.s.

Proof. We first prove existence. Let $L_t = \sup_{s \le t}(-W_s)$ and $X_t = W_t + L_t$. Clearly $X_t \ge 0$. When L_t increases, then $-W_t = L_t$, or $X_t = 0$.

To prove uniqueness, since $X_t \ge 0$, then $L_t \ge L_s \ge -W_s$ if $s \le t$, so $L_t \ge \sup_{s \le t}(-W_s)$. L_t increases only when $X_t = 0$; when this happens, $L_t = -W_t$. Hence we must have $L_t = \sup_{s \le t}(-W_s)$. The same argument applies to L_t'. Therefore $L_t = L_t'$, which implies the theorem. $\square$

We call X_t *reflecting Brownian motion* and L_t the *local time* (at 0) of X_t.

The simplest case of a diffusion in $\mathbb{R}^d$ with reflection, $d \ge 2$, is the following. Let D be the upper-half space, let $Y_t = (Y_t^1, \ldots, Y_t^d)$ be standard d-dimensional Brownian motion, and let L_t be the local time of $|Y_t^d|$. Then $X_t = (Y_t^1, \ldots, Y_t^{d-1}, |Y_t^d|)$ is *reflecting Brownian motion with normal reflection* in D. If $|Y_t^d| = \widetilde{W}_t + L_t$, then X_t solves the stochastic differential equation

$$dX_t = dW_t + \nu(X_t)\,dL_t, \qquad X_t \in \overline{D}, \tag{11.3}$$

where $W_t = (Y_t^1, \ldots, Y_t^{d-1}, \widetilde{W}_t)$ is a d-dimensional Brownian motion, $\nu(x) \equiv (0, \ldots, 0, 1)$ is the inward pointing unit normal vector, and L_t is a continuous nondecreasing process that increases only when X_t is on the boundary of D. The equation (11.3) is an example of what is known as the *Skorokhod equation.*

The process that solves (11.3) is reflecting Brownian motion with normal reflection in D. To consider oblique reflection, we replace ν, the inward pointing normal, by a different vector. Let v be a vector such that $v \cdot \nu > 0$. Thus the vector v started at a point on ∂D points into D. Consider the SDE

$$dX_t = dW_t + v(X_t)\,dL_t, \qquad X_t \in \overline{D}, \tag{11.4}$$

where W_t is a d-dimensional Brownian motion and L_t is a continuous nondecreasing process that grows only when X_t is on ∂D. We call X_t *reflecting Brownian motion with constant oblique reflection.*

It is easy to give an explicit solution to (11.4). Let

$$L_t = \frac{1}{v_d}\sup_{s \le t}(-W_s^d), \quad \text{and} \quad X_t^d = W_t^d + v_d L_t.$$

As we saw in the proof of Theorem 11.1, L_t increases only when X_t^d is at 0, which is when X_t is in ∂D. Also, $X_t^d \geq 0$ for all t, so $X_t \in \overline{D}$. We then set

$$X_t^i = W_t^i + v_i L_t, \qquad i = 1, \ldots, d-1.$$

It is clear that X_t solves (11.4).

We now describe the general Skorokhod equation in C^2 domains. A C^2 *domain* $D \subseteq \mathbb{R}^d$ is one where for each $x \in D$ there exists $r_x > 0$, a C^2 function $\varphi_x : \mathbb{R}^{d-1} \to \mathbb{R}$, and an orthonormal coordinate system CS_x such that

$$D \cap B(x, r_x)$$
$$= \{y = (y_1, \ldots, y_d) \text{ in } CS_x : y_d > \varphi_x(y_1, \ldots, y_{d-1})\} \cap B(x, r_x).$$

In other words, locally the domain D looks like the region above a C^2 function.

Let D be a C^2 domain, σ be matrix-valued, b vector-valued, W_t a standard d-dimensional Brownian motion, and $v(x)$ defined on ∂D such that $v(x) \cdot \nu(x) > 0$ for all $x \in \partial D$. Here $\nu(x)$ is the inward pointing unit normal vector at x. Then the *Skorokhod equation* is the SDE

$$dX_t = \sigma(X_t)\, dW_t + b(X_t)\, dt + v(X_t)\, dL_t, \qquad X_0 = x_0, \qquad (11.5)$$

where $X_t \in \overline{D}$ for all t, $x_0 \in \overline{D}$, and L_t is a continuous nondecreasing process that increases only when $X_t \in \partial D$.

In many cases one can say more about L_t. For example, one can describe it as the local time on the boundary corresponding to a measure on ∂D that is mutually absolutely continuous with respect to surface measure on ∂D. See Bass and Hsu [1] for more details in the case of Brownian motion in Lipschitz domains and Stroock and Varadhan [1] for more general diffusions in C^2 domains. See also Sections II.6 and III.7.

We now make some assumptions on D, σ, b, and v that guarantee a good tightness estimate. Let us assume that b, σ, σ^{-1}, and v are bounded, D is a C^2 domain, and

$$\inf_{x \in \partial D} \nu(x) \cdot v(x) > 0.$$

Let us start with the case where D is the upper-half space H.

(11.2) Proposition. *With $D = H$, the upper-half space, σ, b, and v satisfying the above, and X_t a solution to (11.5), if $\varepsilon, t_0 > 0$, there exists λ such that*

$$\mathbb{P}(\sup_{s \leq t_0} |X_s - X_0| > \lambda) < \varepsilon.$$

Proof. Let λ' and c_1 be positive reals to be chosen in a moment and let

$$B_1 = \{\sup_{s \leq t_0} |X_s - X_0| > \lambda'\},$$

$$B_2 = \{\sup_{s \leq \tau} X_s^d > c_1 \lambda'\},$$

$$B_3 = \{\sup_{s \leq t_0} |X_s - X_0| > 2\lambda\},$$

where τ is the time to exit $B(x_0, \lambda')$. B_2 is the event that X_t^d will be larger than $c_1 \lambda'$ before X_t exits $B(x_0, \lambda')$. We will first show there exists $c_1 < 1/2$ such that if λ' is large enough,

$$\mathbb{P}(B_1 \cap B_2^c) \leq 3/4. \tag{11.6}$$

Let $Y_t = \int_0^t \sigma(X_s)\, dW_s + \int_0^t b(X_s)\, ds$. By Proposition 8.1, there exists λ' such that the probability that $|Y_t|$ exceeds $\lambda'/2$ before time t_0 is less than $1/2$. Suppose the event B_1 holds and also $\sup_{s \leq t_0} |Y_s| \leq \lambda'/2$. Then we must have $|\int_0^\tau v(X_s)\, dL_s|$ greater than or equal to $\lambda'/2$. Since $|v|$ is bounded by a constant c_2, $L_\tau \geq \lambda'/2c_2$. Since v_d is bounded below by a constant c_3,

$$\int_0^\tau v_d(X_s)\, dL_s \geq \frac{c_3}{2c_2} \lambda'.$$

Provided λ' is large enough, the probability that the dth coordinate of Y_t exceeds $c_3 \lambda'/4c_2$ before time t_0 is less than $1/4$. If $\sup_{s \leq t_0} Y_s^d \leq c_3 \lambda'/4c_2$ and $\tau \leq t_0$, the dth coordinate of $X_\tau - x_0$ must be greater than $c_3 \lambda'/4c_2$. Hence, letting $c_1 = (c_3/4c_2) \wedge (1/2)$,

$$\mathbb{P}(B_1 \cap B_2^c) \leq \mathbb{P}(\sup_{s \leq t_0} |Y_s| \geq \lambda'/2) + \mathbb{P}(\sup_{s \leq t_0} Y_s^d > c_3\lambda'/4c_2, \tau \leq t_0)$$

$$\leq 3/4.$$

If $\mathbb{P}(B_1^c) \geq 1/8$, then clearly $\mathbb{P}(B_3^c) \geq 1/8$. Suppose $\mathbb{P}(B_1^c) < 1/8$. Then by (11.6), $\mathbb{P}(B_2) \geq 1/8$. In this case we use Theorem 8.4 with $\varepsilon = c_1 \lambda'/2$ and the strong Markov property at time τ to see that there is positive probability that $\sup_{s \leq t_0} |X_s - X_0| < 2\lambda'$, and hence there exists c_4 such that

$$\mathbb{P}(B_3^c) \geq \mathbb{P}(B_3^c \cap B_2) \geq c_4 \mathbb{P}(B_2) \geq c_4/8. \tag{11.7}$$

Letting $c_5 = (c_4 \wedge 1)/8$, we deduce

$$\mathbb{P}(B_3^c) \geq c_5.$$

We now iterate. Choose n such that $(1 - c_5)^n < \varepsilon$ and set $\lambda = 2n\lambda'$. Let $U_0 = 0$ and let $U_{i+1} = \inf\{t > U_i : |X_s - X_{U_i}| \geq 2\lambda'\}$. What we have shown is that $\mathbb{P}(U_1 < t_0) \leq 1 - c_5$. Note that $X_{U_1+t} - X_{U_1}$ is again a solution to (11.5) starting at X_{U_1}, and recall that $W_{t+U_1} - W_{U_1}$ is independent of $\mathcal{F}_{U_1}$. So the same argument with the same constants shows that

$$\mathbb{P}(U_2 - U_1 < t_0 \mid \mathcal{F}_{U_1}) \leq 1 - c_5.$$

Then

$$\mathbb{P}(U_1 \le U_2 \le t_0) = \mathbb{E}\left[\mathbb{P}(U_2 - U_1 \le t_0 \mid \mathcal{F}_{U_1}); U_1 \le t_0\right]$$
$$\le (1 - c_5)\mathbb{P}(U_1 \le t_0) \le (1 - c_5)^2.$$

Repeating,

$$\mathbb{P}(U_1 \le U_2 \le \cdots \le U_n \le t_0) \le (1 - c_5)^n < \varepsilon,$$

which proves the proposition. $\qquad\square$

A similar argument allows us also to conclude under the above hypotheses that, given $\varepsilon, \lambda > 0$, there exists t_0 such that

$$\mathbb{P}(\sup_{s \le t_0} |X_s - x_0| > \lambda) \le \varepsilon. \qquad (11.8)$$

We now obtain the tightness estimate we want.

(11.3) Proposition. *Suppose $\lambda, \varepsilon > 0$, D is a C^2 domain, X_t solves (11.5), and $\sigma, b,$ and v satisfy the hypotheses above. Then there exists t_0 such that*

$$\mathbb{P}(\sup_{s \le t_0} |X_s - x_0| > \lambda) \le \varepsilon.$$

Proof. Without loss of generality we may take λ smaller so that $\lambda < r_{x_0}$, where r_{x_0} is the radius that arises in the definition of a C^2 domain. We may therefore assume that D is the region above a C^2 function φ_{x_0}. Suppose $x_0 = (x_0^1, \ldots, x_0^d)$. Since $\lambda < r_{x_0}$, if we modify φ_{x_0} outside of the ball in $\mathbb{R}^{d-1}$ of radius r_{x_0} with center at $(x_0^1, \ldots, x_0^{d-1})$ so that it has compact support, there again is no loss of generality.

We now map D onto the upper-half space H by the map

$$y \mapsto (y_1, \ldots, y_{d-1}, y_d - \varphi_{x_0}(y_1, \ldots, y_{d-1})).$$

It is easy to see that X_t is transformed into another process $\widetilde{X}_t$ that satisfies the Skorokhod equation with parameters $\widetilde{\sigma}, \widetilde{b}, \widetilde{v}$, and $\widetilde{L}_t$, and that, moreover, these parameters satisfy bounds of the same type as σ, b, v and L_t. Here it is important that D be a C^2 domain, so that φ_{x_0} is a C^2 function.

By Proposition 11.2, we have an estimate of the type we want for $\widetilde{X}_t$. Since the map taking D onto H has a bounded Jacobian matrix with a bounded inverse, we obtain the estimate we want for X_t also. $\qquad\square$

12. SDEs with reflection: pathwise results

In this section we present some results on pathwise existence and uniqueness of the Skorokhod equation due to Lions and Sznitman [1].

Suppose D is a bounded C^2 domain. Suppose σ and b are Lipschitz, v is C^2 on ∂D, and $v(x) \cdot \nu(x) > 0$ for all $x \in \partial D$.

(12.1) Theorem. *There exists a solution to (11.5). If X_t and X_t' are two solutions to (11.5), then $X_t = X_t'$ a.s. for all t.*

To give an idea of how the proof goes, we will prove uniqueness in the special case that $b \equiv 0$ and $v(x) = \nu(x)$ for all $x \in \partial D$. We will also suppose $\varphi : \mathbb{R}^{d-1} \to \mathbb{R}$ is a bounded C^2 function with compact support, and that

$$D = \{(y_1, \ldots, y_d) : y_d > \varphi(y_1, \ldots, y_{d-1})\}. \tag{12.1}$$

We refer the reader to Lions and Sznitman [1] for the general case and the proof of existence.

Let ψ be a C^2 function on $\mathbb{R}$ taking values in $[-2, 2]$, $\psi(x) = x$ if $|x| \leq 1$, and $\psi(x) \neq 0$ if $x \neq 0$. Let $\Phi : \mathbb{R}^d \to \mathbb{R}$ be defined by

$$\Phi(y_1, \ldots, y_d) = \psi(y_d - \varphi(y_1, \ldots, y_{d-1})).$$

(12.2) Proposition. *(a) $\nabla \Phi \cdot \nu$ is bounded below by a constant $c_1 > 0$ on ∂D.*
(b)   There exists c_2 such that if $x \in \partial D$ and $y \in \overline{D}$, then

$$(y - x) \cdot \nu(x) + c_2 |x - y|^2 \geq 0. \tag{12.2}$$

Proof. Both (a) and (b) follow from our assumptions on ν and the fact that D is the region above a bounded C^2 function. $\qquad\square$

Let us suppose we have two continuous semimartingales X_t, X_t' and

$$Y_t = x_0 + \int_0^t \sigma(X_s)\, dW_s + \int_0^t \nu(Y_s)\, dL_s, \qquad Y_t \in \overline{D} \tag{12.3}$$

$$Y_t' = x_0 + \int_0^t \sigma(X_s')\, dW_s + \int_0^t \nu(Y_s')\, dL_s', \qquad Y_t' \in \overline{D},$$

where L_s and L_s' are continuous nondecreasing processes that increase only when $Y_t \in \partial D$ and $Y_t' \in \partial D$, respectively.

The key to uniqueness and also an important step in existence is the following.

(12.3) Proposition. *Let $t_0 \geq 0$. There exists c_1 such that if $t \leq t_0$,*

$$\mathbb{E} \sup_{s \leq t} |Y_s - Y_s'|^4 \leq c_1 \int_0^t \mathbb{E} \sup_{s \leq r} |X_s - X_s'|^4\, dr.$$

Proof. Let $V_t = \exp(-c_2 \Phi(Y_t) - c_2 \Phi(Y_t'))$, where c_2 will be chosen later. By the product formula and Itô's formula,

$$V_t |Y_t - Y_t'|^2$$

$$= 2 \int_0^t V_s (Y_s - Y_s') \cdot d(Y_s - Y_s') + \int_0^t V_s \, d\langle Y - Y' \rangle_s$$

$$- c_2 \int_0^t V_s |Y_s - Y_s'|^2 \, d(\Phi(Y) + \Phi(Y'))_s$$

$$+ \frac{1}{2} c_2^2 \int_0^t V_s |Y_s - Y_s'|^2 \, d\langle \Phi(Y) + \Phi(Y') \rangle_s$$

$$= I_1(t) + I_2(t) + I_3(t) + I_4(t).$$

Since Φ is bounded, then V_t is bounded above and below. We have

$$I_1(t) = 2 \int_0^t V_s (Y_s - Y_s') [\sigma(X_s) - \sigma(X_s')] \, dW_s$$

$$+ 2 \int_0^t V_s (Y_s - Y_s') \cdot \nu(Y_s) \, dL_s - 2 \int_0^t V_s (Y_s - Y_s') \cdot \nu(Y_s') \, dL_s'$$

$$= 2 I_{11}(t) + 2 I_{12}(t) + 2 I_{13}(t),$$

and by Doob's inequality,

$$\mathbb{E} \sup_{s \le t} I_{11}(s)^2 \le c_3 \mathbb{E} \, I_{11}(t)^2 \le c_4 \mathbb{E} \int_0^t |Y_s - Y_s'|^2 |X_s - X_s'|^2 \, ds$$

$$\le c_5 \int_0^t \mathbb{E} \, |Y_s - Y_s'|^4 \, ds + c_5 \int_0^t \mathbb{E} \, |X_s - X_s'|^4 \, ds. \tag{12.4}$$

Next,

$$I_2(t) = \int_0^t V_s [\sigma(X_s) - \sigma(X_s')] \, [\sigma(X_s) - \sigma(X_s')]^T \, ds,$$

so

$$\mathbb{E} \sup_{s \le t} I_2(s)^2 \le c_6 \mathbb{E} \left(\int_0^t |X_s - X_s'|^2 \, ds \right)^2$$

$$\le c_7 t_0 \mathbb{E} \int_0^t |X_s - X_s'|^4 \, ds. \tag{12.5}$$

Looking at the third term, let $D^2 \Phi$ be the matrix whose i, j entry is $\partial_{ij} \Phi$. Then

$$I_3(t) = -c_2 \int_0^t V_s |Y_s - Y_s'|^2 \nabla\Phi(Y_s)\sigma(X_s)\, dW_s$$

$$- c_2 \int_0^t V_s |Y_s - Y_s'|^2 \nabla\Phi(Y_s)\cdot \nu(Y_s)\, dL_s$$

$$- c_2 \int_0^t V_s |Y_s - Y_s'|^2 \operatorname{trace}\left(\sigma(X_s)^T D^2\Phi(Y_s)\sigma(X_s)\right) ds$$

$$- c_2 \int_0^t V_s |Y_s - Y_s'|^2 \nabla\Phi(Y_s')\sigma(X_s')\, dW_s$$

$$- c_2 \int_0^t V_s |Y_s - Y_s'|^2 \nabla\Phi(Y_s')\cdot \nu(Y_s')\, dL_s'$$

$$- c_2 \int_0^t V_s |Y_s - Y_s'|^2 \operatorname{trace}\left(\sigma(X_s')^T D^2\Phi(Y_s')\sigma(X_s')\right) ds$$

$$= -c_2 I_{31}(t) - c_2 I_{32}(t) - c_2 I_{33}(t) - c_2 I_{34}(t) - c_2 I_{35}(t) - c_2 I_{36}(t),$$

and similarly to (12.4) and (12.5),

$$\mathbb{E}\sup_{s\leq t}[I_{31}(s)^2 + I_{33}(s)^2 + I_{34}(s)^2 + I_{36}(s)^2]$$

$$\leq c_8 \int_0^t \mathbb{E}\,|Y_s - Y_s'|^4\, ds. \tag{12.6}$$

For the last term,

$$I_4(t) = c_2^2 \int_0^t V_s |Y_s - Y_s'|^2 [\nabla\Phi(Y_s) + \nabla\Phi(Y_s')]$$

$$\times [\sigma(X_s) + \sigma(X_s')][\sigma(X_s) + \sigma(X_s')]^T [\nabla\Phi(Y_s) + \nabla\Phi(Y_s')]^T\, ds,$$

so

$$\mathbb{E}\sup_{s\leq t} I_4(s)^2 \leq c_9 \mathbb{E}\int_0^t |Y_s - Y_s'|^4\, ds. \tag{12.7}$$

The key observation is the following. L_t increases only when $Y_s \in \partial D$. So

$$2I_{12}(t) - c_2 I_{32}(t)$$

$$= \int_0^t V_s [2(Y_s - Y_s')\cdot \nu(Y_s) - c_2 |Y_s - Y_s'|^2 \nabla\Phi(Y_s)\cdot \nu(Y_s)]\, dL_s.$$

This is less than or equal to 0 by Proposition 12.2 provided c_2 is taken large enough. Similarly,

$$2I_{13}(t) - c_2 I_{35}(t) \leq 0.$$

Hence by (12.4)–(12.7),

$$\mathbb{E}\sup_{s\leq t} |Y_s - Y_s'|^4 \leq c_{10}\mathbb{E}\left[\sup_{s\leq t}(V_s |Y_s - Y_s'|^2)\right]^2$$

$$\leq c_{11}\mathbb{E}\int_0^t \sup_{s\leq r} |Y_s - Y_s'|^4\, dr$$

$$+ c_{11}\mathbb{E}\int_0^t \sup_{s\leq r} |X_s - X_s'|^4\, dr.$$

The proposition now follows by Gronwall's lemma.   □

Uniqueness follows easily from Proposition 12.3.

(12.4) Theorem. *If X_t and X_t' are two solutions to (11.5), then $X_t = X_t'$ a.s.*

Proof. We have (12.3) holding with $Y_t = X_t$ and $Y_t' = X_t'$. So from Proposition 12.3,

$$\mathbb{E} \sup_{s \leq t} |X_s - X_s'|^4 \leq c_1 \int_0^t \mathbb{E} \sup_{s \leq r} |X_s - X_s'|^4 \, dr,$$

and the conclusion follows by Gronwall's lemma.   □

We remark that if we let $\mathbb{P}^x$ denote the law of X_t when $X_0 = x$, then just as in Section 5, $(\mathbb{P}^x, X_t)$ forms a strong Markov process.

13. Notes

The preliminary material (Section 1) can be found in a large number of places. See, for example, [PTA, Chapter 1], Ikeda and Watanabe [1], or Revuz and Yor [1]. The majority of the rest of the chapter is covered in Ikeda and Watanabe [1] or Protter [1]. Section 5 is from Stroock and Varadhan [2]. Sections 11 and 12 follow Lions and Sznitman [1]. Theorem 4.2 is due to Girsanov; our account is taken from Knight [1]. An argument similar to that of Proposition 11.2 appeared in Kwon [1].

II
REPRESENTATIONS OF SOLUTIONS

This chapter is concerned with giving probabilistic representations of the solutions to PDEs. Throughout we will be assuming that the given PDE has a solution, the solution is unique, and the solution is sufficiently regular. In the next chapter we will use a mixture of PDE and probabilistic techniques – primarily the former – to show that such solutions exist.

Suppose a process X_t is associated to an operator $\mathcal{L}$ as in Section 1.2. The solution to many PDEs involving $\mathcal{L}$ can be written very simply in terms of the expected values of certain functionals of X_t. In Section 1 we discuss Poisson's equation, in Section 2 the Dirichlet problem, in Section 3 the Cauchy problem, and in Section 4 the (real) Schrödinger equation.

For an operator $\mathcal{L}$ given by (1.1), the second-order terms are the key ones. We show in Section 5 how the Girsanov transformation can be used to dispense with first-order terms.

In Section 6 we look at reflecting boundary conditions. Both the Neumann problem and many cases of the oblique derivative problem have solutions that can be represented probabilistically in terms of SDEs with reflections.

Many useful quantities from PDE have probabilistic interpretations. Examples include fundamental solutions and the Green function; see Section 7.

In Section 8 we examine the relationships between fundamental solutions and adjoint operators and between invariant measures and adjoint operators.

1. Poisson's equation

Let

$$\mathcal{L}f(x) = \frac{1}{2}\sum_{i,j=1}^{d} a_{ij}(x)\partial_{ij}f(x) + \sum_{i=1}^{d} b_i(x)\partial_i f(x). \tag{1.1}$$

Throughout this chapter, unless stated otherwise, we assume the a_{ij} and b_i are bounded and at least C^1. We also assume that the operator $\mathcal{L}$ is uniformly strictly elliptic. An operator $\mathcal{L}$ is *strictly elliptic* if for each x there exists $\Lambda(x)$ such that

$$\sum_{i,j=1}^{d} a_{ij}(x)y_iy_j \geq \Lambda(x)\sum_{i=1}^{d} y_j^2 \qquad y = (y_1,\ldots,y_d) \in \mathbb{R}^d. \tag{1.2}$$

The operator $\mathcal{L}$ is *uniformly strictly elliptic* or *uniformly elliptic* if Λ can be chosen to be independent of x. We also call the matrix a strictly elliptic if (1.2) holds and uniformly elliptic if (1.2) holds with $\Lambda(x)$ not depending on x. We also assume throughout that the dimension d is greater than or equal to 3.

We emphasize that the uniform ellipticity of $\mathcal{L}$ is used in Sections 1–4 only to show that the exit times of the domains we consider are finite a.s. For many nonuniformly elliptic operators, it is often the case that the finiteness of the exit times is known for other reasons, and the results of Sections 1–4 then apply to equations involving these operators.

Suppose σ is a matrix such that $a = \sigma\sigma^T$ and each component of σ is bounded and in C^1. Let X_t be the solution to

$$X_t = x + \int_0^t \sigma(X_s)\,dW_s + \int_0^t b(X_s)\,ds. \tag{1.3}$$

We will write $(\mathbb{P}^x, X_t)$ for the strong Markov process corresponding to σ and b.

We consider first Poisson's equation in $\mathbb{R}^d$. Suppose $\lambda > 0$ and f is a C^1 function with compact support. Poisson's equation is

$$\mathcal{L}u(x) - \lambda u(x) = -f(x), \qquad x \in \mathbb{R}^d. \tag{1.4}$$

(1.1) Theorem. *Suppose u is a C^2 solution to (1.4) such that u and its first and second partial derivatives are bounded. Then*

$$u(x) = \mathbb{E}^x \int_0^\infty e^{-\lambda t} f(X_t)\,dt.$$

Proof. Let u be the solution to (1.4). By Itô's formula,

$$u(X_t) - u(X_0) = M_t + \int_0^t \mathcal{L}u(X_s)\,ds,$$

where M_t is a martingale. By the product formula,

$$e^{-\lambda t}u(X_t) - u(X_0) = \int_0^t e^{-\lambda s}dM_s + \int_0^t e^{-\lambda s}\mathcal{L}u(X_s)\,ds$$
$$- \lambda \int_0^t e^{-\lambda s}u(X_s)\,ds.$$

Taking $\mathbb{E}^x$ expectation and letting $t \to \infty$,

$$-u(x) = \mathbb{E}^x \int_0^\infty e^{-\lambda s}(\mathcal{L}u - \lambda u)(X_s)\,ds.$$

Since $\mathcal{L}u - \lambda u = -f$, the result follows. $\qquad\qquad\square$

Let us now let D be a nice bounded domain, e.g., a ball. Poisson's equation in D requires one to find a function u such that $\mathcal{L}u - \lambda u = -f$ in D and $u = 0$ on ∂D, where $f \in C^2(\overline{D})$ and $\lambda \geq 0$. Here we can allow λ to be equal to 0. Recall that by Proposition I.8.2, the time to exit D, namely, $\tau_D = \inf\{t : X_t \notin D\}$, is finite almost surely.

(1.2) Theorem. *Suppose u is a solution to Poisson's equation in a bounded domain D that is C^2 in D and continuous on $\overline{D}$. Then*

$$u(x) = \mathbb{E}^x \int_0^{\tau_D} e^{-\lambda s}f(X_s)\,ds.$$

Proof. The proof is nearly identical to that of Theorem 1.1. By Proposition I.8.2, $\tau_D < \infty$ a.s. Let $S_n = \inf\{t : \text{dist}\,(X_t, \partial D) < 1/n\}$. By Itô's formula,

$$u(X_{t \wedge S_n}) - u(X_0) = \text{ martingale } + \int_0^{t \wedge S_n} \mathcal{L}u(X_s)\,ds.$$

By the product formula,

$$\mathbb{E}^x e^{-\lambda(t \wedge S_n)}u(X_{t \wedge S_n}) - u(x)$$
$$= \mathbb{E}^x \int_0^{t \wedge S_n} e^{-\lambda s}\mathcal{L}u(X_s)\,ds - \mathbb{E}^x \int_0^{t \wedge S_n} e^{-\lambda s}u(X_s)\,ds$$
$$= -\mathbb{E}^x \int_0^{t \wedge S_n} e^{-\lambda s}f(X_s)\,ds.$$

Now let $n \to \infty$ and then $t \to \infty$ and use the fact that u is 0 on ∂D. $\qquad\square$

2. Dirichlet problem

Let D be a ball (or other nice bounded domain) and let us consider the solution to the Dirichlet problem: given f a continuous function on ∂D, find $u \in C(\overline{D})$ such that u is C^2 in D and

$$\mathcal{L}u = 0 \text{ in } D, \qquad u = f \text{ on } \partial D. \tag{2.1}$$

(2.1) Theorem. *The solution to (2.1) satisfies*

$$u(x) = \mathbb{E}^x f(X_{\tau_D}).$$

Proof. By Proposition I.8.2, $\tau_D < \infty$ a.s. Let $S_n = \inf\{t : \operatorname{dist}(X_t, \partial D) < 1/n\}$. By Itô's formula,

$$u(X_{t \wedge S_n}) = u(X_0) + \text{ martingale } + \int_0^{t \wedge S_n} \mathcal{L}u(X_s)\, ds.$$

Since $\mathcal{L}u = 0$ inside D, taking expectations shows

$$u(x) = \mathbb{E}^x u(X_{t \wedge S_n}).$$

We let $t \to \infty$ and then $n \to \infty$. By dominated convergence, we obtain $u(x) = \mathbb{E}^x u(X_{\tau_D})$. This is what we want since $u = f$ on ∂D. $\qquad\square$

There are some further facts that can be deduced from Theorem 2.1. One is the *maximum principle*: if $x \in D$,

$$\sup_{\overline{D}} u \leq \sup_{\partial D} u. \tag{2.2}$$

This follows from

$$u(x) = \mathbb{E}^x f(X_{\tau_D}) \leq \sup_{\partial D} f.$$

There is a sort of converse of Theorem 2.1.

(2.2) Proposition. *Let f be continuous on ∂D and suppose $v(x) = \mathbb{E}^x f(X_{\tau_D})$ is continuous on $\overline{D}$ and C^2 on D. Suppose the coefficients of $\mathcal{L}$ are continuous. Then $\mathcal{L}v = 0$ on D.*

Proof. By the strong Markov property at time $\tau_{B(x,r)}$, the time to exit $B(x,r)$, we have $v(x) = \mathbb{E}^x v(X_{\tau(B(x,r))})$ if r is small enough that $B(x,r) \subseteq D$. By Itô's formula,

$$v(X_{t \wedge \tau(B(x,r))}) = v(X_0) + \text{ martingale } + \int_0^{t \wedge \tau(B(x,r))} \mathcal{L}v(X_s)\, ds.$$

Taking expectations and letting $t \to \infty$,

$$\mathbb{E}^x v(X_{\tau(B(x,r))}) = v(x) + \mathbb{E}^x \int_0^{\tau(B(x,r))} \mathcal{L}v(X_s)\, ds,$$

so $\mathbb{E}^x \int_0^{\tau(B(x,r))} \mathcal{L}v(X_s)\, ds = 0$. Dividing by $\mathbb{E}^x \tau(B(x,r))$, letting $r \to 0$, and using the continuity of $\mathcal{L}v$ implies that $\mathcal{L}v(x) = 0$. $\qquad\square$

We have already supposed that u is a solution to the Dirichlet problem and hence continuous up to the boundary. We will see later on that for domains satisfying an exterior cone condition, we automatically have $\mathbb{E}^x f(X_{\tau_D})$ continuous up to the boundary.

If $\mathcal{L}v = 0$ in D, we say v is $\mathcal{L}$-*harmonic* in D.

3. Cauchy problem

We are primarily interested in elliptic PDEs, but the related parabolic partial differential equation $\partial_t u = \mathcal{L}u$ is often of interest. Here $\partial_t u$ denotes $\partial u / \partial t$.

Suppose for simplicity that the function f is a continuous function with compact support. The Cauchy problem is to find u such that u is bounded, u is C^2 with bounded first and second partial derivatives in x, u is C^1 in t for $t > 0$, and

$$\begin{aligned}
\partial_t u(x,t) &= \mathcal{L}u(x,t), \qquad t > 0, x \in \mathbb{R}^d, \\
u(x,0) &= f(x), \qquad x \in \mathbb{R}^d.
\end{aligned} \tag{3.1}$$

(3.1) Theorem. *The solution to (3.1) satisfies*

$$u(x,t) = \mathbb{E}^x f(X_t).$$

Proof. Fix t_0 and let $M_t = u(X_t, t_0 - t)$. The solution u to (3.1) is known to be C^2 in x and C^1 in t for $t > 0$ (see Friedman [1]). Note $\partial_t[u(x, t_0 - t)] = -(\partial_t u)(x, t_0 - t)$. By Itô's formula on $\mathbb{R}^d \times [0, t_0)$,

$$u(X_t, t_0 - t) = \text{ martingale } + \int_0^t \mathcal{L}u(X_s, t_0 - s)\, ds$$

$$+ \int_0^t (-\partial_t u)(X_s, t_0 - s)\, ds.$$

Since $\partial_t u = \mathcal{L}u$, M_t is a martingale, and $\mathbb{E}^x M_0 = \mathbb{E}^x M_{t_0}$. On the one hand,

$$\mathbb{E}^x M_{t_0} = \mathbb{E}^x u(X_{t_0}, 0) = \mathbb{E}^x f(X_{t_0}),$$

while on the other,

$$\mathbb{E}^x M_0 = \mathbb{E}^x u(X_0, t_0) = u(x, t_0).$$

Since t_0 is arbitrary, the result follows. $\square$

For bounded domains D, the Cauchy problem is to find u such that $\partial_t = \mathcal{L}u$ on D, $u(x,0) = f(x)$ for $x \in D$, and $u(x,t) = 0$ for $x \in \partial D$. The solution is given by

$$u(x,t) = \mathbb{E}^x[f(X_t); t < \tau_D],$$

where τ_D is the exit time of D. The proof is very similar to the case of $\mathbb{R}^d$.

4. Schrödinger operators

We next look at what happens when one adds a potential term, that is, when one considers the operator

$$\mathcal{L}u(x) + q(x)u(x). \tag{4.1}$$

This is known as the *Schrödinger operator*, and $q(x)$ is known as the *potential*. Equations involving the operator in (4.1) are considerably simpler than the quantum mechanics Schrödinger equation because here all terms are real-valued.

If X_t is the diffusion corresponding to $\mathcal{L}$ in the sense of Section I.2, then solutions to PDEs involving the operator in (4.1) can be expressed in terms of X_t by means of the *Feynman-Kac formula*. To illustrate, let D be a nice bounded domain, e.g., a ball, q a C^2 function on $\overline{D}$, and f a continuous function on ∂D; q^+ denotes the positive part of q.

(4.1) Theorem. *Let D, q, f be as above. Let u be a C^2 function on $\overline{D}$ that agrees with f on ∂D and satisfies $\mathcal{L}u + qu = 0$ in D. If*

$$\mathbb{E}^x \exp\left(\int_0^{\tau_D} q^+(X_s)\,ds\right) < \infty,$$

then

$$u(x) = \mathbb{E}^x\left[f(X_{\tau_D})e^{\int_0^{\tau_D} q(X_s)\,ds}\right]. \tag{4.2}$$

We remark that the case when q is a negative constant has been dealt with in Section 1.

Proof. Let $B_t = \int_0^{t\wedge\tau_D} q(X_s)\,ds$. By Itô's formula and the product formula,

$$e^{B(t\wedge\tau_D)}u(X_{t\wedge\tau_D}) = u(X_0) + \text{ martingale } + \int_0^{t\wedge\tau_D} u(X_r)e^{B_r}\,dB_r$$

$$+ \int_0^{t\wedge\tau_D} e^{B_r}\,d[u(X)]_r.$$

Taking $\mathbb{E}^x$ expectation,

$$\mathbb{E}^x e^{B(t\wedge\tau_D)} u(X_{t\wedge\tau_D}) = u(x) + \mathbb{E}^x \int_0^{t\wedge\tau_D} e^{B_r} u(X_r) q(X_r)\, dr$$

$$+ \mathbb{E}^x \int_0^{t\wedge\tau_D} e^{B_r} \mathcal{L}u(X_r)\, dr.$$

Since $\mathcal{L}u + qu = 0$,

$$\mathbb{E}^x e^{B(t\wedge\tau_D)} u(X_{t\wedge\tau_D}) = u(x).$$

If we let $t \to \infty$ and use the exponential integrability of q^+, the result follows. $\qquad\qquad\Box$

The existence of a solution to $\mathcal{L}u + qu = 0$ in D depends on the finiteness of $\mathbb{E}^x e^{\int_0^{\tau_D} q^+(X_s)\, ds}$, an expression that is sometimes known as the *gauge*; see Chung and Zhao [1].

Even in one dimension with $D = (0,1)$ and q a constant function, the gauge need not be finite. By [PTA, (II.4.30)] with $x = 1/2$, $\mathbb{P}^x(\tau_D > t) \geq ce^{-\pi^2 t/2}$ for t sufficiently large. Hence

$$\mathbb{E}^x \exp\left(\int_0^{\tau_D} q\, ds \right) = \mathbb{E}^x e^{q\tau_D}$$

$$= \int_0^\infty q e^{qt} \mathbb{P}^x(\tau_D > t)\, dt;$$

this is infinite if $q \geq \pi^2/2$.

A very similar proof to that of Theorem 4.1 shows that under suitable assumptions on q, g, and D, the solution to $\mathcal{L}u + qu = -g$ in D with boundary condition $u = 0$ on ∂D is given by

$$u(x) = \mathbb{E}^x \left[\int_0^{\tau_D} g(X_s) e^{\int_0^s q(X_r)\, dr}\, ds \right]. \tag{4.3}$$

There is also a parabolic version of Theorem 4.1. The equation $\partial_t u = \mathcal{L}u + qu$ with initial condition $u(x,0) = f(x)$ is solved by

$$u(x,t) = \mathbb{E}^x \left[f(X_t) e^{\int_0^t q(X_s)\, ds} \right]. \tag{4.4}$$

When $q \leq 0$, there is a way of interpreting the right-hand sides of (4.2) through (4.4). We consider (4.4). Let $A_t = \int_0^t (-q)(X_s)\, ds$; this is an additive functional (cf. [PTA, (II.3.38)]). Let Y be a random variable that has a distribution which is exponential with parameter 1 and that is independent of X_t, and let $S = \inf\{t : A_t > Y\}$. Let us change the state space from $\mathbb{R}^d$ to $\mathbb{R}^d \cup \{\Delta\}$, where Δ is an isolated point; we extend any function on $\mathbb{R}^d$ to be 0 at Δ. Let $\widehat{X}_t = X_t$ if $t < S$ and set $\widehat{X}_t = \Delta$ for $t \geq S$. We then can write

$$\mathbb{E}^x\left[f(X_t)e^{-A_t}\right] = \mathbb{E}^x \int_0^\infty f(X_t)e^{-s}1_{(A_t\in ds)}$$

$$= \mathbb{E}^x \int_0^\infty \int_s^\infty e^{-y}\,dy\,f(X_t)1_{(A_t\in ds)}$$

$$= \mathbb{E}^x \int_0^\infty \int_0^y f(X_t)1_{(A_t\in ds)}e^{-y}\,dy$$

$$= \mathbb{E}^x \int_0^\infty \int_0^y f(X_t)1_{(A_t\in ds)}1_{(Y\in dy)}$$

$$= \mathbb{E}^x \int_0^Y f(X_t)1_{(A_t\in ds)}$$

$$= \mathbb{E}^x[f(X_t); A_t < Y] = \mathbb{E}^x f(\widehat{X}_t). \tag{4.5}$$

This is usually phrased by saying the process $\widehat{X}_t$ proceeds until the random clock A_t exceeds Y, at which time $\widehat{X}_t$ is *killed* and is immediately transported to the *cemetery* Δ.

5. Girsanov transformation

Let $\mathcal{L}$ be as in (1.1) and define

$$\mathcal{L}'f(x) = \sum_{i,j=1}^d a_{ij}(x)\partial_{ij}f(x),$$

that is, the operator $\mathcal{L}$ with the first-order terms omitted. Solutions to PDEs involving $\mathcal{L}$ can be written in terms of the diffusion corresponding to $\mathcal{L}'$ (and vice versa). As an example to illustrate this, we consider Poisson's equation in $\mathbb{R}^d$. One can obtain analogous results for the Dirichlet problem and the Cauchy problem.

(5.1) Theorem. *Suppose each coordinate of σ is bounded and in C^2, σ^{-1} is bounded, and $a = \sigma\sigma^T$. Suppose X_t is the solution to*

$$dX_t = \sigma(X_t)\,dW_t,$$

and u is a solution to $\mathcal{L}u - \lambda u = -f$ in $\mathbb{R}^d$ such that u and its first and second partial derivatives are bounded. Then

$$u(x) = \mathbb{E}^x \int_0^\infty e^{-\lambda t}f(X_t)M_t\,dt,$$

where

$$M_t = \exp\left(\int_0^t \rho(X_s)\,dW_s - \frac{1}{2}\int_0^t |\rho(X_s)|^2\,ds\right),$$

and $\rho = b(\sigma^T)^{-1}$.

Proof. Let $N_t = \int_0^t \rho(X_s)\,dW_s$ so that $M_t = \exp(N_t - \langle N\rangle_t/2)$ is a martingale with $M_0 = 1$. Define a new probability measure $\mathbb{Q}$ by setting $d\mathbb{Q}/d\mathbb{P}^x = M_t$ on $\mathcal{F}_t$. By Girsanov's theorem (see Section I.1), if Y_t is a martingale under $\mathbb{P}^x$, then $Y_t - \langle N, Y\rangle_t$ is a martingale under $\mathbb{Q}$, and $\langle Y\rangle_t$ is the same under both probability measures.

We apply this to X_t^i. This process is a martingale under $\mathbb{P}^x$, so $X_t^i - \langle N, X^i\rangle_t$ is a martingale under $\mathbb{Q}$. A calculation shows

$$d\langle N, X^i\rangle_t = \sum_{j=1}^d \rho_j(X_s)\,d\langle W^j, X^i\rangle_t$$

$$= \sum_{j=1}^d \rho_j(X_s)\sigma_{ij}(X_s)\,ds = b_i(X_s)\,ds.$$

The quadratic variation of X^i is the same under both $\mathbb{P}^x$ and $\mathbb{Q}$, and by a polarization argument (cf. [PTA, (I.4.14)]), the mixed quadratic variations $\langle X^i, X^j\rangle$ are as well. If $\widehat{W}_t$ is defined by

$$d\widehat{W}_t = \sigma^{-1}(X_t)\,(dX_t - b(X_t)\,dt),$$

we conclude that $\widehat{W}_t$ is a continuous martingale under $\mathbb{Q}$ with $\langle \widehat{W}^i, \widehat{W}^j\rangle_t = \delta_{ij}\,dt$, hence a Brownian motion under $\mathbb{Q}$ (see Section I.1).

We then can write

$$dX_t^i = \sum_{j=1}^d \sigma(X_t)\,d\widehat{W}_t^j + b_i(X_t)\,dt,$$

and so X_t under $\mathbb{Q}$ is associated to the operator $\mathcal{L}$. Hence by Theorem 1.1,

$$u(x) = \mathbb{E}_{\mathbb{Q}} \int_0^\infty e^{-\lambda t} f(X_t)\,dt = \int_0^\infty e^{-\lambda t}\mathbb{E}_{\mathbb{Q}} f(X_t)\,dt.$$

By the definition of $\mathbb{Q}$, this is

$$\int_0^\infty e^{-\lambda t}\mathbb{E}^x[f(X_t)M_t]\,dt. \qquad \Box$$

6. The Neumann and oblique derivative problems

Suppose D is a bounded smooth domain. The Neumann problem for D is the following: given f a smooth function on ∂D, find $u \in C(\overline{D})$ such that u is C^2 on D and

$$\mathcal{L}u = 0 \text{ in } D, \qquad \partial u/\partial \nu = f \text{ on } \partial D, \tag{6.1}$$

where $\nu(x)$ denotes the inward pointing unit normal vector at $x \in \partial D$. For the Neumann problem to have a solution, side conditions need to be imposed. For example, if $\mathcal{L} = \Delta$, the Laplacian, we must have $\int_{\partial D} f(y)\sigma(dy) = 0$, where $\sigma(dy)$ is surface measure on ∂D. To see this, by Green's first identity ([PTA, Theorem II.3.10]) with $v = 1$,

$$0 = \int_D 1\,\Delta u + \int_D \nabla 1 \cdot \nabla u = \int_D 1\,(\partial u/\partial \nu)\,d\sigma = \int_{\partial D} f\,d\sigma.$$

To avoid dealing with side conditions, let us introduce a smooth compact subset K of D and require instead of (6.1) that

$$\mathcal{L}u = 0 \text{ in } D - K, \qquad \partial u/\partial \nu = f \text{ on } \partial D, \qquad u = 0 \text{ on } K. \qquad (6.2)$$

Suppose X_t satisfies

$$dX_t = \sigma(X_t)\,dW_t + b(X_t)\,dt + \nu(X_t)\,dL_t, \qquad X_t \in \overline{D}, \qquad X_0 = x_0, \qquad (6.3)$$

where W_t is d-dimensional Brownian motion, L_t is a nondecreasing continuous process that increases only when $X_t \in \partial D$, σ and b are smooth, D is a bounded C^2 domain, $a = \sigma\sigma^T$ is uniformly elliptic, and

$$\mathcal{L}u(x) = \frac{1}{2}\sum_{i,j} a_{ij}(x)\partial_{ij}u(x) + \sum_i b_i(x)\partial_i u(x), \qquad x \in D.$$

By Section I.12, we have existence and uniqueness of X_t, and we can construct a strong Markov family $(\mathbb{P}^x, X_t)$, $x \in \overline{D}$.

We will see in Chapter III that the solution to (6.2) is smooth.

(6.1) Theorem. *Suppose T_K, the hitting time to K, is finite a.s. and $\mathbb{E}^x L_{T_K} < \infty$ for all x. The solution to (6.2) satisfies*

$$u(x) = -\mathbb{E}^x \int_0^{T_K} f(X_s)\,dL_s.$$

Proof. By Itô's formula,

$$u(X_{t \wedge T_K}) = u(X_0) + \text{ martingale } + \int_0^{t \wedge T_K} \mathcal{L}u(X_s)\,ds$$

$$+ \int_0^{t \wedge T_K} (\nabla u \cdot \nu)(X_s)\,dL_s.$$

Note $\nabla u \cdot \nu = \partial u/\partial \nu = f$. We take expectations with respect to $\mathbb{P}^x$ and then let $t \to \infty$. Since $u = 0$ on K and $\mathcal{L}u = 0$ in D, we obtain

$$0 = u(x) + \mathbb{E}^x \int_0^{T_K} f(X_s)\,dL_s. \qquad \square$$

The assumptions that $T_K < \infty$ a.s. and $\mathbb{E}^x L_{T_K} < \infty$ actually turn out to be superfluous, but we do not prove that here.

We remark that if f satisfies the appropriate side conditions, we can avoid the introduction of K and write the solution to (6.1) as $u(x) = -\lim_{t\to\infty} \mathbb{E}^x \int_0^t f(X_s)\,dL_s$. See Bass and Hsu [1] for a proof in the case that $\mathcal{L}$ is the Laplacian.

The oblique derivative problem is similar. We consider only a special case: we let v be a smooth vector field on ∂D with $v(y) \cdot \nu(y) > 0$ for all $y \in \partial D$ and we consider the problem

$$\mathcal{L}u = 0 \text{ in } D, \qquad \partial u/\partial v = f \text{ on } \partial D, \qquad u = 0 \text{ on } K, \qquad (6.4)$$

where $\partial u/\partial v$ denotes $\nabla u \cdot v$. (More general boundary conditions can be handled.) We will see in Chapter III that the solution to (6.4) is smooth. We now let X_t be the solution to (6.3) where ν is replaced by v. The same proof as that in Theorem 6.1 proves

(6.2) Theorem. *The solution to (6.4) satisfies*

$$u(x) = -\mathbb{E}^x \int_0^{T_K} f(X_s)\,dL_s.$$

7. Fundamental solutions and Green functions

The function $p(t, x, y)$ is the *fundamental solution* for $\mathcal{L}$ if the solution to

$$\partial_t u = \mathcal{L}u, \qquad u(x, 0) = f(x) \qquad (7.1)$$

is given by

$$u(x, t) = \int p(t, x, y) f(y)\,dy$$

for all continuous f with compact support. We have seen that the solution is also given by $\mathbb{E}^x f(X_t)$. So

$$\int p(t, x, y) f(y)\,dy = \mathbb{E}^x f(X_t) = \int f(y) \mathbb{P}^x(X_t \in dy).$$

Thus the fundamental solution is the same as the transition density for the associated process.

An operator $\mathcal{L}$ in a nice domain D has a *Green function* $G_D(x, y)$ if $G_D(x, y) = 0$ if either x or y is in ∂D and the solution to

$$\mathcal{L}u = f \text{ in } D, \qquad u = 0 \text{ on } \partial D$$

is given by

$$u(x) = -\int G_D(x, y) f(y)\, dy$$

when f is continuous. We have also seen that the solution is given by

$$u(x) = -\mathbb{E}^x \int_0^{\tau_D} f(X_s)\, ds.$$

Thus $G_D(x, y)$ is the same as the occupation time density for X_t. That is, $G_D(x, y)$ is the Radon-Nikodym derivative of the measure $\mu(A) = \mathbb{E}^x \int_0^{\tau_D} 1_A(X_s)\, ds$ with respect to Lebesgue measure. See [PTA, Section II.3] for a discussion of the Laplacian case.

8. Adjoints

The *adjoint* operator to $\mathcal{L}$ is the operator

$$\mathcal{L}^* f(x) = \sum_{i,j=1}^d \partial_{ij}\big(a_{ij}(x) f(x)\big) - \sum_{i=1}^d \partial_i\big(b_i(x) f(x)\big). \tag{8.1}$$

The reason for the name is that

$$\int_{\mathbb{R}^d} f(x) \mathcal{L}g(x)\, dx = \int_{\mathbb{R}^d} g(x) \mathcal{L}^* f(x)\, dx,$$

as integrations by parts show, provided f and g satisfy suitable regularity conditions. The adjoint operator corresponds to the process that is the dual of X_t. Roughly speaking, the dual of X_t is the process run backwards: $X_{t_0 - t}$; see Haussmann and Pardoux [1].

Suppose $p(t, x, y)$ is the fundamental solution for $\mathcal{L}$ and let

$$P_t f(x) = \int p(t, x, y) f(y)\, dy.$$

Let $q(t, x, y) = p(t, y, x)$. Let us suppose that the coefficients of $\mathcal{L}$ are smooth and that $p(t, x, y)$ has bounded derivatives in x, y, and t for each $t > 0$.

(8.1) Proposition. *The fundamental solution for $\mathcal{L}^*$ is $q(t, x, y)$.*

Proof. Let g be continuous and nonnegative and let

$$v(x, t) = \int q(t, x, y) g(y)\, dy.$$

So if f is continuous and nonnegative,

$$\int f(x)v(x,t)\,dx = \int\int f(x)p(t,y,x)g(y)\,dy\,dx = \int P_t f(y)g(y)\,dy.$$

When $t = 0$, $P_t f(y) = f(y)$, or

$$\int f(x)v(x,0)\,dx = \int f(x)g(x)\,dx.$$

This implies $g = v(\cdot,0)$ a.e.

By Itô's formula,

$$P_t f(x) - f(x) = \mathbb{E}^x \int_0^t \mathcal{L}f(X_s)\,ds = \int_0^t P_s \mathcal{L}f(x)\,ds.$$

So $\partial_t P_t f = P_t \mathcal{L}f$, and hence we have

$$\int f(x)\partial_t v(x,t)\,dx = \partial_t\left(\int f(x)v(x,t)\,dx\right)$$
$$= \partial_t\left(P_t f(y)g(y)\,dy\right)$$
$$= \int P_t \mathcal{L}f(y)g(y)\,dy.$$

Now

$$\int P_t h(y)g(y)\,dy = \int\int p(t,y,x)h(x)g(y)\,dx\,dy$$
$$= \int h(x)v(x,t)\,dx.$$

So

$$\int f(x)\partial_t v(x,t)\,dx = \int \mathcal{L}f(x)v(x,t)\,dx$$
$$= \int f(x)\mathcal{L}^* v(x,t)\,dx.$$

Hence $\partial_t v(x,t) = \mathcal{L}^* v(x,t)$ for almost every x. We will see later on that the fundamental solution to $\mathcal{L}^*$ is continuous, so we have equality everywhere.
$\square$

By integrating over t from 0 to ∞, provided the Green function exists, then the Green function for $\mathcal{L}^*$ is $G_D(y,x)$.

Examining the proof of [PTA, Proposition II.4.1] lends credence to the assertion that $q(t,x,y) = p(t,y,x)$ is the transition density of X_{t_0-t}, and so the adjoint operator $\mathcal{L}^*$ corresponds to the process X_t run backwards in time. This is not quite true, but something close to it is; see Haussmann and Pardoux [1] for details.

A measure μ is *invariant* for a strong Markov family if $\mathbb{E}^\mu f(X_t) = \int f(x)\,\mu(dx)$ for all t and all bounded and continuous f, where $\mathbb{E}^\mu f(X_t) =$

$\int \mathbb{E}^y f(X_t)\,\mu(dy)$. We continue to assume the same regularity as in the preceding proposition.

(8.2) Proposition. *Suppose there exists a nonnegative solution v to $\mathcal{L}^* v = 0$. Let $\mu(dx) = v(x)\,dx$. Then μ is invariant for the process associated to $\mathcal{L}$.*

Proof. Let f be continuous and let $u(x,t) = \mathbb{E}^x f(X_t)$. Then

$$0 = \int u(x,t)\mathcal{L}^* v(x)\,dx = \int \mathcal{L}u(x,t)v(x)\,dx = \int \partial_t u(x,t)v(x)\,dx.$$

This implies that

$$\mathbb{E}^\mu f(X_t) = \int \mathbb{E}^x f(X_t)\,\mu(dx) = \int u(x,t)v(x)\,dx$$

is a constant function of t. Letting $t \to 0$,

$$\mathbb{E}^\mu f(X_t) = \int \mathbb{E}^x f(X_t)\,\mu(dx) \to \int f(x)\,\mu(dx),$$

so $\mathbb{E}^\mu f(X_t) = \int f(x)\,\mu(dx)$ for all t. $\qquad\square$

9. Notes

For further information see Durrett [1], Dynkin [1], Pinsky [1], and Stroock and Varadhan [2].

There are certain quasilinear elliptic operators that can be interpreted probabilistically. See Dynkin [2], Chen, Williams, and Zhao [1], LeGall [1], and Funaki [1].

III
REGULARITY OF SOLUTIONS

In order to apply the results of Chapter II we need to know that solutions to Poisson's equation, the Dirichlet problem, etc. exist with sufficient smoothness provided we make suitable assumptions on the coefficients of the operator $\mathcal{L}$ and on the domain D. That is the purpose of this chapter.

In addition, in much of this book we will be interested in elliptic operators whose coefficients are not overly regular. We will often obtain them as limits of operators with more regular coefficients. We therefore will need to show that equations whose operators have smooth coefficients have smooth solutions.

Section 1 is an introduction to the method of variation of parameters. Section 2 contains a discussion of the Hölder and weighted Hölder norms and a derivation of some bounds on the second derivatives of potentials. In Section 3 we show that the diffusions we are interested in have a nice regularity property near the boundary of smooth domains.

Sections 4 and 5 establish the regularity of solutions of Poisson's equation and the Dirichlet problem. In Section 4 we use variation of parameters and the estimates of Sections 2 and 3 to study Poisson's equation. Section 5 combines the results of Sections 3 and 4 to obtain the existence of a solution to the Dirichlet problem.

Section 6 points out how the methods of the previous sections extend to cover the cases of higher order derivatives, of two dimensions, and of first and zero order terms.

Section 7 details the modifications necessary to deal with the Neumann problem and the oblique derivative problem.

Section 8 is another application of variation of parameters to Poisson's equation, whereas Section 9 shows how flows can be used to study the Cauchy problem.

1. Variation of parameters

One of the most common means of proving regularity of solutions of PDEs is that of *variation of parameters*. This is also known as the *parametrix method* or the *perturbation method*. The basic idea is simple. If a and b are real numbers, $r_\lambda = (\lambda - a)^{-1}$, $s_\lambda = (\lambda - (a+b))^{-1}$, and $|br_\lambda| < 1$, then

$$s_\lambda = \frac{1}{\lambda - a - b} = \left(\frac{1}{\lambda - a}\right)\left(\frac{1}{1 - b/(\lambda - a)}\right)$$
$$= r_\lambda \frac{1}{1 - br_\lambda} = r_\lambda(1 + br_\lambda + (br_\lambda)^2 + \cdots).$$

Now let $\mathcal{A}$ and $\mathcal{B}$ be linear operators, $R_\lambda = (\lambda - \mathcal{A})^{-1}$. If $\|\mathcal{B}R_\lambda\| < 1$ with respect to some norm $\|\cdot\|$, then the sum

$$S_\lambda = R_\lambda + R_\lambda\mathcal{B}R_\lambda + R_\lambda\mathcal{B}R_\lambda\mathcal{B}R_\lambda + \cdots \tag{1.1}$$

converges with respect to this norm, and formally, if we apply $\lambda - (\mathcal{A} + \mathcal{B})$ to S_λ, we obtain the identity operator, or $S_\lambda = (\lambda - (\mathcal{A} + \mathcal{B}))^{-1}$. The series in (1.1) is known as a Neumann series, after Carl Neumann.

Suppose $\mathcal{A}$ is the infinitesimal generator of a semigroup P_t, that is,

$$\mathcal{A}f = \frac{d(P_t f)}{dt}\bigg|_{t=0}$$

for a suitable class of fs. There is an analogous formula to (1.1) for the semigroup corresponding to $\mathcal{A} + \mathcal{B}$; see Leviatan [1].

Rather than worrying about domains of the operators $\mathcal{A}$ and $\mathcal{A} + \mathcal{B}$, we will instead show in a concrete situation how variation of parameters may be used to obtain regularity results.

In our first and principal application of variation of parameters, we will take $\lambda = 0$ and $\mathcal{A} = (1/2)\Delta$ on a ball B. R_λ becomes the Green operator G_B defined by

$$G_B f(x) = \int f(y) g_B(x, y)\, dy, \tag{1.3}$$

where g_B is the Green function for Brownian motion on B. We will consider $\mathcal{L}$ defined by

$$\mathcal{L}f(x) = \frac{1}{2}\sum_{i,j=1}^{d} a_{ij}(x)\partial_{ij}f(x),$$

where the a_{ij} are Hölder continuous and strictly elliptic, and we will set

$$\mathcal{B} = \mathcal{L} - \frac{1}{2}\Delta.$$

We thus need estimates on $\mathcal{B}G_B f$ in a suitable norm, and these will arise from estimates on $\partial_{ij}G_B f$.

Let $d \geq 3$ and let Uf be the Newtonian potential of f, that is

$$Uf(x) = \int u(x,y)f(y)\,dy, \qquad u(x,y) = c_0|x-y|^{2-d}, \qquad (1.4)$$

where $c_0 = \Gamma((d/2)-1)/(2\pi)^{d/2}$ is chosen so that $(1/2)\Delta Uf = -f$ for smooth f (see [PTA, Section II.3] for a discussion of potentials). We know

$$Uf(x) = \mathbb{E}^x \int_0^\infty f(X_t)\,dt, \qquad G_B f(x) = \mathbb{E}^x \int_0^{\tau_B} f(X_t)\,dt,$$

where τ_B is the first exit time from B. Let us write

$$P_B f(x) = \int f(y)\mathbb{P}^x(X_{\tau_B} \in dy); \qquad (1.5)$$

this is the harmonic extension of f to B. By the strong Markov property,

$$
\begin{aligned}
Uf(x) &= G_B f(x) + \mathbb{E}^x \int_{\tau_B}^\infty f(X_t)\,dt \\
&= G_B f(x) + \mathbb{E}^x \mathbb{E}^{X(\tau_B)} \int_0^\infty f(X_t)\,dt \\
&= G_B f(x) + \mathbb{E}^x Uf(X_{\tau_B}).
\end{aligned}
$$

We then have

$$G_B f(x) = Uf(x) - P_B(Uf)(x), \qquad (1.6)$$

and we thus need bounds on $\partial_{ij}Uf$ and $\partial_{ij}P_B(Uf)$.

2. Weighted Hölder norms

In order to apply variations of parameters, we first need to find a suitable norm $\|\cdot\|$ to work with. The main one we use is the weighted Hölder norm $\|\cdot\|_{WH}$.

Our strategy is first to consider the C^α norm and to obtain estimates on $\partial_{ij}Uf$. Then we introduce what we call the scaled Hölder norm and again examine $\partial_{ij}Uf$. We next estimate $P_B f$ under the C^α and scaled Hölder norms. From these and (1.6), we deduce bounds on $\partial_{ij}G_B f$ with respect to the scaled Hölder norm. In (2.6) we define the weighted Hölder norm. Theorem 2.4 contains the main result of this section, a bound on the weighted Hölder norm of $\partial_{ij}G_B f$. The remainder of the section discusses some related results that will be needed later in this chapter.

We start with the C^α norm. If f is a real-valued function on a Borel set B, define

$$\|f\|_{C^\alpha(B)} = \sup_{x \in B} |f(x)| + \sup_{x,y \in B} \frac{|f(x) - f(y)|}{|x - y|^\alpha}. \tag{2.1}$$

We make the convention that $0/0 = 0$ in a ratio such as the last term of (2.1). When $B = \mathbb{R}^d$, we write $\|f\|_{C^\alpha}$.

Recall (see [PTA, Theorem II.3.14]) that if $d \geq 3$, $f \in C^\alpha$, the support of f is contained in $B(0,2)$, and Uf is the Newtonian potential of f, then for all i, j we have that $\partial_{ij} Uf \in C^\alpha$ and there exists a constant c_1 independent of f such that

$$\|\partial_{ij} Uf\|_{C^\alpha(B(0,1))} \leq c_1 \|f\|_{C^\alpha}. \tag{2.2}$$

Let us introduce another norm that we will use only temporarily, the scaled Hölder norm. Fix α and define

$$\|f\|_{SH(x_0,R)} = \sup_{x \in B(x_0,R)} |f(x)| + R^\alpha \sup_{x,y \in B(x_0,R)} \frac{|f(x) - f(y)|}{|x - y|^\alpha}.$$

(2.1) Proposition. *Suppose $R \leq 1$, the support of f is contained in $B(x_0, 2R)$, and f is C^α on $B(x_0, 2R)$. There exists c_1 independent of f and R such that*

$$\|\partial_{ij} Uf\|_{SH(x_0,R)} \leq c_1 \|f\|_{SH(x_0,2R)}.$$

Proof. This follows from (2.2) by a scaling argument. Suppose without loss of generality that $x_0 = 0$. Let $g(x) = f(xR)$. Then the support of g is contained in $B(0,2)$, $\|g\|_\infty = \|f\|_\infty$, and

$$|g(x) - g(y)| = |f(xR) - f(yR)| \leq \|f\|_{C^\alpha} |x - y|^\alpha R^\alpha.$$

Note that

$$Ug(x) = c_0 \int |x - y|^{2-d} g(y)\, dy = R^{-d} c_0 \int |x - (z/R)|^{2-d} f(z)\, dz$$

$$= R^{-2} c_0 \int |xR - z|^{2-d} f(z)\, dz = R^{-2} Uf(xR).$$

Then $Uf(x) = R^2 Ug(x/R)$ and hence $\partial_{ij} Uf(x) = \partial_{ij} Ug(x/R)$. If $u = x/R$,

$$|\partial_{ij} Uf(x)| = |\partial_{ij} Ug(u)| \leq c_2 \|g\|_{C^\alpha(B(0,1))} \leq c_2 \|f\|_{SH(x_0,R)}.$$

If $v = y/R$,

$$R^\alpha |\partial_{ij} Uf(x) - \partial_{ij} Uf(y)| = R^\alpha |\partial_{ij} Ug(u) - \partial_{ij} Ug(v)|$$
$$\leq c_2 R^\alpha |u - v|^\alpha \|g\|_{C^\alpha(B(0,1))}$$
$$\leq c_2 |x - y|^\alpha \|f\|_{SH(x_0,2R)}. \qquad \square$$

If $B = B(0,1)$ is the unit ball, let $P_B f$ denote the harmonic extension of f (with respect to the Laplacian) on B. That is, if f is a continuous function on ∂B, then $P_B f$ is the function in B such that $\Delta P_B f = 0$ in B and $P_B f$ agrees with f on the boundary. See [PTA, Section II.1].

(2.2) Proposition. *Suppose $f \in C^\alpha$ on $B(0,2)$. Then $P_B f \in C^\alpha$ on B and there exists c_1 independent of f such that*

$$\|P_B f\|_{C^\alpha(B)} \le c_1 \|f\|_{C^\alpha}.$$

Proof. Clearly $|P_B f(x)| \le \|f\|_\infty$. $P_B f$ is actually C^∞ in B ([PTA, Proposition II.1.3]), so the difficulty is obtaining an estimate on the C^α norm. Let $\varepsilon > 0$ and let f_ε be a C^1 function such that

$$\|f - f_\varepsilon\|_\infty \le c_2 \varepsilon^\alpha \|f\|_{C^\alpha}, \qquad \|\nabla f_\varepsilon\|_\infty \le c_2 \varepsilon^{\alpha-1}\|f\|_{C^\alpha}. \tag{2.3}$$

We will construct such an f_ε in a moment.
Then

$$|P_B f(x) - P_B f(y)| \le |P_B f(x) - P_B f_\varepsilon(x)| + |P_B f(y) - P_B f_\varepsilon(y)|$$
$$+ |P_B f_\varepsilon(x) - P_B f_\varepsilon(y)|.$$

Observe that

$$|P_B f(x) - P_B f_\varepsilon(x)| = |P_B(f - f_\varepsilon)(x)| \le \|f - f_\varepsilon\|_\infty \le c_2 \varepsilon^\alpha \|f\|_{C^\alpha},$$

and similarly with x replaced by y. Also,

$$\begin{aligned}
|P_B f_\varepsilon(x) - P_B f_\varepsilon(y)| &= \int_0^{|y-x|} \partial_t P_B f_\varepsilon\left(x + t\frac{y-x}{|y-x|}\right) dt \\
&\le |y - x|\,\|\nabla P_B f_\varepsilon\|_\infty = |y - x|\,\|P_B(\nabla f_\varepsilon)\|_\infty \\
&\le |y - x|\,\|\nabla f_\varepsilon\|_\infty \le c_2 |y - x|\varepsilon^{\alpha-1}\|f\|_{C^\alpha},
\end{aligned}$$

since $P_B(\partial_i g) = \partial_i(P_B g)$ inside B for all i. If we take $\varepsilon = |y - x|$, combining gives the required estimate.
It remains to construct f_ε. Let $\varphi(x)$ be C^∞ with compact support, nonnegative, and with integral 1. Let $\varphi_\varepsilon(x) = \varepsilon^{-d}\varphi(x/\varepsilon)$ and $f_\varepsilon = f * \varphi_\varepsilon$. Then

$$\begin{aligned}
|f(x) - f_\varepsilon(x)| &= \left|\int [f(x) - f(x - y)]\varphi_\varepsilon(y)\,dy\right| \\
&= \left|\int [f(x) - f(x - \varepsilon y)]\varphi(y)\,dy\right| \\
&\le \varepsilon^\alpha \|f\|_{C^\alpha} \int |y|^\alpha \varphi(y)\,dy \le c_3 \varepsilon^\alpha \|f\|_{C^\alpha}.
\end{aligned}$$

By integration by parts and the fact that φ has compact support, $\int \partial_i \varphi(y)\,dy = 0$. So

$$\begin{aligned}
|\partial_i f_\varepsilon(x)| &= \left| \int f(x-y)\partial_i \varphi_\varepsilon(y)\,dy \right| \\
&= \left| \int [f(x-y) - f(x)]\partial_i \varphi_\varepsilon(y)\,dy \right| \\
&= \varepsilon^{-1} \left| \int [f(x-\varepsilon y) - f(x)]\partial_i \varphi(y)\,dy \right| \\
&\leq \varepsilon^{\alpha-1} \|f\|_{C^\alpha} \int |y|^\alpha \partial_i \varphi(y)\,dy \leq c_3 \varepsilon^{\alpha-1}\|f\|_{C^\alpha},
\end{aligned}$$

which is the other half of (2.3). $\qquad\square$

Let $P_{B(x_0,R)} f$ denote the harmonic extension to the interior of $B(x_0, R)$ of a function f on the boundary. As in Proposition 2.2, if f is C^α on $B(x_0, R)$, then $P_{B(x_0,R)} f$ is C^α on $B(x_0, R)$, and

$$\|P_{B(x_0,R)} f\|_{SH(x_0,R)} \leq c_1 \|f\|_{SH(x_0,R)}. \tag{2.4}$$

Let $G_B f$ be the Green potential of f with respect to the domain B. By (1.6)

$$G_B f = Uf - P_B(Uf). \tag{2.5}$$

(2.3) Proposition. *If f is C^α on $B(x_0, R)$, then $G_{B(x_0,R)} f$ is in $C^{2+\alpha}$ on $B(x_0, R)$ and there exists c_1 independent of f such that*

$$\|\partial_{ij} G_{B(x_0,R)} f\|_{SH(x_0,R)} \leq c_1 \|f\|_{SH(x_0,R)}.$$

Proof. Observe that $G_{B(x_0,R)} f(x) = \mathbb{E}^x \int_0^{\tau_{B(x_0,R)}} f(W_s)\,ds$, where W_t is a Brownian motion and $\tau_{B(x_0,R)}$ is the exit time from $B(x_0, R)$; see [PTA, Section II.3]. Therefore $G_{B(x_0,R)} f$ depends only on the values of f in $B(x_0, R)$; hence without loss of generality we may assume the support of f is contained in $B(x_0, 2R)$ and $\|f\|_{SH(x_0,2R)} \leq c_2 \|f\|_{SH(x_0,R)}$. Since

$$\partial_{ij} G_{B(x_0,R)} f = \partial_{ij} Uf - P_{B(x_0,R)}(\partial_{ij} Uf),$$

the result follows from (2.4) and Proposition 2.1. $\qquad\square$

We now introduce the weighted Hölder norm we are interested in. Let us write d_x for $\mathrm{dist}\,(x, \partial B)$. Define

$$\|f\|_{WH} = \sup_{x \in B} d_x^2 |f(x)| + \sup_{x,y \in B} [d_x \wedge d_y]^{2+\alpha} \frac{|f(x) - f(y)|}{|x-y|^\alpha}. \tag{2.6}$$

If $\|f\|_{WH} < \infty$, we will say $f \in WH$.

(2.4) Theorem. *There exists c_1 such that*

$$\|\partial_{ij} G_B f\|_{WH} \leq c_1 \|f\|_{WH}.$$

Proof. Let $x_0 \in B$ and let $R = d_{x_0}/6$. For $x \in B(x_0, R/2)$, analogously to (1.6),

$$G_B f(x) = G_{B(x_0,R)} f(x) + P_{B(x_0,R)}(G_B f)(x).$$

Then

$$|\partial_{ij} G_B f(x)| \le |\partial_{ij} G_{B(x_0,R)} f(x)| + |\partial_{ij} P_{B(x_0,R)}(G_B f)(x)|$$

$$\le c_2 \|f\|_{SH(x_0,R)}$$

$$= c_2 \sup_{x' \in B(x_0,R)} |f(x')|$$

$$+ c_2 R^\alpha \sup_{x',y' \in B(x_0,R)} \frac{|f(x') - f(y')|}{|x' - y'|^\alpha}.$$

So

$$d_x^2 |\partial_{ij} G_B f(x)| \le c_2 \|f\|_{WH}.$$

Similarly,

$$[d_x \wedge d_y]^{2+\alpha} \frac{|\partial_{ij} G_B f(x) - \partial_{ij} G_B f(y)|}{|x - y|^\alpha} \le c_3 \|f\|_{WH}. \qquad \square$$

We will also need the following proposition.

(2.5) Proposition. *If $f \in C^\alpha$ and $g \in WH$, then*

$$\|fg\|_{WH} \le \|f\|_{C^\alpha} \|g\|_{WH}.$$

Proof. This follows easily from the inequality

$$|f(x)g(x) - f(y)g(y)| \le |f(x)| \, |g(x) - g(y)| + |g(y)| \, |f(x) - f(y)|. \qquad \square$$

Besides the inequality in Theorem 2.4 for the weighted Hölder norm, we also have an analogous inequality for the C^α norm.

(2.6) Proposition. *If $f \in C^\alpha(\overline{B})$, then $\partial_{ij} G_B f \in C^\alpha(\overline{B})$ and there exists c_1 independent of f such that*

$$\|\partial_{ij} G_B f\|_{C^\alpha(\overline{B})} \le c_1 \|f\|_{C^\alpha(\overline{B})}.$$

Proof. We can extend f to have support in $B(0,2)$ such that $\|f\|_{C^\alpha} \le c_2 \|f\|_{C^\alpha(\overline{B})}$. By (2.2) and Proposition 2.2, $\partial_{ij} U f \in C^\alpha$, hence $\partial_{ij}(P_B(U f)) \in C^\alpha(\overline{B})$, and

$$\|\partial_{ij} U f\|_{C^\alpha} \le c_3 \|f\|_{C^\alpha}, \qquad \|\partial_{ij} P_B(U f)\|_{C^\alpha(\overline{B})} \le c_4 \|f\|_{C^\alpha}.$$

Taking the difference and using (2.5) proves our result. $\qquad \square$

Finally, we will need an estimate on harmonic extensions.

(2.7) Proposition. *If φ is bounded on ∂B, then $\partial_{ij} P_B \varphi \in WH$ and there exists a constant c_1 independent of φ such that*

$$\|\partial_{ij} P_B \varphi\|_{WH} \leq c_1 \|\varphi\|_{L^\infty(\partial B)}.$$

Proof. Let $x \in B$, $R = d_x/6$. It suffices to consider the case where $d_y \geq d_x$ and $y \in B(x, R)$. It is well known ([PTA, Corollary II.1.4]) that in $B(x, R)$,

$$\|h\|_{L^\infty(B(x,R))} \leq \frac{c_2}{R^2} \|\varphi\|_{L^\infty(B(x,R))}, \qquad \|\nabla h\|_\infty \leq \frac{c_2}{R^3} \|\varphi\|_\infty,$$

where $h = \partial_{ij} P_B \varphi$. Then

$$|h(y) - h(x)| \leq \left(\sup_{B(x,R)} |\nabla h| \right) |y - x|$$

$$\leq \frac{c_2}{R^3} R^{1-\alpha} |y - x|^\alpha \|\varphi\|_\infty$$

$$= c_2 R^{-(2+\alpha)} |y - x|^\alpha \|\varphi\|_\infty.$$

The proposition follows easily from this. $\qquad\qquad\square$

3. Regularity of hitting distributions

Our main goal in this section is Theorem 3.4, which says that under suitable regularity conditions on X_t and D, the function $H(x) = \mathbb{E}^x \varphi(X_{\tau_D})$ is continuous on $\overline{D}$ if φ is continuous on ∂D. Here τ_D is the time to exit D.

We begin by studying regularity of the boundary. Our first two results lead to a condition in terms of cones for a point x of ∂D to be such that starting at x the process leaves D immediately. A *cone* will be a translate and rotation of the open set $\{(x_1, \ldots, x_d) : x_1^2 + \cdots + x_{d-1}^2 < \alpha x_d^2\}$ for some α.

(3.1) Proposition. *Suppose X_t satisfies (I.2.3), σ, b, and σ^{-1} are bounded, D is a domain, $x \in \partial D$, and there exists a cone V contained in D^c with vertex x. Then $\mathbb{P}^x(\tau_D = 0) > 0$.*

Proof. Without loss of generality, take $x = 0$. Let ψ be a curve starting at 0 and entering the interior of V by time 1, and take ε small enough so that $B(\psi(1), \varepsilon) \subseteq V$. By the support theorem, $\mathbb{P}(X_1 \in V) \geq c_1$.

We now use scaling. Let $X_t^a = a X_{t/a^2}$. By Proposition I.8.6, X_t^a satisfies an SDE of the same form as (I.2.3), and if σ_a and b_a are the corresponding coefficients, then σ_a and b_a satisfy the same bounds as σ and b provided $a > 1$. So by the support theorem (with the same ψ and ε), $\mathbb{P}(X_t^a$ enters V before time 1) $\geq c_1$. This implies, since a cone is invariant under scaling, that

$$\mathbb{P}(\tau_D < a^{-2}) \geq \mathbb{P}(X_t \text{ enters } V \text{ before time } a^{-2}) \geq c_1.$$

Finally, we let $a \to \infty$. $\square$

Let us now suppose in addition that σ and b are smooth enough so that the solutions to (I.2.3) form a strong Markov process and $\mathbb{P}^{x_n}$ converges weakly to $\mathbb{P}^x$ whenever $x_n \to x$. A sufficient condition is that $\sigma(x)$ and $b(x)$ be Lipschitz in x: by Theorem I.10.1, $X(x_n, t)$ converges to $X(x, t)$ almost surely, and weak convergence follows. Later, in Chapter VI, we will see that in fact it is sufficient that b be bounded and σ be bounded, strictly elliptic, and continuous.

A domain D satisfies the *external cone condition* if for all $x \in \partial D$, there exists a cone V whose vertex is at x and which lies in D^c.

(3.2) Corollary. *Suppose $(\mathbb{P}^x, X_t)$ is a strong Markov process, where σ and b are as above. For all x, $\mathbb{P}^x(\tau_D = 0) = 1$.*

Proof. Since $\mathbb{P}^{x_n}$ converges weakly to $\mathbb{P}^x$ whenever $x_n \to x$, then

$$P_t f(x_n) = \mathbb{E}^{x_n} f(X_t) \to \mathbb{E}^x f(X_t) = P_t f(x)$$

if f is bounded and continuous. So if f is bounded and continuous, then $P_t f$ is also. This and the proofs in [PTA, Section I.3] (see in particular Theorem I.3.4, Proposition I.3.5, and Corollary I.3.6 of that book) show that the Blumenthal 0-1 law ([PTA, Corollary I.3.6)] holds for $(\mathbb{P}^x, X_t)$, that is, sets in $\mathcal{F}_{0+}$ must have probability 0 or 1. By Proposition 3.1, $\mathbb{P}^x(\tau_D = 0) > 0$, and hence the probability must be 1. $\square$

The map $X_t(\omega) \to \varphi(X_{\tau_D}(\omega))$ is not necessarily continuous on $C[0, \infty)$, even when φ is continuous. To handle such functionals, we need the following theorem from Billingsley [1].

(3.3) Theorem. *Suppose $\mathbb{P}_n$ converges weakly to $\mathbb{P}$, where $\mathbb{P}_n$ and $\mathbb{P}$ are probability measures on a metric space S. Suppose h maps S to another metric space S' and $E = \{x \in S : h$ is not continuous at $x\}$. If $\mathbb{P}(E) = 0$, then $\mathbb{P}_n h^{-1}$ converges weakly to $\mathbb{P} h^{-1}$.*

Proof. Let F be closed in S'. It suffices to show ([PTA, Theorem I.7.2]) that

$$\limsup_n \mathbb{P}_n h^{-1}(F) \leq \mathbb{P} h^{-1}(F). \tag{3.1}$$

Suppose $x \notin E$. If x is a limit point of $h^{-1}(F)$, then since h is continuous at x, we see that $h(x)$ is a limit point of F. Because F is closed, $h(x) \in F$, or $x \in h^{-1}(F)$. This shows that

$$\overline{h^{-1}(F)} \subseteq E \cup h^{-1}(F).$$

Since $\mathbb{P}_n$ converges weakly to $\mathbb{P}$ and $\mathbb{P}(E) = 0$,

$$\limsup_n \mathbb{P}_n(h^{-1}(F)) \leq \limsup_n \mathbb{P}_n(\overline{h^{-1}(F)}) \leq \mathbb{P}(\overline{h^{-1}(F)})$$

$$\leq \mathbb{P}(h^{-1}(F)) + \mathbb{P}(E) = \mathbb{P}(h^{-1}(F)),$$

which is (3.1). $\qquad\square$

The application we have in mind is the following.

(3.4) Theorem. *Let X_t, D be as above. Suppose φ is a continuous function on ∂D and $H(x) = \mathbb{E}^x \varphi(X_{\tau_D})$. Then H is continuous on $\overline{D}$ and agrees with φ on ∂D.*

Proof. That H agrees with φ on ∂D follows from Corollary 3.2. Suppose $x_n, x \in \overline{D}$ with $x_n \to x$. Let Ω be the space of continuous paths, let $X_t(\omega) = \omega(t)$, and define a mapping $h : \Omega \to \mathbb{R}$ by $h(\omega) = \varphi(X_{\tau_D}(\omega))$. Observe that h is not continuous at ω only if X_t hits ∂D but does not immediately thereafter enter $(\overline{D})^c$, that is, if $T_{(\overline{D})^c} \circ \theta_{T_{\partial D}} > 0$, where θ_t denotes the shift operators defined in Section I.5. By the strong Markov property,

$$\mathbb{P}^x(T_{(\overline{D})^c} \circ \theta_{T_{\partial D}} > 0) = \mathbb{E}^x \mathbb{P}^{X(T_{\partial D})}(T_{(\overline{D})^c} > 0).$$

By Corollary 3.2, $\mathbb{P}^y(T_{(\overline{D})^c} > 0) = 0$ for $y \in \partial D$. Hence h is continuous on Ω except for a set of probability 0. By Theorem 3.3, $\mathbb{P}^{x_n} h^{-1}$ converges weakly to $\mathbb{P}^x h^{-1}$, that is, if f is a continuous bounded function on $\mathbb{R}$, then

$$\int f(y)\, \mathbb{P}^{x_n} h^{-1}(dy) \to \int f(y)\, \mathbb{P}^x h^{-1}(dy). \tag{3.2}$$

Note

$$\mathbb{P}^x h^{-1}(A) = \mathbb{P}^x(h^{-1}(A)) = \int 1_{h^{-1}(A)}(y)\, \mathbb{P}^x(dy)$$

$$= \int 1_A(h(y))\, \mathbb{P}^x(dy).$$

By a limit argument,

$$\int f(y)\, \mathbb{P}^x h^{-1}(dy) = \int f(h(y))\, \mathbb{P}^x(dy),$$

and similarly with $\mathbb{P}^x$ replaced by $\mathbb{P}^{x_n}$. Let f be a continuous bounded function that is equal to the identity on the range of φ. Then by the definition of h,

$$H(x_n) = \mathbb{E}^{x_n} \varphi(X_{\tau_D}) = \mathbb{E}^{x_n} h(X.) = \int f(h(y))\, \mathbb{P}^{x_n}(dy).$$

This converges by (3.2) to $\int f(h(y))\, \mathbb{P}^x(dy)$, which as in the above, is equal to $H(x)$. $\qquad\square$

4. Schauder estimates

We consider elliptic operators in *nondivergence form*

$$\mathcal{L}f(x) = \frac{1}{2}\sum_{i,j=1}^{d} a_{ij}(x)\partial_{ij}f(x) + \sum_{i=1}^{d} b_i(x)\partial_i f(x). \qquad (4.1)$$

We assume that $\mathcal{L}$ is uniformly strictly elliptic.

Recall $\delta_{ij} = 1$ if $i = j$ and 0 otherwise. Let K be the larger of the c_1 in Theorem 2.4 and the c_1 in Proposition 2.6. Let $\varepsilon_0 = 1/(Kd^2)$. In this section let us assume the following.

Assumption 4.1. (i) *The b_i are identically zero;*
(ii) $a_{ij} \in C^\alpha$ *with* $\|a_{ij}(x) - \delta_{ij}\|_{C^\alpha} < \varepsilon_0$.

In the next sections we will weaken these assumptions.

Let B be the unit ball. We use variation of parameters as discussed at the end of Section 1 and Theorem 2.4 to obtain the following key result.

(4.2) Theorem. *Suppose Assumption 4.1 holds. Suppose that $f \in WH$, $f \in C^\alpha(\overline{B})$, and $f = 0$ on ∂B. Then there exists $v \in WH$ such that v is continuous on $\overline{B}$, $\mathcal{L}v = f$, and $v = 0$ on ∂B. Moreover, there exists a constant c_1 independent of f such that for each i and j*

$$\|\partial_{ij}v\|_{WH} \leq c_1\|f\|_{WH}.$$

Proof. Define an operator $\mathcal{B}$ by

$$\mathcal{B}f(x) = \frac{1}{2}\sum_{i,j=1}^{d}(a_{ij}(x) - \delta_{ij})\partial_{ij}f(x). \qquad (4.2)$$

Then using Proposition 2.5,

$$\begin{aligned}
\|\mathcal{B}G_B h\|_{WH} &\leq \frac{1}{2}\sum_{i,j=1}^{d}\|(a_{ij}(x) - \delta_{ij})\partial_{ij}G_B h\|_{WH} \\
&\leq \frac{1}{2}d^2 \sup_{i,j}\|a_{ij} - \delta_{ij}\|_{C^\alpha}\|\partial_{ij}G_B h\|_{WH} \\
&\leq \frac{1}{2}\varepsilon_0 d^2 K\|h\|_{WH} \\
&= \frac{1}{2}\|h\|_{WH}.
\end{aligned}$$

Hence $\mathcal{B}G_B$ maps WH into WH with a norm bounded by $1/2$. Similarly, $\mathcal{B}G_B$ maps $C^\alpha(\overline{B})$ into $C^\alpha(\overline{B})$ with norm bounded by $1/2$.

Let $g = \sum_{m=0}^{\infty}(\mathcal{B}G_B)^n f$. Then $g \in WH$. Let $v = -G_B g$, so for each i and j we have $\partial_{ij}v \in WH$, and

$$\|\partial_{ij}v\|_{WH} \leq c_2\|g\|_{WH} \leq c_3\|f\|_{WH}.$$

Also, $g \in C^\alpha(\overline{B})$, so in particular g is bounded. Hence v is continuous on $\overline{B}$ and equals 0 on ∂B.

Since $\partial_{ij}v \in C^\alpha$ for each i and j, $\mathcal{L}v$ makes sense. $v = -G_B g$ with $g \in C^\alpha$, so $\Delta v/2 = g$. Then

$$\mathcal{L}v = \Delta v/2 + \mathcal{B}v = g - \mathcal{B}G_B g$$
$$= \sum_{m=0}^{\infty}(\mathcal{B}G_B)^m f - \sum_{m=1}^{\infty}(\mathcal{B}G_B)^m f = f. \qquad \square$$

In preparation for discussing the Dirichlet problem we need to obtain some additional estimates of weighted Hölder norms.

(4.3) Theorem. *Suppose Assumption 4.1 holds. There exists c_1 with the following property: if φ is in $C^3(\mathbb{R}^d)$, there exists $w \in WH$ such that $\mathcal{L}w = 0$ in B, $w = \varphi$ on ∂B, and for each i and j*

$$\|\partial_{ij}w\|_{WH} \leq c_1\|\varphi\|_{L^\infty(\partial B)}.$$

Proof. Let $h = P_B(\varphi)$ and let $f = -\mathcal{L}h$. By Proposition 2.7, $\partial_{ij}h \in WH$ for all i and j and $\sup_{i,j}\|\partial_{ij}h\|_{WH} \leq c_2\|\varphi\|_{L^\infty(\partial B)}$, so $f \in WH$ and $\|f\|_{WH} \leq c_3\|\varphi\|_{L^\infty(\partial B)}$. Construct v as in Theorem 4.2 so that $\mathcal{L}v = f$. Then $w = v + h$ is continuous on $\overline{B}$, agrees with φ on ∂B, and

$$\mathcal{L}w = \mathcal{L}v + \mathcal{L}h = f - f = 0$$

inside of B. Finally, for each i and j,

$$\|\partial_{ij}w\|_{WH} \leq \|\partial_{ij}v\|_{WH} + \|\partial_{ij}h\|_{WH} \leq c_4\|\varphi\|_{L^\infty(\partial B)}. \qquad \square$$

Let us continue to assume Assumption 4.1. Now we in addition assume that $a = \sigma\sigma^T$, where σ is symmetric and uniformly elliptic, and that σ is continuous in x if a is continuous in x. If a is Lipschitz in x, we require σ to be also. Given $a(x)$, one can construct $\sigma(x)$ using Taylor series as follows. Let $\sum_{i=0}^{\infty} b_i(1-x)^i$ be the Taylor series for $x^{1/2}$. This expansion will be valid provided $|1 - x| < 1$. Let $A = \sup_{x,i,j}|a_{ij}(x)|$. Since a is uniformly elliptic, then

$$\sup_x\left|1 - \frac{a(x)}{A}\right| < 1,$$

and we set

$$\sigma(x) = A^{1/2}\sum_{i=0}^{\infty} b_i\left(1 - \frac{a(x)}{A}\right)^i.$$

It follows that σ will be continuous (respectively, Lipschitz) when a is continuous (respectively, Lipschitz).

We now assume that $(\mathbb{P}^x, X_t)$ is a strong Markov family of solutions to (I.2.3) and that the hypotheses of Theorem 3.4 are satisfied.

(4.4) Theorem. *Suppose Assumption 4.1 holds. Let φ be a continuous function on ∂B. If we set $u(x) = \mathbb{E}^x \varphi(X_{\tau_B})$, then $\partial_{ij} u \in WH$ for each i and j, u is continuous on $\overline{B}$, u agrees with φ on the boundary of B, and $\mathcal{L}u = 0$ in B.*

Proof. That u is continuous on $\overline{B}$ and agrees with φ on ∂B follows from Theorem 3.4. Suppose first that φ is the restriction to ∂B of a smooth function. Let w be defined as in Theorem 4.3. By Theorem II.2.1, $u(x) = w(x)$, so $\mathcal{L}u = \mathcal{L}w$; by Theorem 4.3, for each i and j

$$\|\partial_{ij}u\|_{WH} = \|\partial_{ij}w\|_{WH} \leq c_1\|\varphi\|_\infty. \tag{4.3}$$

Now let φ_n be a collection of C^3 functions on $\mathbb{R}^d$ whose restriction to ∂B converges uniformly to φ. Let $u_n(x) = \mathbb{E}^x \varphi_n(X_{\tau_B})$. Then $u_n(x)$ converges to $u(x) = \mathbb{E}^x \varphi(X_{\tau_B})$ for each x. By Theorem 3.4, u is continuous on $\overline{B}$ and agrees with φ on ∂B. By (4.3), for each i and j

$$\sup_n \|\partial_{ij}u_n\|_{WH} \leq c_2 \sup_n \|\varphi_n\|_{L^\infty(\partial B)} < \infty.$$

So the $\partial_{ij}u_n$ are equicontinuous on compact subsets of B. It follows easily that

$$\|\partial_{ij}u\|_{WH} \leq \limsup_n \|\partial_{ij}u_n\|_{WH} \leq c_3 \limsup_n \|\varphi_n\|_{L^\infty(\partial B)}$$
$$= c_3\|\varphi\|_{L^\infty(\partial B)}. \qquad \square$$

(4.5) Corollary. *There is at most one function w that is continuous on $\overline{B}$, $\partial_{ij}w \in WH$ for each i and j, w agrees with φ on ∂B, and $\mathcal{L}w = 0$ in B.*

Proof. By Theorem II.2.1, any solution agrees with $u(x) = \mathbb{E}^x \varphi(X_{\tau_B})$. $\qquad \square$

5. Dirichlet problem

We are now ready to solve the Dirichlet problem in a ball. The main work yet to be done is to eliminate the use of Assumption 4.1.

Suppose that the a_{ij} are bounded, strictly elliptic, and in C^α. Let σ be a positive definite square root of a, so that $a = \sigma\sigma^T$, and let X_t be the solution to (I.2.3). We suppose that $(\mathbb{P}^x, X_t)$ forms a strong Markov family and $\mathbb{P}^{x_n}$ converges weakly to $\mathbb{P}^x$ whenever $x_n \to x$. (Supposing that $a(x)$ is Lipschitz is sufficient to guarantee the strong Markov property and this weak convergence condition. In Chapter VI we will find that we have these two properties if the $a(x)$ are continuous in x.)

(5.1) Theorem. *Let φ be continuous on ∂B. Let $u(x) = \mathbb{E}^x \varphi(X_{\tau_B})$. Then u is $C^{2+\alpha}$ in B, continuous on $\overline{B}$, agrees with φ on ∂B, and $\mathcal{L}u = 0$ in B.*

Proof. We will show that u is $C^{2+\alpha}$ in B and that $\mathcal{L}u = 0$ there; the remainder of the assertions follows by Theorem 3.4.

Let $x_0 \in B$. Let A be the symmetric positive definite square root of $a(x_0)$ and let $Y_t = AX_t$. Each component of Y_t is a linear combination of the components of X_t and so is a continuous martingale. We have

$$d\langle Y^i, Y^j\rangle_t = \sum_{k,\ell} A_{ik} A_{j\ell}\, d\langle X^k, X^\ell\rangle_t = \sum_{k,\ell} A_{ik} A_{j\ell} a_{k\ell}(X_t)\, dt,$$

so if $\mathcal{L}'$ is the operator associated to Y_t, then $\mathcal{L}' = (1/2)\sum_{i,j=1}^d a'_{ij}\partial_{ij}$, where $a'_{ij}(y) = Aa(A^{-1}y)A^T$. In particular, $a'(Ax_0)$ equals the identity. Note that the a' are still bounded, strictly elliptic, and in C^α.

Let $B' = \{Ax : x \in B\}$, $\varphi'(y) = \varphi(A^{-1}y)$, and $v(y) = \mathbb{E}^y \varphi'(Y_{\tau(B')})$. Then

$$\begin{aligned}
u(x) &= \mathbb{E}^x \varphi(X_{\tau_B}) = \mathbb{E}^x \varphi(A^{-1}AX_{\tau_B}) \\
&= \mathbb{E}^{Ax} \varphi'(Y_{\tau(B')}) = v(Ax).
\end{aligned}$$

Clearly $u \in C^{2+\alpha}$ and $\mathcal{L}u = 0$ in B if and only if $v \in C^{2+\alpha}$ and $\mathcal{L}'v = 0$ in B'. Setting $y_0 = Ax_0$, we may thus consider $Y_t, \mathcal{L}'$, and v with $a'(y_0)$ being the identity.

Choose r small enough so that $B(y_0, r) \subseteq B$, $|a'_{ij}(y) - \delta_{ij}| < \varepsilon_0$ for $y \in B(y_0, 2r)$, and $r^\alpha \|a'\|_{C^\alpha} < \varepsilon_0$, where ε_0 is the ε_0 of Assumption 4.1. Let $Z_t = r^{-1}Y_{r^2 t}$. By Propositions I.8.6 and I.2.1, Z_t is associated to the operator $\mathcal{L}''$ whose coefficients are $a''(z) = a'(rz)$. The a'' are bounded and strictly elliptic and satisfy the same bounds as the a' do. If $z_0 = r^{-1}y_0$, then $a''(z_0)$ is the identity. For $z \in B(z_0, 2)$,

$$|a''_{ij}(z) - \delta_{ij}| < \varepsilon_0$$

for each i and j. For $w, z \in B(z_0, 1)$ and for each i and j,

$$|a''_{ij}(z) - a''_{ij}(w)| = |a'_{ij}(rz) - a'_{ij}(rw)| \le r^\alpha |z - w|^\alpha \|a'\|_{C^\alpha}.$$

Hence $\|a''_{ij}(z) - \delta_{ij}\|_{C^\alpha(B(z_0,1))} < \varepsilon_0$.

Let $w(z) = \mathbb{E}^z \varphi''(Z_{\tau(B(z_0,1))})$, where $\varphi''(z) = \varphi'(rz)$. Since $Y_{r^2 t}$ is a time change of Y_t, the exit distributions from $B(y_0, r)$ for $Y_{r^2 t}$ and Y_t will be the same. Using this, we see similarly to the above that to show $v \in C^{2+\alpha}$ and $\mathcal{L}'v = 0$ at y_0, it suffices to show $w \in C^{2+\alpha}$ and $\mathcal{L}''w = 0$ at z_0. The proof is now complete, since Theorem 4.4 and a translation of the coordinate systems imply $w \in C^{2+\alpha}$ and $\mathcal{L}''w = 0$ in $B(z_0, 1)$. $\square$

6. Extensions

Smooth solutions. Suppose f is in $C^{k+\alpha}$ with compact support, that is, the kth partial derivatives are all in C^α. Then

$$\partial_{ij}(\partial_{i_1\cdots i_k}Uf) = \partial_{ij}U(\partial_{i_1\cdots i_k}f)$$

by translation invariance. It follows by (2.2) that $Uf \in C^{k+2+\alpha}$. With this fact and corresponding facts for $P_B(f)$, we can deduce much as in Sections 4 and 5 that if the a_{ij} are in $C^{k+\alpha}$ and strictly elliptic, then the solution to $\mathcal{L}v = 0$ in B, $v = \varphi$ on ∂B, is in $C^{k+2+\alpha}$ in the interior of B. Similarly, if $f \in C^{k+\alpha}$, the solution to $\mathcal{L}v = f$ in B, $v = 0$ on ∂B, is in $C^{k+2+\alpha}$.

Two dimensions. Our analysis has not included the case of two dimensions because Uf does not exist in this case. An easy way to deal with the two-dimensional case is by using projection. Define $\widetilde{X}_t = (X_t, W_t)$, where W_t is a one-dimensional Brownian motion independent of the two-dimensional diffusion X_t. It is not hard to see that $\widetilde{X}_t$ corresponds to $\widetilde{a}(x)$, where $\widetilde{a}_{ij}(x_1, x_2, x_3)$ equals $a_{ij}(x_1, x_2)$ if $i, j \leq 2$ and equals δ_{ij} if at least one of i, j equals 3. Given φ continuous on ∂B (in $\mathbb{R}^2$), define $\widetilde{\varphi}$ on ∂C by $\widetilde{\varphi}(x_1, x_2, x_3) = \varphi(x_1, x_2)$, where $C = \{(x_1, x_2, x_3) : (x_1, x_2) \in B\}$. If we let $\widetilde{u}(x) = \mathbb{E}^x\widetilde{\varphi}(\widetilde{X}_{\tau_C})$, then $\widetilde{u}$ does not depend on x_3. By the techniques of Section 5 (with minor modifications because C is an unbounded set), $\widetilde{u}$ is smooth inside C.

First order terms. Suppose we no longer assume that the b_i are identically 0. Let us look at solutions in the neighborhood of x_0, which, without loss of generality, we take to be 0. As in Section 5, let us assume $a_{ij} = \delta_{ij}$, and by a rotation, we may assume that $b_1(0) \neq 0$, but all the other $b_i(0) = 0$. If we perform the transformation $(x_1, \ldots, x_d) \to (e^{-2b_1(0)x_1} - 1, x_2, \ldots, x_d)$, the operator $\mathcal{L}$ transforms to a new operator $\mathcal{L}'$, where $b'(0) = 0$ and the new a'_{ij} are strictly elliptic.

Since $\partial_{ij}Uf \in C^\alpha$ if $f \in C^\alpha$ with compact support, it is easy to see that the same will be true for $\partial_i Uf$. We can then obtain suitable estimates on the WH norm of $\partial_i Uf$. Since we have reduced the problem to the case where $b(0) = 0$, we may imitate the theory of Sections 4 and 5, with

$$\mathcal{B}g(x) = \frac{1}{2}\sum_{i,j=1}^{d}(a_{ij}(x) - a_{ij}(x_0))\partial_{ij}g(x) + \sum_{i=1}^{d}b_i(x)\partial_i g(x)$$

to obtain smoothness of the solution to the Dirichlet problem in a neighborhood of x_0.

Zero order terms. Suppose we want to consider regularity of solutions to equations involving $\mathcal{L}f(x) - q(x)f(x)$, where $q \geq 0$. By a localization argument as in Section 5, we may suppose $\|q(x) - q(x_0)\|_{C^\alpha}$ is small. We proceed as in Sections 4 and 5, except that we write

$$\mathcal{L}f(x) = \frac{1}{2}\Delta f(x) - q(x_0)f(x) + \mathcal{B}f(x).$$

We now need estimates on $(q(x_0) - \Delta/2)^{-1}$, but these may be obtained in a straightforward way.

Poisson's equation. The solution to $(\mathcal{L} - \lambda)u = f$ on $\mathbb{R}^d$ is given by

$$u(x) = \mathbb{E}^x \int_0^\infty e^{-\lambda t} f(X_t)\, dt.$$

To derive properties of u, we use an argument similar to that used in Section 5, the principal difficulty being in showing $u \in C^{2+\alpha}$. Let us consider for simplicity the case where $|X_t| \to \infty$ almost surely, $\lambda = 0$, and f has compact support. Then for $r > 0$

$$u(x) = \mathbb{E}^x \int_0^{\tau_{B(x,r)}} f(X_s)\, ds + \mathbb{E}^x \mathbb{E}^{X_{\tau_{B(x,r)}}} \int_0^\infty f(X_s)\, ds.$$

The latter term is of the form $\mathbb{E}^x u(X_{\tau_{B(x,r)}})$, and hence is in $C^{2+\alpha}$ by Theorem 4.4. Let v be the solution to $\mathcal{L}v = -f$ in $B(x_0, r)$ with $v = 0$ on the boundary; then $v \in C^{2+\alpha}$, and by Theorem II.1.2, $v(x) = \mathbb{E}^x \int_0^{\tau_{B(x,r)}} f(X_s)\, ds$.

The case $\lambda \neq 0$ can be handled by similar ideas. See Gilbarg and Trudinger [1] for the analytic estimates needed.

Elliptic and harmonic measure. The solution to the Dirichlet problem $\mathcal{L}u = 0$ in D, $u = f$ on ∂D is, for nice f, given by

$$u(x) = \mathbb{E}^x f(X_{\tau_D}).$$

Let $\omega(x, dy) = \mathbb{P}^x(X_{\tau_D} \in dy)$. ω is called *elliptic measure* or *$\mathcal{L}$-harmonic measure* relative to x. When X_t is Brownian motion and so $\mathcal{L} = (1/2)\Delta$, then we know from [PTA, Proposition II.3.11] that

$$\omega(x, dy) = c_1 \frac{\partial g_D(x, \cdot)}{\partial \nu}(y)\, \sigma(dy),$$

where g_D is the Green function for D for Brownian motion, ν is the inward pointing unit normal vector, and σ is surface measure on ∂D. If D is a C^2 domain, then $\nu(y)$ is a C^2 function of y, while $g_D(x, \cdot)$ solves the Dirichlet problem in $D - B(x, r)$ for r small with boundary values 0 on ∂D, and $g_D(x, y)$ on $\partial B(x, r)$. By Theorem 5.1, $g_D(x, \cdot)$ is C^2 in a neighborhood of ∂D, and so $(\partial g_D(x, \cdot)/\partial \nu)(y) = \nabla_y g_D(x, y) \cdot \nu(y)$ is C^1 on ∂D. This provides another proof of [PTA, Theorem III.5.2].

7. Neumann and oblique derivative problem

Let D be a bounded C^2 domain and let $\nu(x)$ be the inward pointing unit normal vector at $x \in \partial D$. Suppose $v(x)$ is a C^2 vector field on ∂D (i.e., each component of v is a C^2 function on ∂D) such that $\inf_{\partial D} v \cdot \nu > 0$. Let K be a compact subset of D with smooth boundary. Let us suppose that the a_{ij} are strictly elliptic and C^2 and the b_i are C^2. Let f be a C^2 function on ∂D. In this section we want to show that the solutions to the Neumann problem

$$\mathcal{L}u = 0 \text{ in } D, \qquad \partial u / \partial \nu = f \text{ on } \partial D, \qquad u = 0 \text{ on } K \qquad (7.1)$$

and to the oblique derivative problem

$$\mathcal{L}u = 0 \text{ in } D, \qquad \nabla u \cdot v = f \text{ on } \partial D, \qquad u = 0 \text{ on } K \qquad (7.2)$$

have C^2 solutions in $\overline{D}$. The procedure is very similar to what we did for the Dirichlet problem, so we only sketch the argument, referring the reader to Gilbarg and Trudinger [1] for further details.

Suppose $d \geq 3$. We first consider the half space H and assume that v is constant on ∂D with $|v| = 1$. Let $u(x)$ be the Newtonian potential for $\mathbb{R}^d$ ([PTA, (II.3.1)]) and define

$$G(x, y) = u(x - y) - u(x - \overline{y}) - 2v_d \int_0^\infty \partial_d u(x - \overline{y} + vs) \, ds, \qquad (7.3)$$

where $x, y \in H$ and $\overline{y} = (y_1, \ldots, y_{d-1}, -y_d)$ if $y = (y_1, \ldots, y_d)$. Observe that G is harmonic in x and y for $x \neq y$, and a calculation shows that $(\nabla G(\cdot, y) \cdot v)(x) = 0$ if $x_d = 0$. $G(x, y)$ is thus the Green function for reflecting Brownian motion with constant oblique reflection.

Substituting for u in the integral in (7.3),

$$G(x, y) = u(x, y) - u(x - \overline{y}) \qquad (7.4)$$

$$- c_1 v_d |x - \overline{y}|^{2-d} \int_0^\infty \frac{w_d + v_d s}{(1 + 2(w \cdot v)s + s^2)^{d/2}} \, ds,$$

where $w = (x - \overline{y})/|x - \overline{y}|$. Let $V(x, y)$ denote the last term in (7.4). Since $v \cdot \nu \geq c_2 > 0$, then $1 + 2(w \cdot v)s + s^2$ is bounded away from 0. A calculation shows that

$$\partial_{x_i} V(x, y) = -\partial_{y_i} V(x, y), \qquad i = 1, \ldots, d-1, \qquad (7.5)$$

$$\partial_{x_d} V(x, y) = \partial_{y_d} V(x, y),$$

$$|\partial_{x_i} V(x, y)| \leq c_3 |x - \overline{y}|^{1-d}, \qquad i = 1, \ldots, d,$$

$$|\partial_{x_i x_j} V(x, y)| \leq c_4 |x - \overline{y}|^{-d}, \qquad i, j = 1, \ldots, d.$$

If $|v| \neq 1$, we replace v_d by $v_d/|v|$ in the preceding estimates.

We look at

$$\int_{H \cap B(0,2)} G(x,y)h(y)\,dy \tag{7.6}$$

$$= \int_{H \cap B(0,2)} [u(x-y) - u(x-\overline{y})]h(y)\,dy$$

$$+ \int_{H \cap B(0,2)} V(x,y)h(y)\,dy.$$

Let us extend h to the lower half space by reflection over the hyperplane $\{x_d = 0\}$. If $w(x) = \int_{H \cap B(0,2)} V(x,y)h(y)\,dy$, we can write

$$\partial_{ij}w(x) = \int_{H \cap B(0,2)} \partial_{ij}V(x,y)[h(\overline{y}) - h(x)]\,dy$$

$$- h(x) \int_{H \cap \partial B(0,2)} \partial_i V(x,y)\widehat{\nu}_j(y)\,\sigma(dy)$$

by Green's identity, where $\widehat{\nu}$ is the outward pointing unit normal vector on $\partial B(0,2)$ and σ is surface measure on $\partial B(0,2)$ (cf. [PTA, Proposition II.3.13]). As in [PTA, Theorem II.3.14], we derive the estimates

$$\|\partial_{ij}w\|_{C^\alpha(H \cap B(0,1))} \le c_5 \|h\|_{C^\alpha(H \cap B(0,2))}. \tag{7.7}$$

Since we have the analogous estimate for the first term on the right of (7.6), we have

$$\|\partial_{ij}Gh\|_{C^\alpha(H \cap B(0,1))} \le c_6 \|h\|_{C^\alpha(H \cap B(0,2))}, \tag{7.8}$$

where $Gh(x) = \int_{H \cap B(0,2)} G(x,y)h(y)\,dy$.

With only minor changes, we can use the proofs in Sections 4 and 5 to obtain the existence of $u \in C^2(\overline{H})$ with $\mathcal{L}u = 0$ in $H - K$ and $\nabla u \cdot v = f$ on ∂D, $u = 0$ on K, where now v is a function of x and no longer constant.

Our results are of a local nature. So if we consider $x \in \partial D$, we can map $D \cap B(x,r)$ smoothly onto $H \cap B(y,s)$ if r is sufficiently small, where $y \in \partial H$. The operator $\mathcal{L}$ and reflection vector v are mapped into $\overline{\mathcal{L}}$ and $\overline{v}$, respectively, but both $\overline{\mathcal{L}}$ and $\overline{v}$ satisfy bounds analogous to those of $\mathcal{L}$ and v. The solution to the oblique derivative problem for $\overline{L}$ and $\overline{v}$ is smooth in $\overline{H} \cap B(y,s)$, and taking the inverse image shows that there exists a solution u that is smooth in $\overline{D} \cap B(x,r)$.

The Neumann problem in this formulation is the special case where $v \equiv \nu$. The case $d = 2$ can be handled by a projection argument similar to that in Section 6. We can also handle the analogue of Poisson's equation, etc. See Gilbarg and Trudinger [1] for further details.

8. Calderón-Zygmund estimates

For $1 < p < \infty$ define

$$\|f\|_{W^{2,p}} = \|f\|_p + \sum_{i,j=1}^{d} \|\partial_{ij} f\|_p$$

for $f \in C^2$. Let $W^{2,p}$ be the closure of $C^2 \cap L^\infty$ with respect to this norm. By some well-known estimates of Calderón-Zygmund type, we have $U^\lambda : L^p \to W^{2,p}$ and

$$\|U^\lambda f\|_{W^{2,p}} \leq c_1 \|f\|_p, \tag{8.1}$$

where c_1 is independent of f.

This is easily checked when $p = 2$. The function $\partial_{ij} f$ has Fourier transform $-\xi_i \xi_j \widehat{f}(\xi)$, so Δf has Fourier transform $-|\xi|^2 \widehat{f}(\xi)$. Since $U^\lambda = (\lambda - \Delta/2)^{-1}$, then $U^\lambda f$ has Fourier transform $\widehat{f}(\xi)/(\lambda + |\xi|^2/2)$. Finally, $\partial_{ij} U^\lambda f$ has Fourier transform $-\xi_i \xi_j \widehat{f}(\xi)/(\lambda + |\xi|^2/2)$. By Plancherel's theorem, there exists c_2 independent of f such that $\|\partial_{ij} U^\lambda f\|_2 = c_2 \|(\partial_{ij} U^\lambda f)\widehat{\ }\|_2$ and $\|f\|_2 = c_2 \|\widehat{f}\|_2$. Since $|(\partial_{ij} U^\lambda f)\widehat{\ }| \leq 2|\widehat{f}(\xi)|$, the assertion follows when $p = 2$. For $p \neq 2$, this estimate is considerably harder; see [PTA, Corollary IV.3.9] for a probabilistic proof.

If $f \in W^{2,p}$, we can make sense of $\partial_{ij} f$ and hence of $\mathcal{L}f$, at least up to almost everywhere equivalence. Let us suppose that the a_{ij} are strictly elliptic, the b_i are identically 0, and $|a_{ij}(x) - \delta_{ij}| < \varepsilon_0$, where $\varepsilon_0 = 1/(c_1 d^2)$. We use variation of parameters with respect to the norm of L^p.

(8.1) Theorem. *Suppose $f \in L^p$. Then there exists $v \in W^{2,p}$ such that $\mathcal{L}v - \lambda v = f$.*

Proof. Recall the definition of $\mathcal{B}$ in (4.2). Let $v = U^\lambda \sum_{m=0}^{\infty} (\mathcal{B}U^\lambda)^m f$. Note

$$\|\mathcal{B}U^\lambda g\|_p \leq \frac{d^2}{2} (\sup_{i,j} |a_{ij}(x) - \delta_{ij}|) \|\partial_{ij} U^\lambda g\|_p$$

$$\leq \frac{c_1 \varepsilon_0 d^2}{2} \|g\|_p = \|g\|_p/2,$$

where $\mathcal{B}$ is defined by (4.2). So the L^p norm of $\mathcal{B}U^\lambda$ is bounded by $1/2$, and $h = \sum_{i=0}^{\infty} (\mathcal{B}U^\lambda)^i f \in L^p$. Hence $U^\lambda h \in W^{2,p}$. As in Theorem 4.2, $\mathcal{L}v - \lambda v = f$. $\qquad\square$

9. Flows

Another approach to the Cauchy problem is via flows. Here the assumption of strict ellipticity is not absolutely essential (although we will include it), but additional smoothness is.

(9.1) Proposition. *Suppose the coefficients of $\mathcal{L}$ are in C^3 and are bounded and that $\mathcal{L}$ is strictly elliptic. Let X_t be the associated diffusion. If $f \in C^2$ with bounded* *first and second partial derivatives, then $u(x,t) = \mathbb{E}^x f(X_t)$ solves $\partial_t u = \mathcal{L}u$ in $(0,\infty) \times \mathbb{R}^d$ with $u(x,0) = f(x)$.*

Proof. We first show that u satisfies the appropriate smoothness conditions. By the remarks following Proposition I.10.4, $X(x,t)$ is C^2 in the variable x almost surely, so $u(x,t) = \mathbb{E} f(X(x,t))$ is also C^2 in x. By Itô's formula,

$$f(X_t) - f(X_s) = \text{ martingale } + \int_s^t \mathcal{L}f(X_r)\,dr,$$

which implies

$$u(x,t) - u(x,s) = \mathbb{E}^x \int_s^t \mathcal{L}f(X_r)\,dr.$$

Since $\mathcal{L}f$ is bounded and continuous, this means that u is differentiable in t.

   We now show $\partial_t u = \mathcal{L}u$. Let $t_0 > 0$. Applying Itô's formula to $u(X_t, t_0 - t)$, we obtain

$$u(X_t, t_0 - t) = u(X_0, t_0) + \text{ martingale } - \int_0^t \partial_t u(X_s, t_0 - s)\,ds$$

$$+ \int_0^t \mathcal{L}u(X_s, t_0 - s)\,ds.$$

As in Theorem II.3.1,

$$u(X_t, t_0 - t) = \mathbb{E}^{X_t} f(X_{t_0 - t}) = \mathbb{E}^x[f(X_{t_0}) \mid \mathcal{F}_t]$$

is a martingale. Therefore $\int_0^t (\mathcal{L}u - \partial_t u)(X_s, t_0 - s)\,ds$ is a martingale. A continuous martingale of bounded variation that is 0 at 0 must be identically 0 ([PTA, Proposition I.4.19)]; hence $\int_0^t (\mathcal{L}u - \partial_t u)(X_s, t_0 - s)\,ds$ is identically 0. Thus $(\mathcal{L}u - \partial_t u)(X_s, r) = 0$ a.s. If $(\mathcal{L}u - \partial_t u)(y, r) > 0$ for some $y \in \mathbb{R}^d$ and some $r > 0$, then by continuity it will be positive in a neighborhood of y. By the support theorem, there is positive probability that X_r is in this neighborhood, a contradiction. The case where $\mathcal{L}u - \partial_t u$ takes negative values is the same. Therefore $\partial_t u = \mathcal{L}u$. $\qquad\square$

10. Notes

The bulk of the material in this chapter is based on material in Gilbarg and Trudinger [1]. Section 3 imitates the Brownian motion case in [PTA, Section II.1]. More information along the lines of Section 8 can be found in Stroock and Varadhan [2]. For more on flows see Ikeda and Watanabe [1].

IV

ONE-DIMENSIONAL DIFFUSIONS

The one-dimensional diffusions provide good examples for understanding some of the phenomena that can occur for higher-dimensional diffusions.

Under very mild regularity conditions, every one-dimensional diffusion arises from first time changing a one-dimensional Brownian motion and then making a transformation of the state space.

In Section 1 we study the transformation of the state space by means of scale functions, whereas in Section 2 we investigate the time changes of Brownian motion using speed measures. Section 3 considers the solutions to the SDEs of Chapter I in terms of scale functions and speed measures. In Section 4 we consider diffusions that have boundaries, and Section 5 derives an eigenvalue expansion for the transition densities.

1. Natural scale

Throughout this chapter we suppose that we have a continuous process $(\mathbb{P}^x, X_t)$ defined on an interval I which may be finite or infinite and open or closed (or a combination of the two) and that $(\mathbb{P}^x, X_t)$ is strong Markov with respect to a right continuous filtration $\mathcal{F}_t$. We call such a process a *diffusion* on I. Writing T_y for $\inf\{t : X_t = y\}$, we also assume that every point can be hit from every other point:

For all x, y in the interior of I, $\mathbb{P}^x(T_y < \infty) = 1$. $\hspace{2em}$ (1.1)

If (1.1) holds, we say the diffusion is *regular*.

When X_t is a Brownian motion, it is well known ([PTA, Proposition I.4.9]) that the distribution of X_t upon exiting $[a, b]$ is

$$\mathbb{P}^x(X(\tau_{[a,b]}) = a) = \frac{b - x}{b - a}, \qquad \mathbb{P}^x(X(\tau_{[a,b]}) = b) = \frac{x - a}{b - a}. \tag{1.2}$$

We say that a regular diffusion X_t is on *natural scale* if (1.2) holds for every interval $[a, b]$ contained in the interior of I. In this section we will show that given a regular diffusion, there exists a *scale function* s on I that is continuous, strictly increasing, and such that $s(X_t)$ is on natural scale.

If X_t is regular and x is in the interior of I, then the process started at x must leave x immediately. To see this, let $\varepsilon > 0$ be such that $[x - \varepsilon, x + \varepsilon]$ is contained in the interior of I, $S = \inf\{t : X_t \neq x\}$, and $U = \inf\{t : |X_t - x| \geq \varepsilon\}$. By the regularity of X_t, $\mathbb{E}^x e^{-U} > 0$. By the strong Markov property at time S,

$$\mathbb{E}^x e^{-U} = \mathbb{E}^x(e^{-S}\mathbb{E}^{X_S} e^{-U}) = \mathbb{E}^x e^{-S}\mathbb{E}^x e^{-U},$$

since $X_S = x$. The only way this can happen is if $\mathbb{E}^x e^{-S} = 1$, which implies $S = 0$ a.s.

Let J be a subinterval $[a, b]$ of the interior of I. We define

$$p(x) = p_J(x) = \mathbb{P}^x(X_{\tau_J} = b). \tag{1.3}$$

(1.1) Proposition. *Let $J = [a, b]$ be a finite interval contained in the interior of I. Then $p(X_{t \wedge \tau_J})$ is a regular diffusion on $[0, 1]$ on natural scale.*

Proof. First we show that p is increasing. To get to the point b starting from x, the process must first hit every point between x and b. If $x < y < b$, by the strong Markov property at time T_y, $p(x) \leq p(y)$. There must be a positive probability that the process starting from x hits a before y; otherwise by the strong Markov property, $\mathbb{P}^x(T_a < \infty) = 0$. So by the strong Markov property at T_y,

$$p(x) = \mathbb{P}^x(T_y < T_a)p(y).$$

Since we argued that $\mathbb{P}^x(T_y < T_a) = 1 - \mathbb{P}^x(T_a < T_y)$ is strictly less than 1, then p is strictly increasing.

Next we show that p is continuous. We show continuity from the right; the proof of continuity from the left is similar. Suppose $x_n \downarrow x$. The process X_t has continuous paths, so given ε we can find t small enough so that $\mathbb{P}^x(T_a < t) < \varepsilon$. By the Blumenthal zero-one law, $\mathbb{P}^x(T_{(x,b]} = 0)$ is zero or one (see [PTA, Corollary I.3.6] for a proof in the Brownian motion case; the same proof works for $(\mathbb{P}^x, X_t)$ since X_t was assumed to be Markov with respect to $\mathcal{F}_t$ and $\mathcal{F}_t$ is right continuous, and hence $\mathcal{F}_{0+} = \mathcal{F}_0$). If it is zero, the process immediately moves to the left from x a.s., and by the strong Markov property at T_x, it never hits b, a contradiction. The probability must therefore be one. Thus for n large enough, $\mathbb{P}^x(T_{x_n} < t) \geq 1 - \varepsilon$. Hence with probability at least $1 - 2\varepsilon$, X_t hits x_n before a. Since

$$p(x) = \mathbb{P}^x(T_{x_n} < T_a)\,p(x_n) \geq (1 - 2\varepsilon)p(x_n),$$

we see that $p(x) \geq \liminf_{n \to \infty} p(x_n)$. Since p is nondecreasing, $p(x_n)$ decreases, and therefore $p(x) = \lim p(x_n)$.

Finally, we show $p(X_t)$ is on natural scale. Let $[e, f] \subseteq (0, 1)$ and let

$$r(y) = \mathbb{P}^y(X_t \text{ hits } p^{-1}(f) \text{ before hitting } p^{-1}(e)).$$

Note that

$$\begin{aligned}
\mathbb{P}^x(p(X_t) \text{ hits } f \text{ before } e) &= \mathbb{P}^{p^{-1}(x)}(X_t \text{ hits } p^{-1}(f) \text{ before } p^{-1}(e)) \\
&= r(p^{-1}(x)). \tag{1.4}
\end{aligned}$$

For $y \in [p^{-1}(a), p^{-1}(b)]$, the strong Markov property tells us that

$$\begin{aligned}
p(y) &= \mathbb{P}^y\big(X_t \text{ hits } p^{-1}(f) \text{ before } p^{-1}(e)\big)p\big(p^{-1}(f)\big) \tag{1.5} \\
&\quad + \mathbb{P}^y\big(X_t \text{ hits } p^{-1}(e) \text{ before } p^{-1}(f)\big)p\big(p^{-1}(e)\big) \\
&= r(y)f + (1 - r(y))e.
\end{aligned}$$

Solving for $r(y)$, we obtain $r(y) = (p(y) - e)/(f - e)$. Substituting in (1.4),

$$\begin{aligned}
\mathbb{P}^x(p(X_t) \text{ hits } f \text{ before } e) &= (p(p^{-1}(x)) - e)/(f - e) \\
&= (x - e)/(f - e),
\end{aligned}$$

as desired. $\qquad\square$

Note that if X_t is on natural scale, then so is $c_1 X_t + c_2$ for any constants $c_1 > 0, c_2 \in \mathbb{R}$.

(1.2) Theorem. *There exists a strictly continuous increasing function s such that $s(X_t)$ is on natural scale on $s(I)$.*

Proof. Let J_n be closed subintervals of the interior of I increasing up to the interior of I. Pick two points in J_1; for concreteness let us suppose without loss of generality that they are the points 0 and 1. Choose A_n and B_n so that if $s_n(x) = A_n p_{J_n}(x) + B_n$, then $s_n(0) = 0$ and $s_n(1) = 1$.

We will show that if $n \geq m$, then $s_n = s_m$ on J_m. Once we have that, we can set $s(x) = s_n(x)$ on J_n, set $s(\sup I) = \sup\{s(x) : x \text{ is in the interior of } I\}$, similarly define $s(\inf I)$, and the theorem will be proved.

Suppose $J_m = [e, f]$. By Proposition 1.1, both $s_m(X_t)$ and $s_n(X_t)$ are on natural scale. For all $x \in J_m$,

$$\begin{aligned}
\frac{s_m(x) - s_m(e)}{s_m(f) - s_m(e)} &= \mathbb{P}^{s_m(x)}\big(s_m(X_t) \text{ hits } s_m(f) \text{ before } s_m(e)\big) \\
&= \mathbb{P}^x(X_t \text{ hits } f \text{ before } e).
\end{aligned}$$

We have a similar equation with s_m replaced everywhere by s_n. It follows that

$$\frac{s_m(x) - s_m(e)}{s_m(f) - s_m(e)} = \frac{s_n(x) - s_n(e)}{s_n(f) - s_n(e)}$$

for all x, which implies that $s_n(x) = C s_m(x) + D$ for some constants C and D. Since s_n and s_m are equal at both $x = 0$ and $x = 1$, then C must be 1 and D must be 0. $\square$

2. Speed measures

Suppose that $(\mathbb{P}^x, X_t)$ is a regular diffusion on an open interval I on natural scale. (We will consider boundaries in Section 4.) If $(a, b) \subseteq I$, define

$$G_{a,b}(x, y) = \begin{cases} (2(x - a)(b - y))/(b - a), & a < x \leq y < b \\ (2(y - a)(b - x))/(b - a), & a < y \leq x < b \end{cases} \tag{2.1}$$

and set $G_{a,b}(x, y) = 0$ if x or y is not in (a, b). A measure $m(dx)$ is the *speed measure* for the diffusion if

$$\mathbb{E}^x \tau_{(a,b)} = \int G_{a,b}(x, y)\, m(dy) \tag{2.2}$$

whenever $(a, b) \subseteq I$ and $x \in I$. As (2.2) indicates, the speed measure governs how quickly the diffusion moves through intervals.

As an example, let us argue that the speed measure for Brownian motion is Lebesgue measure. To see this, recall that $X_t^2 - t$ is a martingale, so

$$\mathbb{E}^x(\tau_{(a,b)} \wedge t) = \mathbb{E}^x(X(\tau_{(a,b)} \wedge t) - x)^2.$$

Letting $t \to \infty$ and using monotone convergence on the left and dominated convergence on the right,

$$\mathbb{E}^x \tau_{(a,b)} = \mathbb{E}^x(X_{\tau_{(a,b)}} - x)^2 \tag{2.3}$$

$$= (b - x)^2 \mathbb{P}^x(X(\tau_{(a,b)}) = b) + (x - a)^2 \mathbb{P}^x(X(\tau_{(a,b)}) = a).$$

Since Brownian motion is on natural scale, substituting (1.2) in (2.3) gives

$$\mathbb{E}^x \tau_{(a,b)} = (x - a)(b - x) = \int G_{a,b}(x, y)\, dy.$$

Corollary 2.4 below will imply that $G_{a,b}$ is in fact the Green function for Brownian motion killed on exiting (a, b).

We will show in this section that a regular diffusion on natural scale has one and only one speed measure, that the law of the diffusion is determined by the speed measure, and that there exists a diffusion with a given speed measure.

We first want to show that any speed measure must satisfy $0 < m(a, b) < \infty$ if $[a, b] \subseteq I$. To start we have the following lemma.

(2.1) Lemma. *If $[a,b] \subseteq I$, then $\sup_x \mathbb{E}^x \tau_{(a,b)}^k < \infty$ for each positive integer k.*

Proof. Pick $y \in (a,b)$. Since X_t is a regular diffusion, $\mathbb{P}^y(T_a < \infty) = 1$, and hence there exists t_0 such that $\mathbb{P}^y(T_a > t_0) < 1/2$. Similarly, taking t_0 larger if necessary, $\mathbb{P}^y(T_b > t_0) \leq 1/2$. If $a < x \leq y$, then

$$\mathbb{P}^x(\tau_{(a,b)} > t_0) \leq \mathbb{P}^x(T_a > t_0) \leq \mathbb{P}^y(T_a > t_0) \leq 1/2,$$

and similarly, $\mathbb{P}^x(\tau_{(a,b)} > t_0) \leq 1/2$ if $y \leq x < b$. By the Markov property,

$$\mathbb{P}^x(\tau_{(a,b)} > (n+1)t_0) = \mathbb{E}^x[\mathbb{P}^{X(nt_0)}(\tau_{(a,b)} > t_0); \tau_{(a,b)} > nt_0]$$
$$\leq (1/2)\mathbb{P}^x(\tau_{(a,b)} > nt_0),$$

and by induction, $\mathbb{P}^x(\tau_{(a,b)} > nt_0) \leq 2^{-n}$. The lemma is now immediate. $\quad\square$

(2.2) Proposition. *If $(\mathbb{P}^x, X_t)$ has a speed measure m and $[a,b] \subseteq I$, then $0 < m(a,b) < \infty$.*

Proof. If $m(a,b) = 0$, then for $x \in (a,b)$, we have

$$\mathbb{E}^x \tau_{(a,b)} = \int G_{a,b}(x,y)\, m(dy) = 0,$$

which implies $\tau_{(a,b)} = 0$, $\mathbb{P}^x$-a.s., a contradiction to the continuity of the paths of X_t. Pick (e,f) such that $[a,b] \subseteq (e,f) \subseteq [e,f] \subseteq I$. There exists a constant c_1 such that for $x,y \in (a,b)$, $G_{e,f}(x,y)$ is bounded below by c_1, so

$$m(a,b) \leq c_1^{-1} \int_e^f G_{e,f}(x,y)\, m(dy) = c_1^{-1} \mathbb{E}^x \tau_{(e,f)} < \infty$$

by Lemma 2.1. $\hfill\square$

(2.3) Theorem. *A regular diffusion on natural scale in an open interval I has one and only one speed measure.*

Proof. Suppose first that $I = (e,f)$ is a finite interval. For $n > 1$ let $x_i = e + i(f-e)/2^n$, $i = 0, 1, 2, \cdots, 2^n$. Let

$$m_n(dx) = 2^n \sum_{i=1}^{2^n-1} B(x_i)\delta_{x_i}, \tag{2.4}$$

where $B(x_i) = \mathbb{E}^{x_i} \tau_{(x_{i-1}, x_{i+1})}$. We first want to show that if $[a,b]$ is a subinterval of I with a,b each equal to some x_i and x is also equal to some x_i, then

$$\mathbb{E}^x \tau_{(a,b)} = \int G_{a,b}(x,y)\, m_n(dy). \tag{2.5}$$

To see this, let $S_0 = 0$ and $S_{j+1} = \inf\{t > S_j : |X_t - X_{S_j}| = 2^{-n}\} \wedge \tau_{(a,b)}$; note that X_{S_j} is a simple symmetric random walk up until exiting (a,b) because

X is on natural scale. Let $J(x) = (x - 2^{-n}, x + 2^{-n})$. By repeated use of the strong Markov property,

$$\mathbb{E}^x \tau_{(a,b)} = \sum_{j=0}^{\infty} \mathbb{E}^x (S_{j+1} - S_j)$$

$$\mathbb{E}^x \sum_{j=0}^{\infty} \mathbb{E}^{X(S_j)} \tau_{J(X(S_j))} = \mathbb{E}^x \sum_{j=0}^{\infty} B(X_{S_j}).$$

So

$$\mathbb{E}^x \tau_{(a,b)} = \mathbb{E}^x \sum_{i=1}^{2^n - 1} B(x_i) N_i, \tag{2.6}$$

where N_i is the number of times the random walk X_{S_j} hits x_i before exiting (a, b). $\mathbb{E}^x N_i$ must equal 0 when $x = a$ or $x = b$ and satisfies the equation

$$\mathbb{E}^{x_j} N_i = \delta_{ij} + (\mathbb{E}^{x_{j+1}} N_i + \mathbb{E}^{x_{j-1}} N_i)/2. \tag{2.7}$$

This implies that

$$\mathbb{E}^x N_i = 2^n G_{a,b}(x, x_i). \tag{2.8}$$

This, (2.6), and (2.4) prove (2.5).

By the proof of Proposition 2.2 and (2.5), $m_n(a, b)$ is bounded above by a constant independent of n whenever $[a, b] \subseteq I$. By a diagonalization procedure, there exists a subsequence n_k such that m_{n_k} converges weakly to m, where m is a measure that is finite on every subinterval (a, b) such that $[a, b] \subseteq I$. By the continuity of $G_{a,b}$,

$$\mathbb{E}^x \tau_{(a,b)} = \int G_{a,b}(x, y) \, m(dy) \tag{2.9}$$

whenever a, b, and x are of the form $e + i(f - e)/2^n$ for some i and n.

We now remove this last restriction. If a, b are not of this form, take a_r, b_r of this form such that $(a_r, b_r) \uparrow (a, b)$. Then $\tau_{(a_r, b_r)} \uparrow \tau_{(a,b)}$, and by the continuity of $G_{a,b}$ in a, b, x, and y, we have (2.9) for all a and b. Take $x'_r \uparrow x$, $x''_r \downarrow x$ such that x'_r and x''_r are of the form $e + i(f - e)/2^n$ for some i and n. By the strong Markov property,

$$\mathbb{E}^x \tau_{(a,b)} = \mathbb{E}^x \tau_{(x'_r, x''_r)} + \mathbb{E}^{x'_r} \tau_{(a,b)} \mathbb{P}^x (X(\tau_{(x'_r, x''_r)}) = x'_r)$$

$$+ \mathbb{E}^{x''_r} \tau_{(a,b)} \mathbb{P}^x (X(\tau_{(x'_r, x''_r)}) = x''_r).$$

By the continuity of $G_{a,b}$ in x, and the fact that $\mathbb{E}^x \tau_{(x'_r, x''_r)} \to 0$ as $r \to \infty$, we obtain (2.9) for all x.

Next we show uniqueness for the case of finite intervals. If m_1 and m_2 are two speed measures, then

$$\int G_{a,b}(x, y) \, m_1(dy) = \int G_{a,b}(x, y) \, m_2(dy) \tag{2.10}$$

for all x, a, and b. If $f \in C^2(a,b)$, it is not hard to see that $f(z) = -\int G_{a,b}(z,x) f''(x)\, dx$. So multiplying (2.10) by $-f''(x)$ and integrating over x, we see that $\int f\, dm_1 = \int f\, dm_2$ for $f \in C^2(a,b)$, which implies $m_1 = m_2$.

Finally, if I is infinite, let I_n be finite subintervals increasing up to I. Let m_n be the speed measure for X_t on the interval I_n. By the uniqueness argument, m_n agrees with m_k on I_k if $I_k \subseteq I_n$. Setting m to be the measure whose restriction to I_n is m_n gives us the speed measure. $\square$

The speed measure completely characterizes occupations times.

(2.4) Corollary. *Suppose X_t is a diffusion on natural scale on a finite interval $I = (a,b)$. If f is bounded and measurable,*

$$\mathbb{E}^x \int_0^{\tau(a,b)} f(X_s)\, ds = \int G_{a,b}(x,y) f(y)\, m(dy). \tag{2.11}$$

Proof. Suppose that f is continuous and bounded on I. Let $x_i, S_j, B(x_i), N_i$, and m_n be as in the proof of Theorem 2.3. Let $\varepsilon_n = \sup\{|f(x) - f(y)| : |x - y| \leq 2^{-n}\}$. Note that

$$\mathbb{E}^x \int_0^{\tau(a,b)} f(X_s)\, ds = \sum_{j=0}^{\infty} \mathbb{E}^x \int_{S_j}^{S_{j+1}} f(X_s)\, ds \tag{2.12}$$

and

$$\mathbb{E}^x \sum_{j=0}^{\infty} f(X_{S_j})(S_{j+1} - S_j) = \mathbb{E}^x \sum_{j=0}^{\infty} f(X_{S_j}) \mathbb{E}^{X_{S_j}} S_1 \tag{2.13}$$

$$= \sum_{i=1}^{2^n - 1} f(x_i) B(x_i) \mathbb{E}^x N_i.$$

Moreover, the right-hand side of (2.12) differs from the left-hand side of (2.13) by at most $\varepsilon_n \mathbb{E}^x \tau_{(a,b)}$. By (2.8) the right-hand side of (2.13) is equal to

$$\sum_{i=1}^{2^n - 1} 2^n f(x_i) B(x_i) G_{a,b}(x, x_i) = \int G_{a,b}(x, x_i) f(x_i)\, m_n(dx).$$

So by weak convergence along an appropriate subsequence, the left-hand side and the right-hand side of (2.11) differ by $\limsup_n \varepsilon_n \mathbb{E}^x \tau_{(a,b)}$, which is zero. A limit argument then shows that (2.11) holds for all bounded f. $\square$

We next turn to showing that the speed measure characterizes the law of a diffusion.

(2.5) Theorem. *If $(\mathbb{P}_i^x, X_t)$, $i = 1, 2$, are two diffusions on natural scale on an open interval I with the same speed measure m, then $\mathbb{P}_1^x = \mathbb{P}_2^x$ on $\mathcal{F}_{\tau_I}$.*

Proof. Let I_n be open finite intervals increasing up to I. Clearly it is enough to show uniqueness on each I_n. Hence we may assume that I is finite and $m(I) < \infty$. Suppose $I = (a, b)$.

Define the operator G_i^λ by

$$G_i^\lambda f(x) = \mathbb{E}_i^x \int_0^{\tau_{(a,b)}} e^{-\lambda t} f(X_t)\, dt, \qquad \lambda \geq 0. \tag{2.14}$$

We show first that $G_1^0 = G_2^0$. Let $[c, d] \subseteq (a, b)$, and choose δ positive but small so that $(c - \delta, d + \delta) \subseteq (a, b)$. Let

$$U_0(\delta) = 0, \qquad S_{j+1}(\delta) = \inf\{t > U_j(\delta) : X_t \in [c, d]\} \wedge \tau_{(a,b)},$$

$$U_j(\delta) = \inf\{t > S_j(\delta) : X_t \notin (c - \delta, d + \delta)\} \wedge \tau_{(a,b)}.$$

Because X_t is on natural scale under both $\mathbb{P}_1^x$ and $\mathbb{P}_2^x$, the law of $X_{S_j(\delta)}$ is the same under both. The speed measure is the same under both probabilities, and hence

$$\mathbb{E}_i^x(U_j(\delta) - S_j(\delta)) = \mathbb{E}_i^x \mathbb{E}_i^{X(S_j(\delta))} U_1(\delta)$$

$$= \mathbb{E}_i^x \int G_{c-\delta, d+\delta}(X_{S_j(\delta)}, y)\, m(dy)$$

does not depend on i. The quantities $\sum_{j=0}^\infty (U_j(\delta) - S_j(\delta))$ are bounded by $\tau_{(a,b)}$ and decrease to $\int_0^{\tau_{(a,b)}} 1_{[c,d]}(X_s)\, ds$ as $\delta \to 0$. By dominated convergence, $G_1^0 1_{[c,d]} = G_2^0 1_{[c,d]}$. By linearity and a limit argument, this implies $G_1^0 f = G_2^0 f$ for all bounded and measurable f.

Since $X_{t \wedge \tau_{(a,b)}}$ is a Markov process,

$$G_i^\lambda G_i^\mu f(x) = \mathbb{E}_i^x \int_0^\infty e^{-\lambda t} \mathbb{E}_i^{X_t} \int_0^\infty e^{-\mu s} f(X_s)\, ds\, dt \tag{2.15}$$

$$= \mathbb{E}_i^x \int_0^\infty e^{-\lambda t} \int_0^\infty e^{-\mu s} f(X_{s+t})\, ds\, dt$$

$$= \mathbb{E}_i^x \int_0^\infty e^{-(\lambda - \mu)t} \int_t^\infty e^{-\mu s} f(X_s)\, ds\, dt$$

$$= \mathbb{E}_i^x \int_0^\infty e^{-\mu s} f(X_s) \int_0^s e^{-(\lambda - \mu)t}\, dt\, ds$$

$$= \frac{1}{\lambda - \mu} [G_i^\mu f(x) - G_i^\lambda f(x)].$$

Then

$$G_i^\lambda f(x) = G_i^\mu f(x) - (\lambda - \mu) G_i^\lambda G_i^\mu f(x). \tag{2.16}$$

Iterating, if $\lambda > \mu$ and $|\lambda - \mu| \leq 1/(2\|G_i^\mu\|_\infty)$, then

$$G_i^\lambda = G_i^\mu - (\lambda - \mu)(G_i^\mu)^2 + (\lambda - \mu)^2 (G_i^\mu)^3 - \cdots. \tag{2.17}$$

This also holds for $\mu = 0$ by taking a limit. Observe that $G_i^\lambda f(x) \leq \|f\|_\infty \mathbb{E}^x \tau_{(a,b)}$, or G_i^λ is bounded by a quantity independent of λ.

Since $G_1^0 = G_2^0$, for all λ near 0 we have $G_1^\lambda = G_2^\lambda$. Suppose f is a continuous function. By the uniqueness of the Laplace transform and the fact that $G_1^\lambda f(x) = G_2^\lambda f(x)$, we see that $\mathbb{E}_i^x(f(X_t); t < \tau_{(a,b)})$ does not depend on i for almost every t. By the continuity of f and X_t, this is in fact true for every t. By the Markov property, the finite dimensional distributions of $X_{\tau_{(a,b)}}$ are the same under $\mathbb{P}_1^x$ and $\mathbb{P}_2^x$. By the continuity of the paths of X_t, that is enough to show that $\mathbb{P}_1^x$ and $\mathbb{P}_2^x$ agree on $\mathcal{F}_{\tau_{(a,b)}}$. $\square$

We repeat (2.15) and (2.17) for future reference. Note that their proof uses only the Markov property.

(2.6) Corollary. *Suppose X_t is any Markov process and*

$$G^\lambda f(x) = \mathbb{E}^x \int_0^\infty e^{-\lambda t} f(X_t)\, dt.$$

Then

$$G^\lambda G^\mu f(x) = \frac{G^\mu f(x) - G^\lambda f(x)}{\lambda - \mu},$$

and if $|\lambda - \mu| \leq 1/(2\|G^\mu\|_\infty)$,

$$G^\lambda = G^\mu - (\lambda - \mu)(G^\mu)^2 + (\lambda - \mu)^2 (G^\mu)^3 - \cdots.$$

We now want to show that if m is a measure such that $0 < m(a,b) < \infty$ for all intervals $[a,b] \subseteq I$, then there exists a regular diffusion on natural scale on I having m as a speed measure. If $m(dx)$ had a density, say $m(dx) = b(x)\, dx$, we would proceed as follows. Let W_t be one-dimensional Brownian motion and let

$$A_t = \int_0^t b(W_s)\, ds, \qquad B_t = \inf\{u : A_t > u\}, \qquad X_t = W_{B_t}.$$

In other words, we let X_t be a certain time change of Brownian motion. In general, where $m(dx)$ does not have a density, we make use of the *local times* of Brownian motion. The relevant properties are given by the following theorem.

(2.7) Theorem. *There exist a family of nondecreasing processes $L_t^x = L(t,x)$ that are jointly continuous in x and t a.s. such that*
(a) if f is a nonnegative Borel function, then

$$\int_0^t f(W_s)\, ds = \int f(x) L_t^x\, dx, \qquad a.s.,$$

where the null set can be taken independent of f;
(b) $L_t^x \to \infty$ a.s., as $t \to \infty$;

(c) The set of t on which L_t^x increases is precisely the set $\{t : W_t = x\}$;

(d) L_t^x may be defined by the formula

$$|W_t - x| - |W_0 - x| = \int_0^t \operatorname{sgn}(W_s - x)\, dW_s + L_t^x.$$

A proof of these facts may be found in [PTA, Section I.6].

Let

$$A_t = \int L_t^x \, m(dx), \qquad B_t = \inf\{u : A_u > t\}, \qquad X_t = W_{B_t}. \tag{2.18}$$

(2.8) Theorem. *Let $(\mathbb{P}^x, W_t)$ be a Brownian motion and m a measure on an open interval I such that $0 < m(a, b) < \infty$ for every interval (a, b) whose closure is contained in I. Then, under $\mathbb{P}^x$, X_t as defined by (2.18) is a regular diffusion on natural scale with speed measure m.*

Proof. First we show that X_t is a continuous process. By the continuity of L_t^x, we observe that A_t is a continuous process. Fix ω. If $s < u$, pick $t \in (s, u)$; if $x = W_t$, then L_t^x increases at t by Theorem 2.7(c), or $L_u^x - L_s^x > 0$. By the continuity of local times, $L_u^y - L_s^y > 0$ for all y in a neighborhood of x, say $(x - \delta, x + \delta)$. Since $m(x - \delta, x + \delta) > 0$, then $A_u - A_s > 0$. Hence A_t is strictly increasing. This and the continuity of A_t imply that B_t is continuous, and therefore X_t is continuous.

Next we show that X_t is a regular diffusion on natural scale. By monotone convergence and Theorem 2.7(b), $A_t \uparrow \infty$, hence $B_t \uparrow \infty$, so $\tau_{(a,b)}^X < \infty$ a.s., where $\tau_{(a,b)}^X$ denotes the exit time of (a, b) by X_t and $\tau_{(a,b)}^W$ denotes the corresponding exit time of W_t. Moreover,

$$\mathbb{P}^x(X(\tau_{(a,b)}^X) = b) = \mathbb{P}^x(W(\tau_{(a,b)}^W) = b) = \frac{x - a}{b - a},$$

since X_t is a time change of W_t.

We verify the strong Markov property. Let $\mathcal{F}_t' = \mathcal{F}_{B_t}$. Then if T is a stopping time for $\mathcal{F}_t'$, we have

$$\mathbb{E}^x(f(X_{T+t}) \mid \mathcal{F}_T') = \mathbb{E}^x(f(W(B_{T+t})) \mid \mathcal{F}_{B_T}).$$

B_T is easily seen to be a stopping time for $\mathcal{F}_t$ and $B_{T+t} = B_t \circ \theta_{B_T}$ where θ_t are the shift operators, so this is

$$\mathbb{E}^x E^{W(B_T)} f(W_{B_t}) = \mathbb{E}^x \mathbb{E}^{X_T} f(X_t).$$

This suffices to show that X_t is a strong Markov process by the proof in [PTA, Section I.3].

It remains to determine the speed measure of X_t. We have

$$\mathbb{E}^x \tau_{(a,b)} = \mathbb{E}^x \int_0^\infty 1_{(a,b)}(X(s \wedge \tau_{(a,b)}^X))\, ds$$

$$= \mathbb{E}^x \int_0^\infty 1_{(a,b)}(W(B(s \wedge \tau_{(a,b)}^X)))\, ds$$

$$= E^x \int_0^\infty 1_{(a,b)}(W(t \wedge \tau_{(a,b)}^W))\, dA_t$$

$$= \mathbb{E}^x \int \int_0^\infty 1_{(a,b)}(W(t \wedge \tau_{(a,b)}^W))\, dL_t^y\, m(dy)$$

$$= \mathbb{E}^x \int \int_0^{\tau_{(a,b)}^W} dL_t^y\, m(dy) = \int \mathbb{E}^x L(\tau_{(a,b)}^W, y)\, m(dy).$$

By Theorem 2.7(d),

$$\mathbb{E}^x L(\tau_{(a,b)}^W, y) = \mathbb{E}^x |W(\tau_{(a,b)}^W) - y| - |x - y|.$$

This is equal to

$$|a - y|\frac{b - x}{b - a} + |b - y|\frac{x - a}{b - a} - |x - y| = G_{a,b}(x, y).$$

We thus have

$$\mathbb{E}^x \tau_{(a,b)}^X = \int G_{a,b}(x, y)\, m(dy),$$

as required. $\qquad\qquad\square$

As a corollary of the proof, we see that a regular diffusion on natural scale is a local martingale, since it is a time change of Brownian motion. We also see, in retrospect, why (1.3) is the proper choice for p_J; if $f(a) = 0$ and $f(b) = 1$, then

$$p_J(x) = \mathbb{P}^x(X_{\tau_J} = b) = \mathbb{E}^x f(X_{\tau_J}).$$

So p_J is harmonic for the process X_t, hence $p_J(X_t)$ is a martingale, and hence $p_J(X_t)$ is a time change of Brownian motion.

3. Diffusions as solutions of SDEs

Suppose X_t is given as the solution to

$$dX_t = \sigma(X_t)\, dW_t + b(X_t)\, dt, \qquad\qquad (3.1)$$

where we assume σ and b are continuous and bounded above and σ is bounded below by a positive constant. The process X_t corresponds to the operator

$$\mathcal{L}f(x) = \frac{1}{2}a(x)f''(x) + b(x)f'(x),$$

where $a(x) = \sigma^2(x)$.

(3.1) Theorem. *The scale function $s(x)$ is the solution to $\mathcal{L}s(x) = 0$, and for some constants c_1, c_2, and x_0 is given by*

$$s(x) = c_1 + c_2 \int_{x_0}^{x} e^{-\int_{x_0}^{y} 2b(w)/a(w)\, dw}\, dy. \tag{3.2}$$

Proof. To solve $\mathcal{L}s(x) = 0$, we write

$$\frac{s''(x)}{s'(x)} = -2\frac{b(x)}{a(x)},$$

or $(\log s'(x))' = -2b(x)/a(x)$, from which (3.2) follows. Since we assumed that σ and b are continuous, $s(x)$ given by (3.2) is C^2. Applying Itô's formula, $s(X_t) - s(X_0) - \int_0^t \mathcal{L}s(X_r)\, dr$ is a martingale. This means that $s(X_t)$ is a martingale, hence a time change of Brownian motion. Therefore the exit probabilities of $s(X_t)$ are the same as those of a Brownian motion. $\square$

By Itô's formula,

$$s(X_t) - s(X_0) = \int_0^t s'(X_r)\sigma(X_r)\, dW_r,$$

and if $Y_t = s(X_t)$, then

$$dY_t = (s'\sigma)(s^{-1}(Y_t))\, dW_t. \tag{3.3}$$

Now suppose that b in (3.1) is 0, or $dX_t = \sigma(X_t)\, dW_t$.

(3.2) Theorem. *The speed measure of X_t is given by*

$$m(dx) = \frac{1}{a(x)}\, dx.$$

Proof. Note $\langle X \rangle_t = \int_0^t a(X_s)\, ds$. To obtain a Brownian motion $\overline{W}_t$ by time changing the martingale X_t, we must time change by $\langle X \rangle_t$. On the other hand, from Theorem 2.8, X_t is the time change of a Brownian motion by B_t, where B_t is given by (2.18). Hence

$$B_t = \langle X \rangle_t = \int_0^t a(X_s)\, ds.$$

The inverse of B_t, namely, A_t, must then satisfy

$$\frac{dA_t}{dt} = \frac{1}{a(X_{A_t})} = \frac{1}{a(W_t)},$$

or

$$A_t = \int_0^t \frac{1}{a(W_s)}\, ds = \int L_t^y \frac{1}{a(y)}\, dy$$

for all t. However, $A_t = \int L_t^y\, m(dy)$ by (2.18). So

$$\int L_t^y \frac{1}{a(y)}\,dy = \int L_t^y\, m(dy).$$

By Theorem 2.7(d), $\mathbb{E}^x L(\tau_{(c,d)}, y) = G_{c,d}(x,y)$. Therefore

$$\int G_{c,d}(x,y)\, m(dy) = \int \mathbb{E}^x L(\tau_{(c,d)}, y)\, m(dy) = \mathbb{E}^x A_{\tau_{(c,d)}}$$

$$= \int \mathbb{E}^x L(\tau_{(c,d)}, y)\frac{1}{a(y)}\,dy$$

$$= \int G_{c,d}(x,y)\frac{1}{a(y)}\,dy$$

for all c, d, and x, which implies $m(dy) = (1/a(y))\,dy$. $\qquad\square$

By using the operator $\mathcal{L}$, we can also find transition densities and first passage times. For simplicity, we suppose that the σ and b are Hölder continuous so that we can use the regularity results of Chapter III; actually the result is true under weaker hypotheses.

(3.3) Theorem. *Suppose f is a C^α function with compact support. Then the Laplace transform of $\mathbb{E}^x f(X_t)$, namely,*

$$u_\lambda(x) = \int_0^\infty e^{-\lambda t}(\mathbb{E}^x f(X_t))\,dt,$$

is the solution to

$$\mathcal{L}u - \lambda u = -f, \qquad u(-\infty) = u(\infty) = 0.$$

Proof. By Section III.6, the solution u is $C^{2+\alpha}$. The result now follows by the Feynman-Kac formula (II.4.3) with $D = (-\infty, \infty)$. $\qquad\square$

(3.4) Theorem. *The first passage time to a point x_0 has a Laplace transform*

$$u(x) = \mathbb{E}^x e^{-\lambda T_{x_0}}$$

that is the solution to

$$\mathcal{L}u(x) = \lambda u(x), \qquad u(x_0) = 1, \quad u(-\infty) = u(\infty) = 0.$$

Proof. This also follows from the Feynman-Kac formula. Let $D = (-\infty, x_0)$ or (x_0, ∞), and let $f = 1$ at x_0 and 0 at infinity in Theorem II.4.1. $\qquad\square$

Let us calculate the scale function and the speed measure for some examples of diffusions.

Brownian motion with constant drift. The solution to $dX_t = dW_t + b\,dt$ corresponds to the operator $(1/2)f'' + bf'$. From Theorem 3.1, $s(x) = \exp(-2bx)$

is the scale function. If $Y_t = s(X_t)$, then $(s'\sigma)(s^{-1}(y)) = -2by$, or Y_t corresponds to the operator $2b^2y^2 f''$, so the speed measure is $(4b^2y^2)^{-1}\,dx$.

Bessel processes. We ignore the boundary conditions here and consider a Bessel process of order ν up until the first hit of 0. Then

$$dX_t = dW_t + \frac{\nu - 1}{2X_t}\,dt$$

corresponds to the operator $(1/2)f'' + (\nu - 1)/(2x)f'$. If $\nu \neq 2$, a calculation shows that $s(x) = x^{2-\nu}$. Then $Y_t = s(X_t)$ satisfies

$$dY_t = (2 - \nu)Y_t^{(1-\nu)/(2-\nu)}\,dW_t,$$

and the speed measure is

$$m(dx) = (2 - \nu)^{-2}x^{(2\nu-2)/(2-\nu)}\,dx, \qquad x > 0.$$

If $\nu = 2$, then $s(x) = \log x$, $Y_t = s(X_t)$ satisfies $dY_t = e^{-Y_t}\,dW_t$, and the speed measure is $m(dx) = e^{2x}\,dx$.

4. Boundaries

Let X_t be a diffusion on natural scale on an interval I. We want to consider what happens if I is not the entire real line and is closed on one or both of its endpoints. To be specific, let us suppose $I = [0, \infty)$. We want to see how to assign a value to $m(\{0\})$ so that Theorems 2.3, 2.5, and 2.8 still hold.

Let us describe what happens in a special case first; the proofs of these assertions will be a consequence of the results later in this section. Suppose $m(dx) = dx$ for $x > 0$, so that X_t behaves like Brownian motion on $(0, \infty)$. If we set $m(\{0\}) = \infty$, then X_t will hit 0 and stay there; we call 0 an *absorbing state.* If we set $m(\{0\}) = 0$, we have, as we shall see, reflecting Brownian motion. If $m(\{0\}) = a \in (0, \infty)$, we have what is known as a *sticky boundary.* Upon hitting 0, Brownian motion leaves immediately just as ordinary Brownian motion does, but $\{t : X_t = 0\}$ has positive Lebesgue measure. The set $\{t : X_t = 0\}$ is somewhat analogous to Cantor-like sets of positive Lebesgue measure.

We now turn to general diffusions on natural scale on $[0, \infty)$. The first case to consider is when $m(0, a) = \infty$ for every $a > 0$. An example of this would be if $m(dx) = x^{-1}\,dx$.

(4.1) Lemma. *If X_t is a regular diffusion on natural scale on $[0, \infty)$ and $m(0, a) = \infty$ for all $a > 0$, then $\mathbb{P}^0(T_{(0,\infty)} < \infty) = 0$.*

Proof. A Brownian motion W_t started at 0 leaves 0 immediately and enters $(0, \infty)$. Since the support of L_t^0 is $\{t : W_t = 0\}$, then L_t^0 is positive with

probability one for all $t > 0$. By continuity, L_t^y will be bounded below by a positive number (depending on ω) if y is sufficiently close to 0 (how close also depends on the path ω). It follows that A_t defined in (2.18) will be infinite for all $t > 0$. This implies that $B_t = 0$ for all t, and then $X_t = W_{B_t} = W_0 = 0$.
$$\square$$

In this first case we set $m(\{0\}) = \infty$. Since the law of X_t up to T_0 is determined by m and Lemma 4.1 says that X_t started at 0 remains at 0, we see that the law of X_t is determined by the speed measure. Also, Theorem 2.8 is easily seen to hold.

The second case we look at is when $m(0, a) < \infty$ when a is finite but X_t is absorbing at 0, i.e., upon arriving at 0, the process X_t stays there forever. Again we set $m(\{0\}) = \infty$. Starting at 0, L_t^0 is positive for all $t > 0$, hence $A_t = \infty$, and hence $B_t = 0$. So Theorem 2.8 holds, and the law of X_t is determined by m.

The third and last case is when $m(0, a) < \infty$ for all finite a and 0 is not an absorbing state. Let
$$\overline{G}_r(y) = \begin{cases} 2(r - y) & 0 \le y \le r; \\ 0 & y > r. \end{cases}$$

Define $m(\{0\})$ so that
$$\int \overline{G}_r(y)\, m(dy) = \mathbb{E}^0 T_r \tag{4.1}$$

for all r. We shall see in a moment that this definition is independent of r. We say that X_t has speed measure m if (4.1) and (2.2) hold. Note that this definition is consistent with the first two cases in which $m(\{0\}) = \infty$.

We need to verify that this definition of the value of the speed measure at 0 is independent of r. If $r < s$, by the strong Markov property,
$$\mathbb{E}^0 T_s = \mathbb{E}^0 T_r + \mathbb{E}^r T_s = \mathbb{E}^0 T_r + \mathbb{E}^r \tau_{(0,s)} + \mathbb{P}^r(T_0 < T_s)\mathbb{E}^0 T_s.$$

Solving for $\mathbb{E}^0 T_s$,
$$\mathbb{E}^0 T_s = \frac{\mathbb{E}^0 T_r + \mathbb{E}^r \tau_{(0,s)}}{\mathbb{P}^r(T_s < T_0)}.$$

Since the denominator is r/s, the definition in (4.1) will be consistent if
$$(s/r)(\overline{G}_r(y) + G_{0,s}(r, y)) = \overline{G}_s(y)$$

for all y. Substituting in the definitions of $\overline{G}_r$, $\overline{G}_s$, and $G_{0,s}$ shows that this indeed is the case.

(4.2) Theorem. *Suppose $m(\{0\}) < \infty$.*
(a)   If X_t is defined by (2.18), then X_t has speed measure m.
(b)   The law of $(\mathbb{P}^x, X_t)$ is determined by the speed measure.

Proof. (a) is proved in much the same way as Theorem 2.8, except that we need to verify that X_t has continuous paths. The difficulty is that A_t may have flat intervals, in which case B_t has jumps. Suppose B_t has a jump at time v: suppose $u = B_v > \lim_{t \uparrow v} B_t = s$. This means that A_t is constant in the interval (s, u) and takes the value v. By the definition of A_t, the only way this can happen is if $W_t \leq 0$ for $t \in (s, u)$, and for any larger interval we have $W_t \geq 0$ for some t in the larger interval. Since the paths of Brownian motion are continuous, this means $W_s = W_u = 0$, and therefore $\lim_{t \uparrow v} X_t = 0 = X_v$, or X is continuous at time v.

To prove (b), let $M > 0$ and define

$$G_i f(x) = \mathbb{E}_i^x \int_0^{T_M} f(X_s)\, ds, \qquad i = 1, 2,$$

where $\mathbb{P}_1^x$ and $\mathbb{P}_2^x$ are two families of probabilities under which X_t is a diffusion with speed measure m. As in the proof of Theorem 2.5, it suffices to show that $G_i f$ does not depend on i when f is a continuous function. Since

$$\mathbb{E}_i^x \int_0^{T_M} f(X_s)\, ds = \mathbb{E}_i^x \int_0^{\tau(0, M)} f(X_s)\, ds + \mathbb{E}_i^0 \int_0^{T_M} f(X_s)\, ds,$$

and the first term on the right is determined by m by Theorem 2.5, it suffices to show that $\mathbb{E}_i^0 \int_0^{T_M} f(X_s)\, ds$ does not depend on i. If $\varepsilon \in (0, M)$, then

$$\begin{aligned}
\mathbb{E}_i^0 \int_0^{T_M} f(X_s)\, ds &= \mathbb{E}_i^0 \int_0^{T_\varepsilon} f(X_s)\, ds + \mathbb{E}_i^\varepsilon \int_0^{T_M} f(X_s)\, ds \\
&= \mathbb{E}_i^0 \int_0^{T_\varepsilon} f(X_s)\, ds + \mathbb{E}_i^\varepsilon \int_0^{\tau(0, M)} f(X_s)\, ds \\
&\quad + \mathbb{P}^\varepsilon(T_0 < T_M) \mathbb{E}_i^0 \int_0^{T_M} f(X_s)\, ds.
\end{aligned}$$

Since $1 - \mathbb{P}^\varepsilon(T_0 < T_M) = \mathbb{P}^\varepsilon(T_0 > T_M) = \varepsilon/M$, solving gives

$$\mathbb{E}_i^0 \int_0^{T_M} f(X_s)\, ds = \frac{\mathbb{E}_i^0 \int_0^{T_\varepsilon} f(X_s)\, ds + \mathbb{E}_i^\varepsilon \int_0^{\tau(0,M)} f(X_s)\, ds}{\varepsilon/M}. \tag{4.2}$$

We write

$$\varepsilon^{-1} \mathbb{E}_i^0 \int_0^{T_\varepsilon} f(X_s)\, ds = \varepsilon^{-1} f(0) \mathbb{E}_i^0 T_\varepsilon \tag{4.3}$$

$$+ \varepsilon^{-1} \mathbb{E}_i^0 \int_0^{T_\varepsilon} [f(X_s) - f(0)]\, ds.$$

The first term on the right of (4.3) is equal to $\varepsilon^{-1} f(0) \int \overline{G}_\varepsilon(y)\, m(dy)$, which converges to $f(0) m(\{0\})$ as $\varepsilon \to 0$, and the second term on the right is bounded by

$$(\sup_{0 \leq y \leq \varepsilon} |f(x) - f(0)|)\varepsilon^{-1} \mathbb{E}_i^0 T_\varepsilon.$$

Since $\varepsilon^{-1} \mathbb{E}_i^0 T_\varepsilon = \varepsilon^{-1} \int \overline{G}_\varepsilon(y)\, m(dy)$ remains bounded for small ε, the second term on the right of (4.3) converges to 0.

By Corollary 2.4,

$$\varepsilon^{-1} \mathbb{E}_i^\varepsilon \int_0^{\tau(0,M)} f(X_s)\, ds = \varepsilon^{-1} \int_0^M G_{0,M}(\varepsilon, y) f(y)\, m(dy)$$

$$\to \int_{(0,M)} 2\frac{M-y}{M} f(y)\, m(dy).$$

Substituting in (4.2) and letting $\varepsilon \to 0$,

$$\mathbb{E}_i^0 \int_0^{T_M} f(X_s)\, ds = M f(0) m(\{0\}) + \int_{(0,M)} \overline{G}_M(y) f(y)\, m(dy)$$

$$= \int \overline{G}_M(y) f(y)\, m(dy). \tag{4.4}$$

This shows that $G_i f$ does not depend on i. $\qquad\square$

Let us now return to our examples involving Brownian motion. If $m(\{0\}) = \infty$, we have Brownian motion absorbed at 0 by our second case. If $m(\{0\}) < \infty$, we have $\mathbb{E}^0 T_\varepsilon = \int \overline{G}_\varepsilon(y)\, m(dy) \to 0$ as $\varepsilon \to 0$, or, starting at 0, X_t must enter $(0, \infty)$ immediately. By (4.4), however,

$$\mathbb{E}^0 \int_0^{T_1} 1_{[0,\varepsilon]}(X_s)\, ds = \int \overline{G}_1(y) 1_{[0,\varepsilon]}(y)\, m(dy) \to 2m(\{0\})$$

as $\varepsilon \to 0$, which means that the amount of time X_t spends at 0 has positive Lebesgue measure if $m(\{0\}) > 0$.

We want to provide justification for calling a diffusion reflecting if $m(\{0\}) = 0$. Let Y_t be Brownian motion and let $X_t = |Y_t|$. Then X_t is a diffusion and the speed measure on $(0, \infty)$ for X_t is clearly Lebesgue measure. Since 0 is not an absorbing state and the Lebesgue measure of the time X_t spends at 0 is the same as the Lebesgue measure of the time Y_t spends at 0, which is 0, then $m(\{0\}) = 0$. We conclude that the speed measure of reflecting Brownian motion is Lebesgue measure on $[0, \infty)$.

5. Eigenvalue expansions

Let $I = [a, b]$ be a closed finite interval, X_t a regular diffusion on natural scale on I with speed measure m and absorption at a and b, and suppose $m(a, b) < \infty$. We can then give an eigenvalue expansion for the transition densities of X_t with respect to m.

(5.1) Theorem. *There exist reals $0 < \lambda_1 < \lambda_2 \le \lambda_3 \le \cdots < \infty$ and a collection of continuous functions φ_i on I that are 0 at a and b such that*
(a)   the sequence $\{\lambda_i\}$ has no subsequential limit point other than ∞;
(b) $\{\varphi_i\}$ forms a complete orthonormal system on $L^2(I, m(dx))$;
(c) $\mathbb{P}^x(X_t \in dy) = p(t, x, y)\, m(dy)$, where

$$p(t, x, y) = \sum_{i=1}^{\infty} e^{-\lambda_i t} \varphi_i(x) \varphi_i(y),$$

and where for each t the convergence is absolute and uniform over $(x, y) \in I \times I$;
(d) $\varphi_1 > 0$ in the interior of I.

Proof. Define $Gf(x) = \int_{(a,b)} G(x, y) f(y)\, m(dy)$, where $G = G_{a,b}$ is defined by (2.1). Since G is bounded, G is a bounded operator on $L^2(I)$. Since G is jointly continuous in x and y, then $\{Gf : \|f\|_\infty \le 1\}$ is an equicontinuous family. Note also that G is symmetric in x and y. Let

$$\langle f, g \rangle = \int_I f(x)\overline{g}(x)\, m(dx).$$

By the Hilbert-Schmidt expansion theorem ([PTA, Theorem II.4.12]), there exists a sequence $\mu_1 \ge \mu_2 \ge \cdots \ge 0$ and a complete orthonormal system $\{\varphi_i\}$ of continuous functions such that

$$Gf(x) = \sum_{i=1}^{\infty} \mu_i \langle f, \varphi_i \rangle \varphi_i(x), \qquad (5.1)$$

the sequence $\{\mu_i\}$ has no subsequential limit point other than 0, and (b) holds.

We show that none of the μ_i is 0. For if $\mu_i = 0$, then $(G)^k \varphi_i = 0$. Using (2.17) with $\mu = 0$, $G^\lambda \varphi_i = 0$ if $\lambda \le 1/(2\|G^0\|_\infty)$. Repeatedly using (2.17), $G^\lambda \varphi_i = 0$ for all λ. Then

$$0 = \lambda G^\lambda \varphi_i(x) = \mathbb{E}^x \int_0^\infty \lambda e^{-\lambda t} \varphi_i(X_{t \wedge \tau_{(a,b)}})\, dt \to \varphi_i(x)$$

as $\lambda \to \infty$. This says that $\varphi_i \equiv 0$, a contradiction.

Set $\lambda_i = \mu_i^{-1}$. We next show that $\sum_{i=1}^{\infty} \lambda_i^{-2} < \infty$. Because

$$\langle G(y, \cdot), \varphi_i \rangle = G\varphi_i(y) = \lambda_i^{-1} \varphi_i(y),$$

then using (5.1) with $f(x) = G(y, x)$ gives

$$Gf(x) = \sum_{i=1}^{\infty} \lambda_i^{-1} \langle f, \varphi_i \rangle \varphi_i(x) = \sum_{i=1}^{\infty} \lambda_i^{-1} G\varphi_i(y) \varphi_i(x)$$

$$= \sum_{i=1}^{\infty} \lambda_i^{-2} \varphi_i(x) \varphi_i(y).$$

Since G is bounded,

$$\infty > \int_I Gf(y)\, m(dy) = \sum_{i=1}^{\infty} \lambda_i^{-2} \int_I \varphi_i^2(y)\, m(dy) = \sum_{i=1}^{\infty} \lambda_i^{-2}. \tag{5.2}$$

Define $p(t,x,y) = \sum_{i=1}^{\infty} e^{-\lambda_i t}\varphi_i(x)\varphi_i(y)$. We have $\varphi_i(x) = \lambda_i G\varphi_i(x)$ and

$$|G\varphi_i(x)| = \left| \int G(x,y)\varphi_i(y)\, m(dy) \right|$$
$$\leq \left(\int G(x,y)^2\, m(dy) \right)^{1/2} \left(\int \varphi_i(y)^2\, m(dy) \right)^{1/2},$$

which implies that $|\varphi_i(x)| \leq c_1 \lambda_i$. If we let $c_2 = \sup_{\lambda \geq 0} \lambda^2 e^{-\lambda t/2}$, then

$$\sum_i e^{-\lambda_i t}|\varphi_i(x)\varphi_i(y)| \leq c_1^2 \sum_i \lambda_i^2 e^{-\lambda_i t} \leq c_1^2 c_2 e^{-\lambda_i t/2}.$$

This is finite by comparing with (5.2), so the convergence of the series defining $p(t,x,y)$ is absolute and uniform.

We now show that $p(t,x,y)$ is the transition density. Note that

$$\int p(t,x,y)\varphi_i(y)\, m(dy) = e^{-\lambda_i t}\varphi_i(x).$$

Therefore

$$\int_0^{\infty} e^{-\lambda t} \int p(t,x,y)\varphi_i(y)\, m(dy)\, dt = (\lambda + \lambda_i)^{-1}\varphi_i(x).$$

On the other hand, $(G)^k \varphi_i(x) = \lambda_i^{-k}\varphi_i(x)$, so $G^\lambda \varphi_i(x) = (\lambda + \lambda_i)^{-1}\varphi_i(x)$ for λ small by (2.17). Comparing the two expressions, the uniqueness of the Laplace transform shows that $\int p(t,x,y)\varphi_i(y)\, m(dy) = \mathbb{E}^x \varphi_i(X_{t \wedge \tau_{(a,b)}})$ for almost every t; by the continuity of both sides in t, this holds for every t. Since the $\{\varphi_i\}$ form a complete orthonormal system, this shows that $p(t,x,y)$ is the transition density.

We show next that $\varphi_1 > 0$ in the interior of I. Recall that the way φ_1 is defined in the proof of the Hilbert-Schmidt expansion theorem is that we take a sequence of functions f_n with $\|f_n\|_2 = 1$ and $\langle f_n, Gf_n \rangle \to \sup\{\langle f, Gf \rangle : \|f\|_2 = 1\}$, we take a subsequential limit point g of Gf_n, and we set $\varphi_1 = c_3 Gg$ for a suitable constant c_3. Since G is nonnegative, $\langle |f_n|, G|f_n| \rangle \geq \langle f_n, Gf_n \rangle$, so we may as well take all the $f_n \geq 0$. Then $Gf_n \geq 0$, and so $g \geq 0$, and finally $\varphi_1 \geq 0$. Since $G(x,y) > 0$ if x,y are in the interior of I, then $\varphi_1 = \lambda_1 G\varphi_1$ will be strictly positive in the interior of I.

It remains to show that $\lambda_2 > \lambda_1$. Suppose instead that $\lambda_2 = \lambda_1$. We have

$$|\varphi_2| = \lambda_2 |G\varphi_2| \leq \lambda_2 G(|\varphi_2|). \tag{5.3}$$

By the symmetry of G,

$$\langle \varphi_1, |\varphi_2| \rangle \leq \lambda_2 \langle \varphi_1, G(|\varphi_2|) \rangle = \lambda_2 \langle G\varphi_1, |\varphi_2| \rangle = \langle \varphi_1, |\varphi_2| \rangle.$$

We must therefore have equality a.e. in (5.3), or $G(|\varphi_2|) = \lambda_2^{-1}|\varphi_2|$. The function $|\varphi_2|$ is greater than or equal to 0 a.e. and is not identically 0, so must be strictly positive in the interior of I by the argument of the preceding paragraph. Define $\theta(x)$ such that $|\varphi_2(x)| = \varphi_2(x)e^{-i\theta(x)}$. Then

$$G|\varphi_2(x)| = \lambda_2^{-1}|\varphi_2(x)| = \lambda_2^{-1}e^{-i\theta(x)}\varphi_2(x) = e^{-i\theta(x)}G\varphi_2(x),$$

or

$$\int G(x,y)|\varphi_2(y)|\, m(dy) = \int G(x,y)e^{-i\theta(x)}\varphi_2(y)\, m(dy). \qquad (5.4)$$

The real part of $G(x,y)e^{-i\theta(x)}\varphi_2(y)$ is less than or equal to $G(x,y)|\varphi_2(y)|$, yet the integrals in (5.4) are equal, so

$$G(x,y)|\varphi_2(y)| = G(x,y)\mathrm{Re}\,(e^{-i\theta(x)}\varphi_2(y)), \qquad \text{a.e.}$$

By the continuity of φ_2 and the positivity of G, $|\varphi_2(y)| = \mathrm{Re}\,(e^{-i\theta(x)}\varphi_2(y))$. Since

$$\left| e^{-i\theta(x)}\varphi_2(y) \right| = |\varphi_2(y)| = \mathrm{Re}\,(e^{-i\theta(x)}\varphi_2(y)),$$

$\mathrm{Im}\,(e^{-i\theta(x)}\varphi_2(y)) = 0$, or $|\varphi_2(y)| = e^{-i\theta(x)}\varphi_2(y)$. This implies that $\theta(x)$ must be a constant, say θ, and $|\varphi_2| = e^{-i\theta}\varphi_2$. But then $\langle |\varphi_2|, \varphi_1 \rangle = e^{-i\theta}\langle \varphi_2, \varphi_1 \rangle = 0$, a contradiction to $|\varphi_2|$ and φ_1 being positive in the interior of I. $\square$

6. Notes

For further information see Breiman [1], Rogers and Williams [1], and especially Itô and McKean [1]. In some accounts the definition of speed measure differs from ours by a factor of 2. Our proof of Theorem 2.5 is based on Blumenthal and Getoor [1], Chapter V. Section 5 is based on McKean [1]. The proof that $\lambda_2 > \lambda_1$ in Theorem 5.1 is from Krein and Rutman [1].

V
NONDIVERGENCE FORM OPERATORS

In this chapter and the next we consider operators in nondivergence form. This chapter is primarily concerned with the Harnack inequality of Krylov-Safonov and consequences.

Section 1 defines nondivergence form operators and relates them to solutions of SDEs. Section 2 contains some basic estimates and formulas, whereas Section 3 gives some examples of singular behavior.

Section 4 presents an important estimate of Alexandroff-Bakelman-Pucci. This is used in Section 5 to obtain bounds on Green functions.

Section 6 is about an approximation procedure due to Krylov. This allows us to extend results about operators with smooth coefficients to those with nonsmooth coefficients.

Section 7 proves the Harnack inequality of Krylov-Safonov. A key estimate is that processes associated to operators in nondivergence form hit small sets. The result that these processes also spend at least a certain amount of time in small sets is proved in Section 8 and is then applied to finish the discussion of approximation.

1. Definitions

We consider operators in *nondivergence form*, that is, operators of the form

$$\mathcal{L}f(x) = \frac{1}{2} \sum_{i,j=1}^{d} a_{ij}(x) \partial_{ij} f(x) + \sum_{i=1}^{d} b_i(x) \partial_i f(x). \tag{1.1}$$

These operators are sometimes said to be of *nonvariational form*. Operators in divergence form or variational form will be considered in Chapter VII.

We assume throughout this chapter that the coefficients a_{ij} and b_i are bounded and measurable. Unless stated otherwise, we also assume that the operator $\mathcal{L}$ is uniformly elliptic (see (II.1.2)). The coefficients a_{ij} are called the *diffusion* coefficients and the b_i are called the *drift* coefficients. We let $\mathcal{N}(\Lambda_1, \Lambda_2)$ denote the set of operators of the form (1.1) with $\sup_i \|b_i\|_\infty \leq \Lambda_2$ and

$$\Lambda_1 |y|^2 \leq \sum_{i,j=1}^d y_i a_{ij}(x) y_j \leq \Lambda_1^{-1} |y|^2, \qquad y \in \mathbb{R}^d, x \in \mathbb{R}^d. \tag{1.2}$$

We saw in Chapter I that if X_t is the solution to

$$dX_t = \sigma(X_t)\, dW_t + b(X_t)\, dt, \qquad X_0 = x_0, \tag{1.3}$$

where σ is a $d \times d$ matrix, b is a vector, and W_t is a Brownian motion, then X_t is associated to the operator $\mathcal{L}$ with $a = \sigma\sigma^T$. Proposition I.2.1 says that if $f \in C^2$, then

$$f(X_t) - f(X_0) - \int_0^t \mathcal{L}f(X_s)\, ds \tag{1.4}$$

is a local martingale under $\mathbb{P}$.

A very fruitful idea of Stroock and Varadhan is to phrase the association of X_t to $\mathcal{L}$ in terms which use (1.4) as a key element. Let Ω consist of all continuous functions ω mapping $[0, \infty)$ to $\mathbb{R}^d$. Let $X_t(\omega) = \omega(t)$ (cf. [PTA, (I.2.1)]) and let $\mathcal{F}_t$ be the right continuous modification of the σ-field generated by the X_s, $s \leq t$. A probability measure $\mathbb{P}$ is a solution to the *martingale problem for $\mathcal{L}$ started at x_0* if

$$\mathbb{P}(X_0 = x_0) = 1 \tag{1.5}$$

and

$$f(X_t) - f(X_0) - \int_0^t \mathcal{L}f(X_s)\, ds$$

is a local martingale under $\mathbb{P}$ whenever $f \in C^2(\mathbb{R}^d)$. The martingale problem is *well posed* if there exists a solution and this solution is unique.

Uniqueness of the martingale problem for $\mathcal{L}$ is closely connected to weak uniqueness or uniqueness in law of (1.3). Recall that the cylindrical sets are ones of the form $\{\omega : \omega(t_1) \in A_1, \ldots, \omega(t_n) \in A_n\}$ for $n \geq 1$ and $A_1, \ldots, A_n$ Borel subsets of $\mathbb{R}^d$.

(1.1) Theorem. *Suppose $a = \sigma\sigma^T$. Weak uniqueness for (1.3) holds if and only if the solution for the martingale problem for $\mathcal{L}$ started at x_0 is unique. Weak existence for (1.3) holds if and only if there exists a solution to the martingale problem for $\mathcal{L}$ started at x_0.*

Proof. We prove the uniqueness assertion. Let Ω be the continuous functions on $[0, \infty)$ and Z_t the coordinate process: $Z_t(\omega) = \omega(t)$. First suppose the solution to the martingale problem is unique. If (X_t^1, W_t^1) and (X_t^2, W_t^2) are two weak solutions to (1.3), define $\mathbb{P}_1^{x_0}$ and $\mathbb{P}_2^{x_0}$ on Ω by $\mathbb{P}_i^{x_0}(Z. \in A) = \mathbb{P}(X^i \in A)$, $i = 1, 2$, for any cylindrical set A. Clearly $\mathbb{P}_i^{x_0}(Z_0 = x_0) = \mathbb{P}(X_0^i = x_0) = 1$. By Proposition I.2.1, (1.4) is a local martingale under $\mathbb{P}_i^{x_0}$ for each i and each $f \in C^2$. By the hypothesis of uniqueness for the solution of the martingale problem, $\mathbb{P}_1^{x_0} = \mathbb{P}_2^{x_0}$. This implies that the laws of X_t^1 and X_t^2 are the same, or weak uniqueness holds.

Now suppose weak uniqueness holds for (1.3). Let

$$Y_t = Z_t - \int_0^t b(Z_s)\, ds.$$

Let $\mathbb{P}_1^{x_0}$ and $\mathbb{P}_2^{x_0}$ be solutions to the martingale problem. If $f(x) = x_k$, the kth coordinate of x, then $\partial_i f(x) = \delta_{ik}$ and $\partial_{ij} f = 0$, or $\mathcal{L}f(Z_s) = b_k(Z_s)$. Therefore the kth coordinate of Y_t is a local martingale under $\mathbb{P}_i^{x_0}$. Now let $f(x) = x_k x_m$. Computing $\mathcal{L}f$, we see that $Y_t^k Y_t^m - \int_0^t a_{km}(Z_s)\, ds$ is a local martingale. We set

$$W_t = \int_0^t \sigma^{-1}(Z_s)\, dY_s.$$

The stochastic integral is finite since

$$\mathbb{E} \int_0^t \sum_j (\sigma^{-1})_{ij}(Z_s) \sum_k (\sigma^{-1})_{ik}(Z_s)\, d\langle Y^j, Y^k \rangle_s \tag{1.6}$$

$$= \mathbb{E} \int_0^t \sum_{i,k} (a^{-1})_{ik}(Z_s) a_{ik}(Z_s)\, ds = t < \infty.$$

It follows that W_t is a martingale, and a calculation similar to (1.6) shows that $W_t^k W_t^m - \delta_{km} t$ is also a martingale under $\mathbb{P}_i^{x_0}$. So by Lévy's theorem (Section I.1), W_t is a Brownian motion under both $\mathbb{P}_1^{x_0}$ and $\mathbb{P}_2^{x_0}$, and (Z_t, W_t) is a weak solution to (1.3). By the weak uniqueness hypothesis, the laws of Z_t under $\mathbb{P}_1^{x_0}$ and $\mathbb{P}_2^{x_0}$ agree, which is what we wanted to prove.

A similar proof shows that the existence of a weak solution to (1.3) is equivalent to the existence of a solution to the martingale problem. $\qquad\square$

Since pathwise existence and uniqueness imply weak existence and uniqueness, if the σ_{ij} and b_i are Lipschitz, then the martingale problem for $\mathcal{L}$ is well posed for every starting point.

2. Some estimates

It will be handy to have the formula for the radial component of a diffusion.

(2.1) Proposition. *Suppose the drift coefficients of $\mathcal{L}$ are zero and $R_t = |X_t|$. If $\mathbb{P}$ is a solution to the martingale problem for $\mathcal{L}$ started at x_0, then R_t satisfies the following stochastic differential equation up until $T_{\{0\}}$, the hitting time of 0, where $\overline{W}_t$ is a one-dimensional Brownian motion.*

$$R_t = |x_0| + \int_0^t \left[\sum_{i,j} \frac{X_s^i a_{ij}(X_s) X_s^j}{R_s^2}\right]^{1/2} d\overline{W}_t \tag{2.1}$$

$$+ \frac{1}{2} \int_0^t \left[\frac{\operatorname{trace} a(X_s)}{R_s} - \sum_{i,j} \frac{X_s^i a_{ij}(X_s) X_s^j}{R_s^3}\right] ds.$$

Proof. Let σ be a positive definite square root of a. Using Proposition 1.1, we can find W_t so that under $\mathbb{P}$, W_t is a Brownian motion and X_t solves (1.3).

Let $f(x) = |x|$. For $x \neq 0$, $\partial_i f(x) = x_i/|x|$ and $\partial_{ij} f(x) = (\delta_{ij}|x|^2 - x_i x_j)/|x|^3$. If $\varepsilon > 0$, applying Itô's formula to a C^2 function that is equal to f in $B(0,\varepsilon)^c$, we have

$$R_t = R_0 + \int_0^t \sum_{i=1}^d \frac{X_s^i}{R_s} dX_s^i \tag{2.2}$$

$$+ \frac{1}{2} \int_0^t \sum_{i,j=1}^d \frac{\delta_{ij} R_s^2 - X_s^i X_s^j}{R_s^3} d\langle X^i, X^j\rangle_s$$

$$= |x_0| + \int_0^t \sum_{i,j=1}^d \frac{X_s^i}{R_s} \sigma_{ij}(X_s) dW_s^j$$

$$+ \frac{1}{2} \int_0^t \sum_{i,j=1}^d \frac{\delta_{ij} R_s^2 - X_s^i X_s^j}{R_s^3} a_{ij}(X_s) ds$$

for $t < T_{B(0,\varepsilon)}$. (2.2) holds for $t < T_{B(0,\varepsilon)}$ for each $\varepsilon > 0$; hence (2.2) holds for $t < T_{\{0\}}$. Since $\sum_{i,j} \delta_{ij} a_{ij}(x) = \operatorname{trace} a(x)$, the proof will be complete if we identify the martingale term M_t in the last line of (2.2).

M_t is a martingale and its quadratic variation is given by

$$d\langle M\rangle_t = d\Big\langle \sum_{i,j} \frac{X_s^i}{R_s}\sigma_{ij}(X_s) dW^j, \sum_{k,\ell} \frac{X_s^k}{R_s}\sigma_{k\ell}(X_s) dW^\ell \Big\rangle_t$$

$$= \sum_{i,j,k,\ell} \frac{X_s^i X_s^k}{R_s^2}\sigma_{ij}(X_s)\sigma_{k\ell}(X_s) d\langle W^j, W^\ell\rangle_t$$

$$= \sum_{i,j,k} \frac{X_s^i X_s^k}{R_s^2}(\sigma_{ij}\sigma_{jk}^T)(X_s) dt = \sum_{i,k} \frac{X_s^i a_{ik}(X_s) X_s^k}{R_s^2} ds.$$

If we define $\overline{W}_t$ by

$$d\overline{W}_t = \left[\sum i,j \frac{X_s^i a_{ij}(X_s) X_s^j}{R_s^2}\right]^{-1/2} dM_s,$$

then $\overline{W}_t$ is a continuous martingale with quadratic variation equal to t, hence a Brownian motion. The proposition follows. $\square$

Diffusions corresponding to elliptic operators in nondivergence form do not have an exact scaling property as does Brownian motion, i.e., rX_{t/r^2} does not necessarily have the same law as X_t. However, they do have a weak scaling property that is nearly as useful: rX_{t/r^2} is again a diffusion corresponding to another elliptic operator of the same type. See also Proposition I.8.6.

(2.2) Proposition. *Suppose $\mathcal{L}$ is an elliptic operator with zero drift coefficients. Suppose $\mathbb{P}$ is a solution to the martingale problem for $\mathcal{L}$ started at x_0. Then the law of rZ_{t/r^2} is a solution to the martingale problem for $\mathcal{L}_r$ started at rx_0, where*

$$\mathcal{L}_r f(x) = \sum_{i,j=1}^{d} a_{ij}(x/r)\partial_{ij} f(x), \qquad f \in C^2.$$

Proof. It is obvious that rZ_{t/r^2} starts at rx_0 with $\mathbb{P}$ probability one. If $f \in C^2$, let $g(x) = f(rx)$. Setting $V_t = rZ_{t/r^2}$,

$$f(V_t) = g(Z_{t/r^2}) \tag{2.3}$$

$$= g(x_0) + \text{ martingale } + \int_0^t \sum_{i,j} \partial_{ij} g(Z_{s/r^2}) \, d\langle Z^i, Z^j \rangle_{s/r^2}.$$

By the definition of g, $\partial_{ij} g(x) = r^2 \partial_{ij} f(rx)$, so $\partial_{ij} g(Z_{s/r^2}) = r^2 \partial_{ij} f(V_s)$. From the definition of martingale problem applied to the function $x_i x_j$, we see that as in the proof of Theorem 1.1, $Z_t^i Z_t^j - \int_0^t a_{ij}(Z_s)\,ds$ is a local martingale under $\mathbb{P}$, and hence $d\langle Z^i, Z^j \rangle_s = a_{ij}(Z_s)\,ds$ and

$$d\langle Z^i, Z^j \rangle_{s/r^2} = r^{-2} a_{ij}(Z_{s/r^2})\,ds = r^{-2} a_{ij}(V_s/r)\,ds.$$

Substituting in (2.3),

$$f(V_t) = f(V_0) + \text{ martingale } + \int_0^t a_{ij}(V_s/r)\partial_{ij} f(V_s)\,ds.$$

Thus the law of V_t under $\mathbb{P}$ is a solution to the martingale problem for $\mathcal{L}_r$. $\square$

The following elementary bounds on the time to exit a ball will be used repetitively. Recall that τ_A denotes the hitting time of A.

(2.3) Proposition. *Suppose $\mathcal{L} \in \mathcal{N}(\Lambda, 0)$, so that the drift coefficients of $\mathcal{L}$ are 0. Suppose $\mathbb{P}$ is a solution to the martingale problem for $\mathcal{L}$ started at 0.*
(a)   There exists c_1 depending only on Λ such that

$$\mathbb{P}(\tau_{B(0,1)} \leq t) \leq c_1 t.$$

(b)   There exist c_2 and c_3 depending only on Λ such that

$$\mathbb{P}(\tau_{B(0,1)} \geq t) \leq c_2 e^{-c_3 t}.$$

Proof. Write B for $B(0,1)$. Let f be a C^2 function that is zero at 0, one on ∂B, with $\partial_{ij} f$ bounded by a constant c_4. Since $\mathbb{P}$ is a solution to the martingale problem,

$$\mathbb{E}\, f(X_{t \wedge \tau_B}) = \mathbb{E} \int_0^{t \wedge \tau_B} \mathcal{L} f(X_s)\, ds \leq c_5 t,$$

where c_5 depends on c_4 and Λ. Since $f(X_{t \wedge \tau_B}) \geq 1_{(\tau_B \leq t)}$, this proves (a).

To prove (b), look at X_t^1. Since $\mathbb{P}$ is a solution to the martingale problem, taking $f(x) = x_1$ in (1.4) shows that X_t^1 is a local martingale. Taking $f(x) = x_1^2$ in (1.4) shows that $(X_t^1)^2 - \int_0^t a_{11}(X_s)\, ds$ is also a local martingale. So $d\langle X^1 \rangle_t = a_{11}(X_t)\, dt$, and X_t^1 is a nondegenerate time change of a one-dimensional Brownian motion. By the argument of Proposition I.8.2, X_s^1 stays in the interval $[-1,1]$ up until time t only if a Brownian motion stays in the interval $[-1,1]$ up until time $c_6 t$, and this is bounded (see [PTA, (II.4.30)]) by $c_7 e^{-c_8 t}$. If X_s has not exited B by time t, then X_s^1 has not exited $[-1,1]$, and (b) follows. $\qquad\square$

We need an estimate on the modulus of continuity of the paths of X_t, which will be used in the next chapter to establish tightness.

(2.4) Theorem. *Let $\mathcal{L} \in \mathcal{N}(\Lambda_1, \Lambda_2)$ and let $\mathbb{P}$ be a solution to the martingale problem for $\mathcal{L}$ started at x_0. If $\varepsilon > 0$ and $N > 0$, there exists M depending on Λ_1, Λ_2, N, and ε such that*

$$\mathbb{P}\left(\sup_{0 \leq s \leq t \leq N} \frac{|X_t - X_s|}{|t - s|^{1/4}} \geq M \right) < \varepsilon. \tag{2.4}$$

Proof. We will first show that there exists c_1 such that

$$\mathbb{P}\left(\sup_{s \leq u \leq t} |X_u - X_s| \geq \lambda |t - s|^{1/4} \right) \leq 2d e^{-c_1 \lambda^2 / |t-s|^{1/2}}, \tag{2.5}$$

if $0 \leq s \leq t \leq N$ and $\lambda / |t - s|^{3/4}$ is sufficiently large (recall that d is the dimension). Fix i and s for the moment. As in the proof of Proposition 2.3,

$$M_v = X_{s+v}^i - X_s^i - \int_s^{s+v} b_i(X_r)\, dr$$

is a local martingale with quadratic variation $\int_s^{s+v} a_{ii}(X_r)\, dr$, which is bounded by $c_2 v$. So (see [PTA, Exercise I.8.13])

$$\mathbb{P}(\sup_{u \leq v} |M_u| > \gamma) \leq 2e^{-\gamma^2/2c_2 v}.$$

If $\gamma \geq 2\Lambda_2(t - s)$, then

$$\mathbb{P}(\sup_{s \leq u \leq t} |X_u^i - X_s^i| \geq \gamma) \leq \mathbb{P}(\sup_{u \leq t-s} |M_u| \geq \gamma/2) \leq 2e^{-\gamma^2/8c_2(t-s)}.$$

Setting $\gamma = \lambda |t - s|^{1/4}/\sqrt{d}$, then $\gamma \geq 2\Lambda_2(t - s)$ if $\lambda > 2\sqrt{d}\Lambda_2|t - s|^{3/4}$, and then

$$\mathbb{P}(\sup_{s \leq u \leq t} |X_u^i - X_s^i| \geq \lambda |t - s|^{1/4}/\sqrt{d}) \leq 2e^{-\lambda^2/8dc_2|t-s|^{1/2}}. \tag{2.6}$$

Since $|X_t - X_s|$ is greater than $\lambda |t - s|^{1/4}$ only if $|X_t^i - X_s^i|$ is greater than $\lambda |t - s|^{1/4}/\sqrt{d}$ for some i, (2.5) is proved.

To obtain (2.4) from (2.5), note that if $|X_t - X_s| \geq M|t - s|^{1/4}$ for some $0 \leq s \leq t \leq N$, then for some integer $n \geq 1$ and some integer $j \leq N2^n$,

$$\sup_{j2^{-n} \leq u \leq (j+2)2^{-n}} |X_u - X_{j2^{-n}}| \geq M(2^{-n})^{1/4}/2.$$

If M is large enough, $M2^{-n/4}/2 > 4\sqrt{d}\Lambda_2 2^{-3n/4}$. Using the estimate in (2.5),

$$\mathbb{P}\left(\sup_{0 \leq s \leq t \leq N} \frac{|X_t - X_s|}{|t - s|^{1/4}} \geq M\right)$$

$$\leq \sum_{n=1}^{\infty} \sum_{j=0}^{N2^n} \mathbb{P}\left(\sup_{j2^{-n} \leq u \leq (j+2)2^{-n}} |X_u - X_{j2^{-n}}| \geq M2^{-n/4}/2\right)$$

$$\leq c_3 \sum_{n=1}^{\infty} N2^n \exp(-c_4 M^2/2^{-n/2}),$$

which will be less than ε if M is large enough. $\qquad\square$

Finally, we have a support theorem.

(2.5) Theorem. *Suppose $\mathcal{L} \in \mathcal{N}(\Lambda_1, \Lambda_2)$ and $\mathbb{P}$ is a solution to the martingale problem for $\mathcal{L}$ started at x_0. Suppose $\psi : [0, t] \to \mathbb{R}^d$ is continuous with $\psi(0) = x_0$ and $\varepsilon > 0$. There exists c_1 depending only on ε, Λ_1, Λ_2, t, and the modulus of continuity of ψ such that*

$$\mathbb{P}(\sup_{s \leq t} |X_s - \psi(s)| < \varepsilon) > c_1.$$

Proof. As in the proof of Theorem 1.1, there exists a Brownian motion W_t such that X_t is a weak solution to (1.3). The result now follows from Theorem I.8.5. $\qquad\square$

As an example of the use of the support theorem, if $r < 1$, there exists c_1 depending only on r such that

$$\mathbb{P}^x(T_{B(y,r)} < \tau_{B(0,2)}) \geq c_1, \qquad x, y \in B(0,1). \tag{2.7}$$

To see this, let $\varepsilon = r/3$ and let ψ be the line segment connecting x and y. The line segment ψ never comes within ε of $\partial B(0,2)$, and our assertion follows from the support theorem.

3. Examples

In this section we show that the behavior of a diffusion near a point can be quite different from that of a Brownian motion, even if the coefficients are uniformly strictly elliptic. To motivate this, consider approximating a two-dimensional Brownian motion by the following Markov chain. At a point $x \neq 0$, with probability 1/4 each, the chain Y_n jumps $\pm h$ in the radial direction, i.e., $\mathbb{P}^x(Y_1 = x \pm hx/|x|) = 1/4$. With probability 1/4 each, the chain jumps $\pm h$ in the direction orthogonal to the radial direction:

$$\mathbb{P}^x\left(Y_1 = x \pm h\frac{(x_2, -x_1)}{|x|}\right) = 1/4, \qquad x = (x_1, x_2) \neq 0.$$

If the chain moves in the radial direction, $|Y_n|$ behaves like a simple random walk up until first hitting $B(0,h)$. If the chain moves in the angular direction, the radius must increase, from $|x|$ to $(|x|^2 + h^2)^{1/2}$. So the radial component of Y_n looks like a random walk with an outward drift. By changing the angular component to $\pm Ah$ instead of $\pm h$, we can increase or decrease the amount of outward drift, without making the chain degenerate. Since two-dimensional Brownian motion is neighborhood recurrent but not point recurrent, varying the angular component can vary the outward drift of the radial component and drastically affect the recurrence properties of the diffusion.

(3.1) Proposition. *Suppose $d \geq 2$.*

(a)   There exists a uniformly elliptic operator $\mathcal{L}$ with zero drift coefficients such that if $\mathbb{P}$ is a solution to the martingale problem for $\mathcal{L}$ started at $x \neq 0$, then $\mathbb{P}(T_{\{0\}} < \infty) = 1$.

(b)   There exists a uniformly elliptic operator $\mathcal{L}$ with zero drift coefficients such that if $\mathbb{P}$ is a solution to the martingale problem for $\mathcal{L}$ started at $x \neq 0$, then $\mathbb{P}(|X_t| \to \infty) = 1$.

Proof. Let $A \in (-1, \infty)$ and

$$a_{ij}^A(x) = (1 + A)^{-1}(\delta_{ij} + Ax_ix_j/|x|^2), \qquad x \neq 0. \tag{3.1}$$

Let $a^A(0)$ be the identity. If $A > -1$, it is easy to see that a_{ij}^A is strictly positive definite, although not continuous at 0. We also have

$$\sum_{i,j=1}^{d} x_i a_{ij}^A(x)x_j = (1 + A)^{-1}\left[|x|^2 + A\sum_{i,j} x_i^2 x_j^2/|x|^2\right] = |x|^2.$$

Note trace $a^A(x) = (d + A)/(1 + A)$ if $x \neq 0$. Let $\mathcal{L}^A$ be the operator with zero drift coefficients and diffusion coefficients a^A. By Proposition 2.1, if $R_t = |X_t|$,

$$dR_t = d\overline{W}_t + \frac{1}{2R_t}\left[\frac{d + A}{1 + A} - 1\right] dt, \qquad t < T_{\{0\}}.$$

Using Proposition I.7.2, this says that up until the first hit of 0 by R_t, if there is one, R_t has the law of a Bessel process of order $(d + A)/(1 + A)$. If we choose A to be large enough, the order will be less than 2, and so R_t will hit 0 with probability one. If $d > 2$ or $d = 2$ and $A \in (-1, 0)$, then the Bessel process will have order larger than 2, and such processes tend to ∞ with probability 1. $\qquad\square$

Adjusting A suitably, we can arrange matters so that the solution to the martingale problem corresponding to $\mathcal{L}^A$ does not hit 0, but spends significantly more or less time (depending on A) in neighborhoods of 0 than does d-dimensional Brownian motion.

The analytic counterpart of Proposition 3.1 is the following. It shows that some care is needed in describing the Dirichlet problem and in using the maximum principle.

(3.2) Proposition. *If $(d + A)/(1 + A) < 2$, then there exists $h \neq 0$ such that $\mathcal{L}^A h = 0$ in $B(0, 1) - \{0\}$, h is bounded and continuous on $\overline{B(0, 1)}$, h is C^2 in $B(0, 1) - \{0\}$, and $h = 0$ on $\partial B(0, 1)$.*

Proof. Let

$$h(x) = \mathbb{P}^x(T_{\{0\}} < \tau_{B(0,1)}), \qquad h(0) = 1,$$

where $\mathbb{P}^x$ denotes the solution to the martingale problem for $\mathcal{L}^A$ started at x. (Since the a_{ij}^A are Lipschitz away from 0, the martingale problem is well posed as long as we restrict attention to $\mathcal{F}_{T_{\{0\}}}$.) The only assertion that is not immediate from Proposition 3.1 and the results of Chapter III is the one that h is continuous at 0. Note that $|X_t|$ has the law of a Bessel process of order $\nu = (d + A)/(1 + A)$ started at $|x|$. By Itô's formula, $|X_t|^{2-\nu}$ is a martingale, so by [PTA, Corollary I.4.10],

$$\mathbb{P}^x(T_{\{0\}} < \tau_{B(0,1)}) = \mathbb{P}^{|x|}(|X_t| \text{ hits } 0 \text{ before } 1) = 1 - |x|^{2-\nu} \to 1$$

as $|x| \to 0$. $\qquad\square$

If the coefficients of the a_{ij} are sufficiently smooth, then X_t cannot hit points. Let

$$\psi(r) = \sup_{i,j} \sup_{|x-y| \leq r} |a_{ij}(x) - a_{ij}(y)|.$$

(3.3) Theorem. *Let $\mathcal{L} \in \mathcal{N}(\Lambda, 0)$ so that the drift coefficients are zero. If $d \geq 3$ and $\psi(r) \to 0$ as $r \to 0$, then $\mathbb{P}(T_{\{0\}} < \infty) = 0$ for any solution $\mathbb{P}$ to the martingale problem started at $x \neq 0$ for the operator $\mathcal{L}$. If $d = 2$, and $\int_0^1 \psi(r)/r\, dr < \infty$, then the same conclusion holds.*

Proof. If σ is a positive definite square root of a, then the law of $\sigma^{-1}(0)X_t$ under $\mathbb{P}$ will be a solution to the martingale problem for the operator whose diffusion coefficients are $a^{-1}(0)a(x)$ started at $\sigma^{-1}(0)x$. So without loss of generality, we may assume $a_{ij}(0) = \delta_{ij}$. Let $R_t = |X_t|$. By Proposition 2.1, $R_t = M_t + A_t$, where the martingale term M_t and bounded variation term A_t are given by (2.2). Let us perform a time change: let $B_t = \inf\{u : \langle M \rangle_u > t\}$ and let $S_t = R_{B_t}$. M_{B_t} is a martingale with quadratic variation t, hence a Brownian motion. By the ellipticity assumption, $d\langle M \rangle_t / dt$ is bounded above and below by positive constants, so R_t hits 0 if and only if S_t does. By the definition of ψ and our assumption on $a(0)$,

$$\left| \sum_{i,j} \frac{x_i a_{ij}(x) x_j}{|x|^2} - 1 \right| \leq |x|^{-2} \sum_{i,j} |x_i|\,|x_j|\,|a_{ij}(x) - \delta_{ij}| \leq d^2 \psi(|x|)$$

and

$$|\operatorname{trace} a(x) - d| \leq d\psi(|x|).$$

A calculation shows that

$$\frac{dA_{B_t}}{dt} = \frac{1}{2R_{B_t}} \frac{F_t}{G_t}\, dt = \frac{1}{2S_t} D(X_{B_t})\, dt,$$

where

$$F_t = \operatorname{trace} a(X_{B_t}) - \sum_{i,j} \frac{X_{B_t}^i a_{ij}(X_{B_t}) X_{B_t}^j}{R_{B_t}^2},$$

$$G_t = \sum_{i,j} \frac{X_{B_t}^i a_{ij}(X_{B_t}) X_{B_t}^j}{R_{B_t}^2},$$

and $|D(x) - (d-1)| \leq c_1 \psi(|x|)$.

Let Z_t be the solution to the one-dimensional equation

$$dZ_t = dM_{B_t} + \frac{(d-1) - c_1 \psi(Z_t)}{2Z_t}\, dt, \qquad Z_0 = |x|.$$

By the comparison theorem, Theorem I.6.2, S_t is always larger than Z_t. Z_t is a one-dimensional diffusion corresponding to the operator $(1/2)f''(z) + ((d-1) - c_1\psi(z))/2z\, f'(z)$. The scale function for this operator (cf. Theorem IV.3.1) is

$$s(x) = \int_{x_1}^x \exp\left(-\int_{x_1}^y \frac{(d-1) - c_1\psi(z)}{z}\, dz \right) dy,$$

where x_1 is any point in $(0, \infty)$. If $d \geq 3$ and $\psi(z) \to 0$ as $z \to 0$, or if $d = 2$ and $\int_0^1 \psi(z)/z\, dz < \infty$, then $s(x) \to -\infty$ as $x \to 0$. By Section IV.1,

$$\mathbb{P}^z(Z_t \text{ hits } \delta \text{ before } M) = \mathbb{P}^{s(z)}(s(Z_t) \text{ hits } s(\delta) \text{ before } s(M))$$

$$= \frac{s(M) - s(z)}{s(M) - s(\delta)}.$$

Letting $\delta \to 0$ and then $M \to \infty$ shows that $\mathbb{P}^z(Z_t \text{ hits } 0) = 0$. $\qquad\square$

4. Convexity

In this section we will let the a_{ij} be smooth (C^2, say) and strictly elliptic, and assume that the drift coefficients are identically 0. Let D be either $B(0, 1)$ or a unit cube centered at 0.

Suppose u is continuous. The *upper contact set* of u is the set

$$U_u = \{y \in D : \text{ there exists } p \in \mathbb{R}^d \text{ such that}$$
$$u(x) \le u(y) + p \cdot (x - y) \text{ for all } x \in D\}.$$

Here $p \cdot (x - y)$ denotes the inner product. In this definition p will depend on y. A point y is in U_u if there is a hyperplane, namely, $u(y) = u(x) + p \cdot (x - y)$, that lies above the graph of u but touches the graph at $(y, u(y))$. With this interpretation we see that when u is concave (i.e., $-u$ is convex), then $U_u = D$, and conversely, if $U_u = D$, then u is concave.

When $u \in C^1$, for $y \in U_u$ there is only one p such that $u(x) \le u(y) + p \cdot (x - y)$, namely, $p = \nabla u(y)$. For $u \in C^2$ let H_u denote the *Hessian* matrix:

$$(H_u)_{ij}(x) = \partial_{ij} u(x).$$

(4.1) Proposition. *If $u \in C^2$ and $y \in U_u$, then $H_u(y)$ is nonpositive definite.*

Proof. Let h be a unit vector. $y \in U_u$ implies there exists p such that $u(y + \varepsilon h) \le u(y) + \varepsilon p \cdot h$ and $u(y - \varepsilon h) \le u(y) - \varepsilon p \cdot h$. Combining,

$$u(y + \varepsilon h) + u(y - \varepsilon h) - 2u(y) \le 0.$$

Dividing by ε^2 and letting $\varepsilon \to 0$ gives $h^T H_u(y) h \le 0$. $\qquad\square$

Let $S_u(y)$ be the set of slopes of supporting hyperplanes to u at y. That is,

$$S_u(y) = \{p \in \mathbb{R}^d : u(x) \le u(y) + p \cdot (x - y) \text{ for all } x \in D\}.$$

As we noted above, $S_u(y) \ne \emptyset$ if and only if $y \in U_u$, and if $u \in C^1$ and $y \in U_u$, then $S_u(y) = \{\nabla u(y)\}$. Let

$$S_u(A) = \bigcup_{y \in A} S_u(y).$$

Let $|A|$ denote the Lebesgue measure of A and $\det H$ the determinant of H. Recall that if V is a neighborhood in D, $v : D \to \mathbb{R}^d$ is in C^1, and $v(V)$ is the image of V under v, then

$$|v(V)| \le \int_V |\det J_v|, \tag{4.1}$$

where J_v is the Jacobian of v. (We have inequality instead of equality because we are not assuming v is one-to-one.)

(4.2) Proposition. *Suppose u is continuous on $\overline{D}$ and C^2 in D. There exists c_1 not depending on u such that*

$$\sup_D u \le \sup_{\partial D} u + c_1 \left(\int_{U_u} |\det H_u| \right)^{1/d}.$$

Proof. Replacing u by $u - \sup_{\partial D} u$, we may assume $u \le 0$ on ∂D. We first show

$$|S_u(D)| = |S_u(U_u)| \le \int_{U_u} |\det H_u|. \tag{4.2}$$

Since $S_u(y) = \{\nabla u(y)\}$, the Jacobian matrix of the mapping S_u is H_u. For each ε, $H_u - \varepsilon I$ is negative definite in a neighborhood of U_u. We apply (4.1) to $v = S_u - \varepsilon I$ and let $\varepsilon \to 0$, which gives (4.2).

Next suppose u takes a positive maximum at $y \in D$. Let v be the function such that the region below the graph of v is the cone with base D and vertex $(y, u(y))$. More precisely, let G_1 be the smallest convex set in $D \times [0, \infty)$ containing $\partial D \times \{0\}$ and the point $(y, u(y))$; let $v(x) = \sup\{z \ge 0 : (x, z) \in G_1\}$ for $x \in D$.

Suppose $p \in S_v(D)$. We look at the family of hyperplanes $\alpha + p \cdot (x - y)$. If we start with α large and let α decrease to $-\infty$, there is a first hyperplane that touches the graph of u (not necessarily at $(y, u(y))$). Consequently $p \in S_u(D)$. We have thus shown that $S_v(D) \subseteq S_u(D)$.

Let w be the function whose support is $B(y, d)$ (where d is the dimension) and the region below w is the cone with vertex $(y, u(y))$. To be more precise again, let G_2 be the smallest convex set in $\overline{B}(y, d) \times [0, \infty)$ containing $\partial B(y, d) \times \{0\}$ and the point $(y, u(y))$, and let $w(x) = \sup\{z \ge 0 : (x, z) \in G_2\}$ for $x \in \overline{B}(y, d)$. A picture shows that $S_w(D) \subseteq S_v(D)$, and we see then that

$$|S_w(D)| \le |S_v(D)| \le |S_u(D)| \le \int_{U_u} |\det H_u|. \tag{4.3}$$

We now compute $|S_w(\{y\})|$. Note that $w(x) = u(y)(1 - |x - y|/d)$ for $x \in B(y, d)$. If each coordinate of p is between $-u(y)/d$ and $+u(y)/d$, then $p \in S_w(y)$. So

$$|S_w(D)| \ge |S_w(\{y\})| \ge c_2(u(y)/d)^d.$$

Combining with (4.2),

$$u(y)^d \le c_2^{-1} d^d |S_w(D)| \le c_3 \int_{U_u} |\det H_u|. \qquad \square$$

We will use the inequality

$$\frac{1}{d} \sum_{j=1}^d \lambda_j \ge \prod_{j=1}^d \lambda_j^{1/d}, \qquad \lambda_j \ge 0, \quad j = 1, \dots, d. \tag{4.4}$$

One way to prove (4.4) is to let $\Omega = \{1, 2, \ldots, d\}$, let $\mathbb{P}$ assign mass $1/d$ to each point of Ω, let X be the random variable defined by $X(j) = \lambda_j$, and apply Jensen's inequality to the convex function $-\log x$. We then have

$$-\log\left(\sum_{j=1}^{d}\lambda_j\frac{1}{d}\right) \leq \frac{1}{d}\sum_{j=1}^{d}(-\log\lambda_j),$$

which implies (4.4).

We now prove a key estimate due to Alexandroff-Bakelman-Pucci.

(4.3) Theorem. *Suppose* $\mathcal{L} \in \mathcal{N}(\Lambda, 0)$, *the coefficients of* $\mathcal{L}$ *are in* C^2, $u \in C^2$, *and* $\mathcal{L}u = f$ *in* D. *There exists* c_1 *independent of* u *such that*

$$\sup_{D} u \leq \sup_{\partial D} u + c_1\left(\int_{D}|f(x)|^d\,dx\right)^{1/d}.$$

Proof. Fix $y \in U_u$, let $B = -H_u(y)$, and let A be the matrix $a(y)$. Let $\lambda_1, \ldots, \lambda_d$ be the eigenvalues of B. Since H_u is nonpositive definite, $\lambda_j \geq 0$. Let P be an orthogonal matrix and C a diagonal matrix such that $B = P^T C P$. Note $|\det H_u| = \det B = \lambda_1 \cdots \lambda_d$ and

$$(AB)_{ii} = \sum_{j=1}^{d} A_{ij}B_{ji} = -\sum_{j} a_{ij}(y)\partial_{ij}u(y).$$

Then

$$-f(y) = -\sum_{i,j} a_{ij}(y)\partial_{ij}u(y) = \text{trace}\,(AB) \tag{4.5}$$

$$= \text{trace}\,(AP^TCP) = \text{trace}\,(CPAP^T) = \sum_{j=1}^{d}\lambda_j(PAP^T)_{jj}.$$

Since A is uniformly positive definite, there exists c_2 such that $(PAP^T)_{jj} \geq c_2$, so by (4.4),

$$-f(y) \geq \sum_{j} c_2\lambda_j = c_2 d\sum_{j}(\lambda_j/d)$$

$$\geq c_2 d(\prod_{j}\lambda_j)^{1/d} = c_2 d|\det H_u|^{1/d}.$$

Taking dth powers, integrating over U_u, and using Proposition 4.2 completes the proof. $\qquad\square$

5. Green functions

Let $\mathbb{P}$ be a solution to the martingale problem for $\mathcal{L}$ started at x (assuming one exists) and let $\mathbb{E}$ be the corresponding expectation. If D is a domain, a function $G_D(x,y)$ is called a *Green function* for the operator $\mathcal{L}$ in the domain D if

$$\mathbb{E}\int_0^{\tau_D} f(X_s)\, ds = \int_D G_D(x,y)f(y)\, dy \qquad (5.1)$$

for all nonnegative Borel measurable functions f on D. The function $G^\lambda(x,y)$ is called the *λ-resolvent density* if

$$\mathbb{E}\int_0^\infty e^{-\lambda s} f(X_s)\, ds = \int_{\mathbb{R}^d} G^\lambda(x,y)f(y)\, dy \qquad (5.2)$$

for all nonnegative Borel measurable f on $\mathbb{R}^d$.

An immediate consequence of the Alexandroff-Bakelman-Pucci estimate is the following.

(5.1) Theorem. *Suppose $\mathcal{L} \in \mathcal{N}(\Lambda, 0)$ and the diffusion coefficients are in C^2. Then there exists c_1 depending only on Λ such that*

$$\left| \mathbb{E}\int_0^{\tau_{B(0,1)}} f(X_s)\, ds \right| \le c_1 \left(\int_{B(0,1)} |f(y)|^d\, dy \right)^{1/d}.$$

Proof. We prove this inequality for f that are C^2 in $\overline{B(0,1)}$; a limit argument then yields the inequality for arbitrary f. Let $u(y) = \mathbb{E}^y \int_0^{\tau_{B(0,1)}} f(X_s)\, ds$. By Section III.6, u is C^2 in $B(0,1)$, continuous on the closure of $B(0,1)$, and $\mathcal{L}u = -f$. In fact, u is 0 on the boundary of $B(0,1)$. Now apply Theorem 4.3. $\qquad\square$

(5.2) Corollary.
$$G_B(x, \cdot) \in L^{d/(d-1)}(B).$$

Proof. By Theorem 5.1 and (5.1),

$$\left| \int_B G_B(x,y)f(y)\, dy \right| \le c_1 \|f\|_{L^d(B)}$$

for all $f \in L^d(B)$. The result follows by the duality of L^d and $L^{d/(d-1)}$. $\qquad\square$

We also have

(5.3) Theorem. *Suppose $\mathcal{L} \in \mathcal{N}(\Lambda, 0)$ and the diffusion coefficients are in C^2. There exists c_1 not depending on f such that if $f \in L^d$, then*

$$\left| E \int_0^\infty e^{-\lambda t} f(X_t)\, dt \right| \le c_1 \left(\int_{\mathbb{R}^d} |f(y)|\, dy \right)^{1/d}.$$

Proof. By the smoothness of the diffusion coefficients, there is a unique solution to the martingale problem for $\mathcal{L}$ starting at each $x \in \mathbb{R}^d$ (see the remarks at the end of Section 1); we denote it $\mathbb{P}^x$. Moreover, $(\mathbb{P}^x, X_t)$ forms a strong Markov family by Theorem I.5.1.

Let $S_0 = 0$ and $S_{i+1} = \inf\{t > S_i : |X_t - X_{S_i} : |X_t - X_{S_i}| > 1\}$, $i = 0, 1, \ldots$ Then $S_{i+1} = S_i + S_1 \circ \theta_{S_i}$. By Proposition 2.3, there exists t_0 such that $\sup_x \mathbb{P}^x(S_1 \le t_0) \le 1/2$. Then

$$\mathbb{E}^x e^{-\lambda S_1} \le \mathbb{P}^x(S_1 \le t_0) + e^{-\lambda t_0} \mathbb{P}^x(S_1 > t_0)$$
$$= (1 - e^{-\lambda t_0})\mathbb{P}^x(S_1 \le t_0) + e^{-\lambda t_0}.$$

So if $\rho = \sup_x \mathbb{E}^x e^{\lambda S_1}$, then $\rho < 1$. By the strong Markov property,

$$\mathbb{E}^x e^{-\lambda S_{i+1}} = \mathbb{E}^x \left(e^{-\lambda S_i} \mathbb{E}^x (e^{-\lambda S_1 \circ \theta_{S_i}} \mid \mathcal{F}_{S_i}) \right) \le \rho \mathbb{E}^x e^{-\lambda S_i},$$

and by induction $\mathbb{E}^x e^{-\lambda S_i} \le \rho^i$.

We now write

$$\mathbb{E}^x \int_0^\infty e^{-\lambda t} f(X_t)\, dt = \sum_{i=0}^\infty \int_{S_i}^{S_{i+1}} e^{-\lambda t} f(X_t)\, dt. \tag{5.3}$$

By the strong Markov property at time S_i and Theorem 5.1,

$$\left| \mathbb{E}^x \int_{S_i}^{S_{i+1}} e^{-\lambda t} f(X_t)\, dt \right| = \left| \mathbb{E}^x \left(e^{-\lambda S_i} \mathbb{E}^{X_{S_i}} \int_0^{S_1} e^{-\lambda t} f(X_t)\, dt \right) \right|$$
$$\le c_2 \mathbb{E}^x e^{-\lambda S_i} \|f\|_d \le c_2 \rho^i \|f\|_d.$$

Substituting in (5.3) proves the theorem. $\qquad\square$

As in Corollary 4.2, this implies $G^\lambda(x, \cdot) \in L^{d/(d-1)}$.

Using the theory of A_p weights, Fabes and Stroock [1] obtained an improvement of this, namely, that $G^\lambda(x, \cdot) \in L^{d/(d-1)+\varepsilon}$ for some $\varepsilon > 0$.

One disadvantage of Theorems 5.1 and 5.3 is that we required the diffusion coefficients to be smooth. We will remove this restriction in the next section by an approximation procedure due to Krylov. Earlier Krylov [1] had also proved, however, that Theorems 5.1 and 5.3 hold whenever $X_t = x + \int_0^t \sigma_s\, dW_s$, where $\sigma_s(\omega)$ is an adapted, matrix-valued process that is bounded and is uniformly positive definite (that is, there exists c_1 such that $y^T \sigma_s(\omega) y \ge c_1 |y|^2$ for all $y \in \mathbb{R}^d$, where c_1 is independent of s and y).

6. Resolvents

In this section we present a theorem of Krylov on approximating resolvents and then apply it to extend Theorem 5.3 to arbitrary solutions of the martingale problem for an elliptic operator $\mathcal{L}$. We suppose that $\mathcal{L} \in \mathcal{N}(\Lambda, 0)$ for some $\Lambda > 0$, but make no smoothness assumptions on the coefficients. Let $\mathcal{L}$ be defined as in (1.1). Let $\mathbb{P}$ be any solution to the martingale problem for $\mathcal{L}$ started at x_0.

Recall that $f * g(x) = \int f(y)g(x - y)\, dy$. Let φ be a nonnegative radially symmetric C^∞ function with compact support such that $\int_{\mathbb{R}^d} \varphi = 1$ and $\varphi > 0$ on $B(0, r)$ for some r. Let $\varphi_\varepsilon(x) = \varepsilon^{-d}\varphi(x/\varepsilon)$.

(6.1) Theorem. *Let $\lambda > 0$. There exist a_{ij}^ε in C^∞ with the following properties:*
(i)   if $\mathcal{L}^\varepsilon$ is defined by

$$\mathcal{L}^\varepsilon f(x) = \frac{1}{2} \sum_{i,j=1}^{d} a_{ij}^\varepsilon(x)\partial_{ij} f(x), \tag{6.1}$$

then $\mathcal{L}^\varepsilon \in \mathcal{N}(\Lambda, 0)$, and
(ii)   if $\mathbb{P}_\varepsilon^x$ is the solution to the martingale problem for $\mathcal{L}^\varepsilon$ started at x and

$$G_\varepsilon^\lambda h(x) = \mathbb{E}_\varepsilon^x \int_0^\infty e^{-\lambda t} h(X_t)\, dt \tag{6.2}$$

for h bounded, then

$$(G_\varepsilon^\lambda f * \varphi_\varepsilon)(x_0) \to \mathbb{E} \int_0^\infty e^{-\lambda t} f(X_t)\, dt \tag{6.3}$$

whenever f is continuous.

We will see later (Section 8) that $G_\varepsilon^\lambda f$ is equicontinuous in ε, so that in fact $G_\varepsilon^\lambda f(x_0)$ converges to the right-hand side of (6.3).

The a_{ij}^ε depend on $\mathbb{P}$, and different solutions to the martingale problem could conceivably give us different sequences a_{ij}^ε.

Proof. Define a measure μ by

$$\mu(C) = \mathbb{E} \int_0^\infty e^{-\lambda t} 1_C(X_t)\, dt. \tag{6.4}$$

By the support theorem, Theorem 2.5, for each $y \in \mathbb{R}^d$ and $s > 0$, there is positive probability under $\mathbb{P}$ that X_t enters the ball $B(y, s)$ and stays there a positive length of time. So $\mu(B(y, s)) > 0$ for all y and s. Define

$$a_{ij}^\varepsilon(x) = \frac{\int \varphi_\varepsilon(x - y) a_{ij}(y)\, \mu(dy)}{\int \varphi_\varepsilon(x - y)\, \mu(dy)}. \tag{6.5}$$

By our assumptions on φ, the denominator is not zero. It is clear that (i) holds.

Suppose u is in C^2 and bounded. By the product formula and Itô's formula,

$$e^{-\lambda t}u(X_t) = u(X_0) - \int_0^t u(X_s)\lambda e^{-\lambda s}\,ds + \int_0^t e^{-\lambda s}\,d[u(X)]_s$$

$$= u(X_0) - \int_0^t u(X_s)\lambda e^{-\lambda s}\,ds + \text{ martingale}$$

$$+ \int_0^t e^{-\lambda s}\mathcal{L}u(X_s)\,ds.$$

Taking expectations and letting $t \to \infty$,

$$u(x_0) = \mathbb{E}\int_0^\infty e^{-\lambda s}(\lambda u - \mathcal{L}u)(X_s)\,ds = \int (\lambda u - \mathcal{L}u)(x)\,\mu(dx). \tag{6.6}$$

We next apply (6.6) to $u = v * \varphi_\varepsilon$, where v is a bounded and C^2 function. On the left-hand side we have $\int v(x_0 - y)\varphi_\varepsilon(y)\,dy$. Note that

$$\mathcal{L}(v * \varphi_\varepsilon)(z) = \frac{1}{2}\sum_{i,j} a_{ij}(z)\partial_{ij}(v * \varphi_\varepsilon)(z) \tag{6.7}$$

$$= \frac{1}{2}\sum_{i,j} a_{ij}(z)((\partial_{ij}v) * \varphi_\varepsilon)(z)$$

$$= \frac{1}{2}\sum_{i,j}\int a_{ij}(z)\partial_{ij}v(x)\varphi_\varepsilon(x - z)\,dx.$$

However, by (6.5),

$$\int a_{ij}(z)\varphi_\varepsilon(x - z)\,\mu(dz) = a_{ij}^\varepsilon(x)\int \varphi_\varepsilon(x - y)\,\mu(dy). \tag{6.8}$$

Combining (6.6), (6.7), and (6.8),

$$\int v(x_0 - y)\varphi_\varepsilon(y)\,dy = \int [\lambda(v * \varphi_\varepsilon) - \mathcal{L}(v * \varphi_\varepsilon)](x)\,\mu(dx) \tag{6.9}$$

$$= \int\int (\lambda - \mathcal{L}^\varepsilon)v(x)\varphi_\varepsilon(x - y)\,\mu(dy)\,dx.$$

Suppose f is smooth, and let $v(x) = G_\varepsilon^\lambda f(x)$. By Section III.6, v is in C^2 and bounded and $(\lambda - \mathcal{L}^\varepsilon)v = f$. Substituting in (6.9),

$$\int G_\varepsilon^\lambda f(x_0 - y)\varphi_\varepsilon(y)\,dy = \int\int f(x)\varphi_\varepsilon(x - y)\,\mu(dy)\,dx \tag{6.10}$$

$$= \int f * \varphi_\varepsilon(y)\,\mu(dy).$$

By a limit argument, we have (6.10) when f is continuous. Since f is continuous, $f * \varphi_\varepsilon$ is bounded and converges to f uniformly. Hence

$$\int f * \varphi_\varepsilon(y)\,\mu(dy) \to \int f(y)\mu(dy) = \mathbb{E} \int_0^\infty e^{-\lambda t} f(X_t)\,dt. \qquad \square$$

Defining b_i^ε by the analogue of (6.5), there is no difficulty extending this theorem to the case $\mathcal{L} \in \mathcal{N}(\Lambda_1, \Lambda_2)$, $\Lambda_2 > 0$.

(6.2) Theorem. *Let $\mathbb{P}$ be as above. There exists c_1 not depending on f such that*

$$\left| \mathbb{E} \int_0^\infty e^{-\lambda t} f(X_t)\,dt \right| \le c_1 \|f\|_d.$$

Proof. By Theorem 6.1, the left-hand side is the limit of $|G_\lambda^\varepsilon f * \varphi_\varepsilon(x_0)|$ if f is continuous and bounded. The coefficients in $\mathcal{L}^\varepsilon$ are smooth, so by Theorem 5.3 $\|G_\varepsilon^\lambda f\|_\infty \le c_1 \|f\|_d$, c_1 independent of ε. This proves the proposition for f smooth, and the case of general f follows by a limit argument. $\qquad \square$

(6.3) Corollary. *Under the assumptions of Theorem 6.1,*

$$(G_\varepsilon^\lambda f * \varphi_\varepsilon)(x_0) \to \mathbb{E} \int_0^\infty e^{-\lambda t} f(X_t)\,dt,$$

if f is bounded.

Proof. We start with (6.10). By a limit argument, we have (6.10) holding for f bounded. So we need to show that the right-hand side of (6.10) converges to $\int f(y)\,\mu(dy)$. Since f is bounded, $f * \varphi_\varepsilon$ converges to f almost everywhere and boundedly (see [PTA, Theorem IV.1.2] or Stein [1], Chapter 3). By Theorem 6.2 and (6.4), μ is absolutely continuous with respect to Lebesgue measure. Then

$$\int f * \varphi_\varepsilon(y)\,\mu(dy) = \int f * \varphi_\varepsilon(y)(d\mu/dy)\,dy$$

$$\to \int f(y)(d\mu/dy)\,dy = \int f(y)\,\mu(dy)$$

by dominated convergence. $\qquad \square$

7. Harnack inequality

In this section we prove some theorems of Krylov and Safonov concerning positive $\mathcal{L}$-harmonic functions. Recall that a function h is $\mathcal{L}$-harmonic in a domain D if $h \in C^2$ and $\mathcal{L}h = 0$ in D. These results were first proved probabilistically (see Krylov and Safonov [1]) and are a good example of the power of the probabilistic point of view.

In this section we assume that $\mathcal{L} \in \mathcal{N}(\Lambda, 0)$ so that the drift coefficients are 0. We assume that for each $x \in \mathbb{R}^d$ we have a solution to the martingale problem for $\mathcal{L}$ started at x and that $(\mathbb{P}^x, X_t)$ forms a strong Markov family. In Chapter VI we will see that this is not really any restriction on the generality.

Let $Q(x, r)$ denote the cube of side length r centered at x. Our main goal is to show that X_t started at x must hit a set A before exiting a cube with positive probability if A has positive Lebesgue measure and x is not too near the boundary. The first proposition starts things off by handling the case when A nearly fills the cube. Recall that we are using $|A|$ to denote the Lebesgue measure of A.

(7.1) Proposition. *There exist ε and $c_1 = c_1(\varepsilon)$ such that if $x \in Q(0, 1/2)$, $A \subseteq Q(0, 1)$, and $|Q(0, 1) - A| < \varepsilon$, then $\mathbb{P}^x(T_A < \tau_{Q(0,1)}) \geq c_1$.*

Proof. Let us write τ for $\tau_{Q(0,1)}$. By Propositions 2.2 and 2.3, there exist c_2 and c_3 not depending on x such that $\mathbb{E}^x \tau \geq c_2$ and $\mathbb{E}^x \tau^2 \leq c_3$.

Note that $\mathbb{E}^x \int_0^\tau 1_{A^c}(X_s)\, ds = \mathbb{E}^x \int_0^\tau 1_{(Q(0,1)-A)}(X_s)\, ds$. Since

$$\mathbb{E}^x(\tau - (\tau \wedge t_0)) \leq \mathbb{E}^x(\tau; \tau \geq t_0) \leq \mathbb{E}^x \tau^2 / t_0,$$

we can choose t_0 large enough so that $\mathbb{E}^x(\tau - (\tau \wedge t_0)) \leq c_2/4$. Then

$$\mathbb{E}^x \int_0^\tau 1_{(Q(0,1)-A)}(X_s)\, ds \tag{7.1}$$

$$\leq c_2/4 + e^{t_0} \mathbb{E}^x \int_0^{t_0} e^{-s} 1_{(Q(0,1)-A)}(X_s)\, ds$$

$$\leq c_2/4 + e^{t_0} \mathbb{E}^x \int_0^\infty e^{-s} 1_{(Q(0,1)-A)}(X_s)\, ds$$

$$\leq c_2/4 + c_5 e^{t_0} \|1_{Q(0,1)-A}\|_d$$

$$\leq c_2/4 + c_5 e^{t_0} \varepsilon^{1/d}.$$

If ε is chosen small enough, then $\mathbb{E}^x \int_0^\tau 1_{A^c}(X_s)\, ds < c_2/2$.

On the other hand,

$$c_2 \leq \mathbb{E}^x \tau = \mathbb{E}^x(\tau; T_A < \tau) + \mathbb{E}^x \int_0^\tau 1_{A^c}(X_s)\, ds$$

$$\leq (\mathbb{E}^x \tau^2)^{1/2} (\mathbb{P}^x(T_A < \tau))^{1/2} + c_2/2$$

$$\leq c_3^{1/2} (\mathbb{P}^x(T_A < \tau))^{1/2} + c_2/2,$$

and the result follows with $c_1 = c_2^2 / 4c_3$. $\square$

We used Theorem 6.2 because it applies to arbitrary solutions to the martingale problem, whereas Theorem 5.1 requires the a_{ij} to be smooth.

As noted at the end of Section 5, Theorem 5.1 actually holds for arbitrary solutions to the martingale problem; if we used that fact, we then could have obtained the estimate in (7.1) more directly.

Next we decompose $Q(0, 1)$ into smaller subcubes such that a set A fills up a certain percentage of each of the smaller subcubes. If Q is a cube, let $\widehat{Q}$ denote the cube with the same center as Q but side length three times as long.

(7.2) Proposition. *Let $q \in (0, 1)$. If $A \subseteq Q(0, 1)$ and $|A| \leq q$, then there exists D such that (i) D is the union of cubes $\widehat{R}_i$ such that the interiors of the R_i are pairwise disjoint, (ii) $|A| \leq q|D \cap Q(0, 1)|$, and (iii) for each i, $|A \cap R_i| > q|R_i|$.*

Proof. We will do the case $d = 2$; the higher-dimensional case differs only in the notation. We form a collection of subsquares $\mathcal{R} = \{R_i\}$ as follows. Divide $Q(0, 1)$ into four equal squares Q_1, Q_2, Q_3, and Q_4 with disjoint interiors. For $j = 1, 2, 3, 4$, if $|A \cap Q_j| > q|Q_j|$, we let Q_j be one of the squares in $\mathcal{R}$. If not, we split Q_j into four equal subsquares $Q_{j1}, Q_{j2}, Q_{j3}, Q_{j4}$ and repeat; Q_{jk} will be one of the R_i if $|A \cap Q_{jk}| > q|Q_{jk}|$, and otherwise we divide Q_{jk}. To be more precise, let $\mathcal{Q}_n$ be the collection of squares of side lengths 2^{-n} with vertices of the form $[j/2^n, k/2^n]$ for integers j and k. An element Q' of $\mathcal{Q}_n$ will be in $\mathcal{R}$ if $|A \cap Q'| > q|Q'|$ and Q' is not contained in any $Q'' \in \mathcal{Q}_0 \cup \mathcal{Q}_1 \cup \cdots \cup \mathcal{Q}_{n-1}$ with $|A \cap Q''| > q|Q''|$.

We let $D = \cup_i \widehat{R}_i$ where the union is over $R_i \in \mathcal{R}$. Assertions (i) and (iii) are clear and it remains to prove (ii). Recall that almost every point $z \in A$ is a point of density of A, that is, $|B(z, r) \cap A|/|B(z, r)| \to 1$ a.e. for $z \in A$; this follows by the Lebesgue density theorem ([PTA, Theorem IV.1.2], for example). If z is a point of density of A and T_n denotes the element of $\mathcal{Q}_n$ containing z, then $|T_n \cap A|/|T_n| \to 1$ (see [PTA, Exercise II.8.13], for example). If z is a point of density of A and z is not on the boundary of some square in $\mathcal{Q}_n$ for some n, it follows that z must be in some $R_i \in \mathcal{R}$. We conclude that $|A - D| = 0$.

We form a new collection of squares $\mathcal{S}$. We divide $Q(0, 1)$ into four equal subsquares Q_1, Q_2, Q_3, Q_4. If $Q_j \subseteq D$, it will be in $\mathcal{S}$; otherwise split Q_j into four subsquares and continue. More exactly, $Q' \in \mathcal{Q}_n$ will be in $\mathcal{S}$ if $Q' \subseteq D$ but Q' is not contained in any $Q'' \in \mathcal{Q}_0 \cup \cdots \mathcal{Q}_{n-1}$ for which $Q'' \subseteq D$.

Since D is the union of cubes $\widehat{R}_i$, then $|D \cap Q(0, 1)| = \sum_i |S_i|$ where the sum is over $S_i \in \mathcal{S}$. Hence $|A| = \sum_i |S_i \cap A|$. It thus suffices to show that

$$|A \cap S_i| \leq q|S_i| \tag{7.2}$$

for each $S_i \in \mathcal{S}$. We then sum over i and (ii) will be proved.

Consider $S_i \in \mathcal{S}$. If $S_i = Q(0, 1)$, we are done by the hypotheses on A. Otherwise S_i is in $\mathcal{Q}_n$ for some $n \geq 1$ and is contained in a square $Q' \in \mathcal{Q}_{n-1}$. Let C_1, C_2, C_3 denote the other three squares of $\mathcal{Q}_n$ that are contained in Q'. Since $S_i \in \mathcal{S}$, then $Q' = S_i \cup C_1 \cup C_2 \cup C_3$ is not in $\mathcal{S}$. Since $S_i \subseteq D$, at least one of the squares C_1, C_2, C_3 cannot be contained in D. We have $S_i \cup C_1 \cup C_2 \cup C_3 \subseteq \widehat{S}_i$. $\widehat{S}_i$ is not contained in D, which implies that $S_i \notin \mathcal{R}$. We thus have $S_i \cup C_1 \cup C_2 \cup C_3$ is not contained in D but $S_i \notin \mathcal{R}$; this could only happen if $|S_i \cap A| \leq q|S_i|$, which is (7.2). $\square$

(7.3) Lemma. *Let $r \in (0,1)$. Let $y \in Q(0,1)$ with $\mathrm{dist}\,(y, \partial Q(0,1)) > r$, $\mathcal{L}' \in \mathcal{N}(\Lambda, 0)$, and P be a solution to the martingale problem for $\mathcal{L}'$ started at y. If $Q(z,r) \subseteq Q(0,1)$, then $\mathbb{P}(T_{Q(z,r)} < \tau_{Q(0,1)}) \geq \zeta(r)$ where $\zeta(r) > 0$ depends only on r and Λ.*

Proof. This follows easily from the support theorem. $\square$

We now prove the key result, that sets of positive Lebesgue measure are hit with positive probability.

(7.4) Theorem. *There exists a nondecreasing function $\varphi : (0,1) \to (0,1)$ such that if $B \subseteq Q(0,1)$, $|B| > 0$, and $x \in Q(0,1/2)$, then*

$$\mathbb{P}^x(T_B < \tau_{Q(0,1)}) \geq \varphi(|B|).$$

Proof. Again we suppose the dimension d is 2 for simplicity of notation. Set

$$\varphi(\varepsilon) = \inf\{\mathbb{P}^y(T_B < \tau_{Q(z_0,R)}) : z_0 \in \mathbb{R}^d, R > 0, y \in Q(z_0, R/2),$$
$$|B| \geq \varepsilon|Q(z_0,R)|, B \subseteq Q(z_0,R)\}.$$

By Proposition 7.1 and scaling, $\varphi(\varepsilon) > 0$ for ε sufficiently close to 1. Let q_0 be the infimum of those ε for which $\varphi(\varepsilon) > 0$. We suppose $q_0 > 0$, and we will obtain our contradiction.

Choose $q > q_0$ such that $(q + q^2)/2 < q_0$. This is possible, since $q_0 < 1$. Let $\eta = (q - q^2)/2$. Let $\beta = (q \wedge (1 - q))/16$ and let ρ be equal to $\zeta((1 - \beta)/6)$ as defined in Lemma 7.3. There exist $z_0 \in \mathbb{R}^d$, $R > 0$, $B \subseteq Q(z_0, R)$, and $x \in Q(z_0, R/2)$ such that $q > |B|/|Q(z_0, R)| > q - \eta$ and $\mathbb{P}^x(T_B < \tau_{Q(z_0,R)}) < \rho\varphi(q)^2$. Without loss of generality, let us assume $z_0 = 0$ and $R = 1$, and so we have $\mathbb{P}^x(T_B < \tau_{Q(0,1)}) < \rho\varphi(q)^2$.

We next use Proposition 7.2 to construct the set D (with A replaced by B). Since $|B| > q - \eta$ and

$$|B| \leq q|D \cap Q(0,1)|,$$

then

$$|D \cap Q(0,1)| \geq \frac{|B|}{q} > \frac{q - \eta}{q} = \frac{q + 1}{2}.$$

Let $\widetilde{D} = D \cap Q(0, 1 - \beta)$. Then $|\widetilde{D}| > q$. By the definition of φ, this implies that

$$\mathbb{P}^x(T_{\widetilde{D}} < \tau_{Q(0,1)}) \geq \varphi(q).$$

We want to show that if $y \in \widetilde{D}$, then

$$\mathbb{P}^y(T_B < \tau_{Q(0,1)}) \geq \rho\varphi(q). \tag{7.3}$$

Once we have that, we write

$$\mathbb{P}^x(T_B < \tau_{Q(0,1)}) \geq \mathbb{P}^x(T_{\widetilde{D}} < T_B < \tau_{Q(0,1)})$$

$$\geq \mathbb{E}^x(\mathbb{P}^{X(T(\widetilde{D}))}(T_B < \tau_{Q(0,1)}); T_{\widetilde{D}} < \tau_{Q(0,1)})$$

$$\geq \rho\varphi(q)\mathbb{P}^x(T_{\widetilde{D}} < \tau_{Q(0,1)}) \geq \rho\varphi(q)^2,$$

our contradiction.

We now prove (7.3). If $y \in \partial\widetilde{D}$, then $y \in \widehat{R}_i$ for some $R_i \in \mathcal{R}$ and dist $(y, \partial Q(0,1)) \geq 1 - \beta$. Let R_i^* be the cube with the same center as R_i but side length half as long. By Lemma 7.3,

$$\mathbb{P}^y(T_{R_i^*} < \tau_{Q(0,1)}) \geq \rho.$$

By the definition of q and the fact that $R_i \in \mathcal{R}$, then $|B \cap R_i| \geq q|R_i|$. By the definition of $\varphi(q)$, we have $\mathbb{P}^z(T_{B \cap R_i} < \tau_{R_i}) \geq \varphi(q)$ if $z \in R_i^*$. So by the strong Markov property,

$$\mathbb{P}^y(T_B < \tau_{Q(0,1)}) \geq \mathbb{E}^y(\mathbb{P}^{X(T_{R_i^*})}(T_B < \tau_{R_i}); T_{R_i^*} < \tau_{Q(0,1)})$$

$$\geq \rho\varphi(q). \qquad \square$$

Theorem 7.4 is the key estimate. We now proceed to show that $\mathcal{L}$-harmonic functions are Hölder continuous and that they satisfy a Harnack inequality. A function h is $\mathcal{L}$-harmonic in D if $h \in C^2$ and $\mathcal{L}h = 0$ in D. If h is $\mathcal{L}$-harmonic, then by Itô's formula, $h(X_{t \wedge \tau_D})$ is a martingale. There may be very few $\mathcal{L}$-harmonic functions unless the coefficients of $\mathcal{L}$ are smooth, so we will use the condition that $h(X_{t \wedge \tau_D})$ is a martingale as our hypothesis.

(7.5) Theorem. *Suppose h is bounded in $Q(0,1)$ and $h(X_{t \wedge \tau_{Q(0,1)}})$ is a martingale. Then there exist α and c_1 not depending on h such that*

$$|h(x) - h(y)| \leq c_1\|h\|_\infty|x - y|^\alpha, \qquad x, y \in Q(0, 1/2).$$

Proof. Define $\operatorname{Osc}_B h = \sup_{x \in B} h(x) - \inf_{x \in B} h(x)$. To prove the theorem, it suffices to show there exists $\rho < 1$ such that for all $z \in Q(0, 1/2)$ and $r \leq 1/4$,

$$\operatorname*{Osc}_{Q(z,r/2)} h \leq \rho \operatorname*{Osc}_{Q(z,r)} h. \tag{7.4}$$

If we look at $Ch + D$ for suitable constants C and D, we see that it is enough to consider the case where $\inf_{Q(z,r)} h = 0$ and $\sup_{Q(z,r)} h = 1$. Let $B = \{x \in Q(z, r/2) : h(x) \geq 1/2\}$. We may assume that $|B| \geq (1/2)|Q(z,r/2)|$, for if not, we replace h by $1 - h$.

If $x \in Q(z, r/2)$, then $h(x) \leq 1$. On the other hand, since we know $h(X_{t \wedge \tau_{Q(0,1)}})$ is a martingale,

$$h(x) = \mathbb{E}^{x}[h(X(\tau_{Q(z,r)} \wedge T_B))]$$
$$\geq (1/2)\mathbb{P}^{x}(T_B < \tau_{Q(z,r)}) \geq (1/2)\varphi(2^{-(d+1)}),$$

from Theorem 7.4 and scaling. Hence $\mathrm{Osc}_{Q(z,r/2)}\, h \geq 1 - \varphi(2^{-(d+1)})/2$. Setting $\rho = 1 - \varphi(2^{-(d+1)/2})/2$ proves (7.4). $\square$

(7.6) Theorem. *Suppose $\mathcal{L} \in \mathcal{N}(\Lambda, 0)$. There exists c_1 depending only on Λ such that if h is nonnegative, bounded in $Q(0, 16)$, and $h(X(t \wedge \tau_{Q(0,16)}))$ is a martingale, then $h(x) \leq c_1 h(y)$ if $x, y \in Q(0,1)$.*

Proof. If we look at $h + \varepsilon$ and let $\varepsilon \to 0$, we may assume $h > 0$. By looking at Ch, we may assume $\inf_{Q(0,1/2)} h = 1$. By Theorem 7.5, we know that h is Hölder continuous in $Q(0,8)$, so there exists $y \in Q(0,1/2)$ such that $h(y) = 1$. We want to show that h is bounded above by a constant in $Q(0,1)$, where the constant depends only on Λ.

By the support theorem and scaling, if $x \in Q(0,2)$, there exists δ such that
$$\mathbb{P}^{y}(T_{Q(x,1/2)} < \tau_{Q(0,8)}) \geq \delta.$$

By scaling, if $w \in Q(x, 1/2)$, then $\mathbb{P}^{w}(T_{Q(x,1/4)} < \tau_{Q(0,8)}) \geq \delta$. So by the strong Markov property,
$$\mathbb{P}^{y}(T_{Q(x,1/4)} < \tau_{Q(0,8)}) \geq \delta^2.$$

Repeating and using induction,
$$\mathbb{P}^{y}(T_{Q(x,2^{-k})} < \tau_{Q(0,8)}) \geq \delta^k.$$

Then
$$1 = h(y) \geq \mathbb{E}^{y}[h(X_{T(Q(x,2^{-k}))}); T_{Q(x,2^{-k})} < \tau_{Q(0,8)}]$$
$$\geq \delta^k \Big(\inf_{Q(x,2^{-k})} h \Big),$$

or
$$\inf_{Q(x,2^{-k})} h \leq \delta^{-k}. \tag{7.5}$$

By (7.4) there exists $\rho < 1$ such that
$$\mathop{\mathrm{Osc}}_{Q(x,2^{-(k+1)})} h \leq \rho \mathop{\mathrm{Osc}}_{Q(x,2^{-k})} h. \tag{7.6}$$

Take m large so that $\rho^{-m} \geq \delta^{-2}/(\delta^{-1} - 1)$. Let $M = 2^m$. Then
$$\mathop{\mathrm{Osc}}_{Q(x,M2^{-k})} h \geq \rho^{-m} \mathop{\mathrm{Osc}}_{Q(x,2^{-k})} h \geq \frac{\delta^{-2}}{\delta^{-1} - 1} \mathop{\mathrm{Osc}}_{Q(x,2^{-k})} h. \tag{7.7}$$

Take K large so that $\sqrt{d}M2^{-K} < 1/8$. Suppose there exists $x_0 \in Q(y,1)$ such that $h(x_0) \geq \delta^{-K-1}$. We will construct a sequence $x_1, x_2, \ldots$ by induction. Suppose we have $x_j \in Q(x_{j-1}, M2^{-(K+j-1)})$ with $h(x_j) \geq \delta^{-K-j-1}$, $j \leq n$. Since $|x_j - x_{j-1}| < \sqrt{d}M2^{-(K+j-1)}$, $1 \leq j \leq n$, and $|x_0 - y| \leq 1$, then $|x_n - y| < 2$. Since $h(x_n) \geq \delta^{-K-n-1}$ and by (7.5), $\inf_{Q(x_n,2^{-K-n})} h \leq \delta^{-K-n}$,

$$\operatorname*{Osc}_{Q(x_n,2^{-K-n})} h \geq \delta^{-K-n}(\delta^{-1} - 1).$$

So $\operatorname{Osc}_{Q(x_n,M2^{-K-n})} h \geq \delta^{-K-n-2}$, which implies that there exists $x_{n+1} \in Q(x_n, M2^{-K-n})$ with $h(x_{n+1}) \geq \delta^{-K-n-2}$ because h is nonnegative. By induction we obtain a sequence x_n with $x_n \in Q(y,4)$ and $h(x_n) \to \infty$. This contradicts the boundedness of h on $Q(0,8)$. Therefore h is bounded on $Q(0,1)$ by δ^{-K-1}. $\qquad\square$

(7.7) Corollary. *Suppose D is a bounded connected open domain and $r > 0$. There exists c_1 depending only on D, Λ, and r such that if h is nonnegative, bounded in D, and $h(X_{t \wedge \tau_D})$ is a martingale, then $h(x) \leq c_1 h(y)$ if $x, y \in D$ and $\operatorname{dist}(x, \partial D)$ and $\operatorname{dist}(y, \partial D)$ are both greater than r.*

Proof. We form a sequence $x = y_0, y_1, y_2, \ldots, y_m = y$ such that $|y_{i+1} - y_i| < (a_{i+1} \wedge a_i)/32$, where $a_i = \operatorname{dist}(y_i, \partial D)$ and each $a_i < r$. By compactness we can choose M depending only on r so that no more than M points y_i are needed. By scaling and Theorem 7.6, $h(y_i) \leq c_2 h(y_{i+1})$ with $c_2 > 1$. So

$$h(x) = h(y_0) \leq c_2 h(y_1) \leq \cdots \leq c_2^m h(y_m) = c_2^m h(y) \leq c_2^M h(y). \qquad\square$$

8. Equicontinuity and approximation

We first prove an equicontinuity result for $G^\lambda f$. Then we show that X_t spends positive time in sets of positive Lebesgue measure. Finally, we complete the discussion of approximation started in Section 6 by showing that the a_{ij}^ε defined in (6.5) converge to a_{ij} almost everywhere.

For the next chapter we will need a modulus of continuity for $G^\lambda f$.

(8.1) Theorem. *Let $\mathcal{L} \in \mathcal{N}(\Lambda, 0)$ so that the drift coefficients of $\mathcal{L}$ are 0. Suppose for each x, $\mathbb{P}^x$ is a solution to the martingale problem for $\mathcal{L}$ started at x and $(\mathbb{P}^x, X_t)$ is a strong Markov family. Then $G^\lambda f(x) = \mathbb{E}^x \int_0^\infty e^{-\lambda t} f(X_t)\, dt$ is continuous with a modulus of continuity that depends only on λ, $\|f\|_\infty$, and Λ.*

Proof. Fix x_0 and λ. Take R sufficiently small so that

$$\sup_{x \in B(x_0,R)} \mathbb{E}^x \tau_{B(x_0,R)} \leq \varphi(1/2)/4\lambda, \tag{8.1}$$

where φ is defined in Theorem 7.4. This is possible by Theorem 2.3. We will write B for $B(x_0, R)$. Let g be bounded on ∂B, and consider the function

$$h(x) = \mathbb{E}^x[e^{-\lambda \tau_B} g(X_{\tau_B})]. \tag{8.2}$$

Our first goal is to obtain a modulus of continuity for h in $B(x_0, R/2)$. We obtain the Hölder continuity of h as in Theorem 7.5: it suffices to show there exists $\rho < 1$ such that

$$\operatorname*{Osc}_{Q(z,r/2)} h \le \rho \operatorname*{Osc}_{Q(z,r)} h \tag{8.3}$$

for $z \in B(x_0, R/2)$ and $r < R/4d^{1/2}$. Note that $e^{-\lambda(t\wedge\tau_B)} h(X_{t\wedge\tau_B})$ is a martingale, since by the strong Markov property,

$$
\begin{aligned}
e^{-\lambda(t\wedge\tau_B)} & h(X_{t\wedge\tau_B}) \\
&= e^{-\lambda(t\wedge\tau_B)} \mathbb{E}^{X_{t\wedge\tau_B}}[e^{-\lambda\tau_B} g(X_{\tau_B})] \\
&= e^{-\lambda(t\wedge\tau_B)} \mathbb{E}^{x}[e^{-\lambda\tau_B\circ\theta_{t\wedge\tau_B}} g(X_{\tau_B}\circ\theta_{t\wedge\tau_B}) \mid \mathcal{F}_{t\wedge\tau_B}] \\
&= \mathbb{E}^{x}[e^{-\lambda(t\wedge\tau_B)+\tau_B\circ\theta_{t\wedge\tau_B})} g(X_{\tau_B}\circ\theta_{t\wedge\tau_B}) \mid \mathcal{F}_{t\wedge\tau_B}] \\
&= \mathbb{E}^{x}[e^{-\lambda\tau_B} g(X_{\tau_B}) \mid \mathcal{F}_{t\wedge\tau_B}]
\end{aligned}
$$

and $t\wedge\tau_B + \tau_B\circ\theta_{t\wedge\tau_B} = \tau_B$ and $X_{\tau_B}\circ\theta_{t\wedge\tau_B} = X_{\tau_B}$.

To show (8.3) it suffices to show

$$\operatorname*{Osc}_{Q(z,r/2)} (Ch+D) \le \rho \operatorname*{Osc}_{Q(z,r)} (Ch+D),$$

where we choose C and D so that $\sup_{Q(z,r)}(Ch+D) = 1$, $\inf_{Q(z,r)}(Ch+D) = 0$, and $|F| \ge (1/2)|Q(z,r)|$, where $F = \{z \in Q(z,r); (Ch+D) \ge 1/2\}$. If $x \in Q(z,r/2)$,

$$
\begin{aligned}
(Ch+D)(x) &\ge C\mathbb{E}^{x}[e^{-\lambda\tau_F} h(X_{\tau_F}); \tau_F < \tau_{Q(z,r)}] + D \\
&\ge \mathbb{E}^{x}[e^{-\lambda\tau_F}(Ch+D)(X_{\tau_F}); \tau_F < \tau_{Q(z,r)}] \\
&\ge \mathbb{E}^{x}[(Ch+D)(X_{\tau_F}); \tau_F < \tau_{Q(z,r)}] - \mathbb{E}^{x}(1 - e^{-\lambda\tau_F}) \\
&\ge \frac{1}{2}\mathbb{P}^{x}(\tau_F < \tau_{Q(z,r)}) - \lambda E^{x}\tau_F \\
&\ge \frac{1}{2}\varphi(1/2) - \lambda\mathbb{E}^{x}\tau_B \ge \frac{1}{4}\varphi(1/2).
\end{aligned}
$$

We thus obtain (8.3) with $\rho = 1 - \varphi(1/2)/4$.

Now fix x_0 and write

$$
\begin{aligned}
G^{\lambda}f(x) &= \mathbb{E}^{x}\int_0^{\tau_{B(x_0,R)}} e^{-\lambda t} f(X_t)\, dt + \mathbb{E}^{x}\int_{\tau_{B(x_0,R)}}^{\infty} e^{-\lambda t} f(X_t)\, dt \tag{8.4} \\
&= \mathbb{E}^{x}\int_0^{\tau_{B(x_0,R)}} e^{-\lambda t} f(X_t)\, dt + \mathbb{E}^{x} e^{-\lambda\tau_{B(x_0,R)}} G^{\lambda}f(X_{\tau_{B(x_0,R)}}).
\end{aligned}
$$

Let $\varepsilon > 0$ and take R small enough so that (8.1) holds and also

$$\sup_{x\in B(x_0,R)} \mathbb{E}^{x}\tau_{B(x_0,R)} < \varepsilon/3.$$

So the first term on the last line of (8.4) is bounded by $\varepsilon\|f\|_\infty/3$. Note that $\|G^{\lambda}f\|_\infty \le \lambda^{-1}\|f\|_\infty$. By what we showed above,

$$h(x) = \mathbb{E}^{x} e^{-\lambda\tau_{B(x_0,R)}} G^{\lambda}f(X_{\tau_{B(x_0,R)}})$$

is a Hölder continuous function of x with a modulus of continuity depending only on $\|G^{\lambda}f\|_\infty$ for $x \in B(x_0, R/2)$. So there exist c_1 and α such that if

$x, y \in B(x_0, R/2)$, then $|h(x) - h(y)| \leq c_1 |x - y|^\alpha$. Thus if $|x - y| \leq (\varepsilon/3c_1)^\alpha$ and $x, y \in B(x_0, R/2)$,

$$|G^\lambda f(x) - G^\lambda f(y)| \leq \varepsilon \|f\|_\infty.$$

This proves the modulus of continuity result. $\qquad\square$

Combining with Corollary 6.3, we have the following.

(8.2) Theorem. *Let $\lambda > 0$ and let $\mathbb{P}$ be a solution to the martingale problem started at x for an operator $\mathcal{L} \in \mathcal{N}(\Lambda, 0)$. There exist a_{ij}^ε smooth such that if $\mathcal{L}^\varepsilon$ is defined by (6.1) and G_ε^λ is defined by (6.2), then $\mathcal{L}^\varepsilon \in \mathcal{N}(\Lambda, 0)$ and*

$$G_\varepsilon^\lambda f(x) \to \mathbb{E} \int_0^\infty e^{-\lambda t} f(X_t) \, dt$$

whenever f is bounded.

Proof. By the preceding theorem, $G_\varepsilon^\lambda f$ is equicontinuous in ε. Since φ_ε in Corollary 6.3 has compact support,

$$|(G_\varepsilon^\lambda f * \varphi_\varepsilon)(x_0) - G_\varepsilon^\lambda f(x_0)|$$
$$\leq \int |G_\varepsilon^\lambda f(x_0 - \varepsilon y) - G_\varepsilon^\lambda f(x_0)| \varphi(y) \, dy \to 0$$

as $\varepsilon \to 0$. Combining with Corollary 6.3 proves the convergence. $\qquad\square$

We have not yet proved that the a_{ij}^ε in Theorem 8.2 converge almost everywhere to the a_{ij}. In order to do so, we need to show that X_t spends positive time in sets of positive Lebesgue measure.

(8.3) Lemma. *Suppose $r > 1$ and let W be a cube in $Q(0, 1)$. Let W^* be the cube with the same center as W but side length half as long. Let V be a subset of W with the property that there exists δ such that*

$$\mathbb{E}^y \int_0^{\tau_W} 1_V(X_s) \, ds \geq \delta \mathbb{E}^y \tau_W, \qquad y \in W^*.$$

Then there exists $\zeta(\delta)$ depending on δ, r, and Λ such that

$$\mathbb{E}^y \int_0^{\tau_{Q(0,r)}} 1_V(X_s) \, ds \geq \zeta(\delta) \mathbb{E}^y \int_0^{\tau_{Q(0,r)}} 1_W(X_s) \, ds, \qquad y \in Q(0, 1).$$

Proof. Let S be the cube in $Q(0, r)$ with the same center as W and side length $r \wedge 2^{1/d}$ as long. Let $T_1 = \inf\{t : X_t \in W\}$, $U_1 = \inf\{t > T_1 : X_t \notin S\}$, $T_{i+1} = \inf\{t > U_i : X_t \in W\}$, and $U_{i+1} = \inf\{t > T_{i+1} : X_t \notin S\}$. Then

$$\mathbb{E}^{\,y} \int_0^{\tau_{Q(0,r)}} 1_W(X_s)\,ds = \sum \mathbb{E}^{\,y}\left[\int_{T_i}^{U_i} 1_W(X_s)\,ds; T_i < \tau_{Q(0,r)}\right],$$

$$= \sum \mathbb{E}^{\,y}\left[\mathbb{E}^{\,X(T_i)} \int_0^{\tau_S} 1_W(X_s)\,ds; T_i < \tau_{Q(0,r)}\right],$$

with a similar expression with W replaced by V. So it suffices to show there exists $\zeta(\delta)$ such that

$$\mathbb{E}^{\,w} \int_0^{\tau_S} 1_V(X_s)\,ds \geq \zeta(\delta)\mathbb{E}^{\,w} \int_0^{\tau_S} 1_W(X_s)\,ds, \qquad w \in W.$$

By the support theorem, there exists c_1 depending only on r and Λ such that

$$\mathbb{P}^w(T_{W^*} < \tau_S) \geq c_1, \qquad w \in W.$$

So if $w \in W$, by the strong Markov property,

$$\mathbb{E}^{\,w} \int_0^{\tau_S} 1_V(X_s)\,ds \geq \mathbb{E}^{\,w}\left[\int_0^{\tau_S} 1_V(X_s)\,ds; T_{W^*} < \tau_S\right]$$

$$= \mathbb{E}^{\,w}\left[\mathbb{E}^{\,X(T(W^*))} \int_0^{\tau_S} 1_V(X_s)\,ds; T_{W^*} < \tau_S\right]$$

$$\geq c_1 \inf_{z\in W^*} \mathbb{E}^{\,z} \int_0^{\tau_S} 1_V(X_s)\,ds$$

$$\geq c_1 \inf_{z\in W^*} \mathbb{E}^{\,z} \int_0^{\tau_W} 1_V(X_s)\,ds.$$

By our hypothesis, if $z \in W^*$,

$$\mathbb{E}^{\,z} \int_0^{\tau_W} 1_V(X_s)\,ds \geq \delta \mathbb{E}^{\,z}\tau_W.$$

By Proposition 2.3 and scaling,

$$\mathbb{E}^{\,z}\tau_W \geq c_2 \sup_{v\in S} \mathbb{E}^{\,v}\tau_S \geq c_2 \mathbb{E}^{\,w} \int_0^{\tau_S} 1_W(X_s)\,ds.$$

We now take $\zeta(\delta) = c_1 c_2 \delta$. $\qquad\qquad\qquad\qquad\qquad\qquad\qquad\square$

(8.4) Lemma. *There exist c_1 and ε such that if $|B| \subseteq Q(0,1)$, $x \in Q(0,1/2)$, and $|Q(0,1) - B| < \varepsilon$, then*

$$\mathbb{E}^{\,x} \int_0^{\tau_{Q(0,1)}} 1_B(X_s)\,ds \geq c_1.$$

Proof. By Proposition 2.3, there exists c_2 such that $\mathbb{E}^{\,x}\tau_{Q(0,1)} \geq c_2$. As in the proof of (7.1), if ε is small enough, then

$$\mathbb{E}^{\,x} \int_0^{\tau_{Q(0,1)}} 1_{(Q(0,1)-B)}(X_s)\,ds \leq c_2/2.$$

We thus have the lemma with $c_1 = c_2/2$. $\qquad\qquad\qquad\qquad\qquad\qquad\square$

(8.5) Theorem. *There exists a nondecreasing function* $\psi : (0,1) \to (0,1)$ *such that if* $\mathbb{P}$ *is a solution to the martingale problem for an operator in* $\mathcal{N}(\Lambda, 0)$ *started at an* $x \in Q(0, 1/2)$ *and* $B \subseteq Q(0,1)$, *then*

$$\mathbb{E} \int_0^{\tau Q(0,1)} 1_B(X_s)\, ds \geq \psi(|B|).$$

Proof. Let

$$\psi(\varepsilon) = \inf\Big\{ \mathbb{E}^y \int_0^{\tau Q(z_0, R)} 1_B(X_s)\, ds : z_0 \in \mathbb{R}^d, R > 0, B \subseteq Q(z_0, R),$$

$$|B| \geq \varepsilon |Q(z_0, R)|, y \in Q(z_0, R/2) \Big\}.$$

Lemma 8.4 and scaling tell us that $\psi(\varepsilon) > 0$ if ε is sufficiently close to 1. We need to show $\psi(\varepsilon) > 0$ for all $\varepsilon > 0$.

Let q_0 be the infimum of those ε for which $\psi(\varepsilon) > 0$. We suppose $q_0 > 0$ and we will obtain a contradiction. As in the proof of Theorem 7.4, choose $q > q_0$ such that $(q + q^2)/2 < q_0$ and let $\eta = (q - q^2)/2$. Let β be a number of the form 2^{-n} with

$$(\eta \wedge q \wedge (1 - q))/32d \leq \beta < (\eta \wedge q \wedge (1 - q))/16d.$$

There exist $z_0 \in \mathbb{R}^d, R > 0, B_1 \subseteq Q(z_0, R)$, and $x \in Q(z_0, R/2)$ such that $q > |B_1|/|Q(z_0, R)| > q - \eta/2$ and

$$\mathbb{E}^x \int_0^{\tau Q(z_0, R)} 1_{B_1}(X_s)\, ds < \zeta(\psi(q))\psi(q),$$

where ζ is defined in Lemma 8.3. Without loss of generality, we can suppose $z_0 = 0$ and $R = 1$, and so

$$\mathbb{E}^x \int_0^{\tau Q(0,1)} 1_{B_1}(X_s)\, ds < \zeta(\psi(q))\psi(q).$$

Let $B = B_1 \cap Q(0, 1 - \beta)$. Then

$$\mathbb{E}^x \int_0^{\tau Q(0,1)} 1_B(X_s)\, ds < \zeta(\psi(q))\psi(q), \tag{8.5}$$

and by our choice of β, $q > |B| > q - \eta$.

As in the proof of Theorem 7.4, we use Proposition 7.2 to construct D consisting of the union of cubes with $|D \cap Q(0,1)| \geq (q + 1)/2$. Let $\widetilde{D} = D \cap Q(0,1)$, and as in the proof of Theorem 7.5, $|\widetilde{D}| > q$. Since $|\widetilde{D}| > q > q_0$,

$$\mathbb{E}^x \int_0^{\tau Q(0,1)} 1_{\widetilde{D}}(X_s)\, ds > \psi(q).$$

D consists of the union of cubes $\widehat{R}_i$ such that the R_i have pairwise disjoint interiors, where R_i is the cube with the same center as $\widehat{R}_i$ but one-third the side length. Let $V_i = \widehat{R}_i \cap Q(0, 1 - \beta)$. We have by our construction $|B \cap R_i| \geq q|R_i|$. We will show that for each i,

$$\mathbb{E}^x \int_0^{\tau_{Q(0,1)}} 1_{B \cap R_i}(X_s)\, ds \geq \zeta(\psi(q)) \mathbb{E}^x \int_0^{\tau_{Q(0,1)}} 1_{V_i}(X_s)\, ds. \qquad (8.6)$$

Once we have (8.6), we sum and we have

$$\mathbb{E}^x \int_0^{\tau_{Q(0,1)}} 1_B(X_s)\, ds \geq \sum_i \int_0^{\tau_{Q(0,1)}} 1_{B \cap R_i}(X_s)\, ds$$

$$\geq \zeta(\psi(q)) \sum_i \mathbb{E}^x \int_0^{\tau_{Q(0,1)}} 1_{V_i}(X_s)\, ds$$

$$\geq \zeta(\psi(q)) \mathbb{E}^x \int_0^{\tau_{Q(0,1)}} 1_{\widetilde{D}}(X_s)\, ds$$

$$\geq \zeta(\psi(q))\psi(q),$$

our contradiction.

We now prove (8.6). Fix i. By our definition of β, if V_i is not empty, then V_i is contained in a cube W_i that is itself contained in $Q(0, 1 - \beta)$ and $|W_i| \leq 3^d |R_i|$. Let R_i^* be the cube with the same center as R_i but side length half as long. By the definition of ψ,

$$\mathbb{E}^y \int_0^{\tau_{R_i}} 1_{B \cap R_i}(X_s)\, ds \geq \psi(q) \mathbb{E}^y \tau_{R_i}$$

if $y \in R_i^*$. (8.6) now follows from Lemma 8.3 and scaling. $\qquad \square$

(8.6) Corollary. *If* $|B| > 0$, *then* $\mathbb{E}^x \int_0^\infty e^{-\lambda t} 1_B(X_t)\, dt > 0$.

Proof. We select a unit cube Q such that $|Q \cap B| > 0$. Let Q^* be the cube with the same center as Q but side length half as long. By the strong Markov property and support theorem, there exists c_1 such that

$$\mathbb{E}^x \int_0^\infty e^{-\lambda t} 1_B(X_t)\, dt \geq c_1 \inf_{y \in Q^*} \mathbb{E}^y \int_0^\infty e^{-\lambda t} 1_{Q \cap B}(X_t)\, dt.$$

By Proposition 2.3, there exists c_2 such that if $y \in Q^*$,

$$\mathbb{E}^y(\tau_Q - (\tau_Q \wedge t)) = \mathbb{E}^y(\mathbb{E}^{X_t} \tau_Q; t < \tau_Q)$$

$$\leq c_2 \mathbb{P}^y(t < \tau_Q) \leq c_2 \mathbb{E}^x \tau_Q^2 / t^2.$$

Using Proposition 2.3 again and taking t_0 large enough, we have

$$\sup_{y \in Q^*} \mathbb{E}^y(\tau_Q - (\tau_Q \wedge t_0)) \leq \psi(|Q \cap B|)/2.$$

Then

$$\mathbb{E}^y \int_0^\infty e^{-\lambda t} 1_{Q \cap B}(X_t)\, dt$$

$$\geq \mathbb{E}^y \int_0^{\tau_Q \wedge t_0} e^{-\lambda t} 1_{Q \cap B}(X_t)\, dt$$

$$\geq e^{-\lambda t_0} \mathbb{E}^y \int_0^{\tau_Q \wedge t_0} 1_{Q \cap B}(X_t)\, dt$$

$$\geq e^{-\lambda t_0} \left[\mathbb{E}^y \int_0^{\tau_Q} 1_{Q \cap B}(X_t)\, dt - \mathbb{E}^y(\tau_Q - \tau_Q \wedge t_0) \right]$$

$$\geq e^{-\lambda t} (\psi(|Q \cap B|) - \psi(|Q \cap B|)/2) > 0. \qquad \square$$

(8.7) Proposition. *Let $\mathbb{P}$ be a solution to the martingale problem for an operator in $\mathcal{N}(\Lambda, 0)$ started at x. If $|C| > 0$, then $\mathbb{E} \int_0^\infty e^{-\lambda t} 1_C(X_t)\, dt > 0$.*

Proof. This follows from Corollary 8.6 and Theorem 8.2 with $f = 1_C$. $\qquad \square$

Finally, we show that the a_{ij}^ε defined in (6.5) converge.

(8.8) Theorem. *Let $a_{ij}^\varepsilon(x)$ be defined by (6.5). Then for each i and j, the a_{ij}^ε converge to a_{ij} almost everywhere.*

Proof. By Theorem 6.2,

$$\left| \int f(y)\, \mu(dy) \right| = \left| \mathbb{E} \int_0^\infty e^{-\lambda t} f(X_t)\, dt \right| \leq c_1 \|f\|_d.$$

Hence $\mu(dy)$ has a density $m(y)\, dy$, and by a duality argument, $m \in L^{d/(d-1)}$. If $C = \{y : m(y) = 0\}$, then

$$\mathbb{E} \int_0^\infty e^{-\lambda t} 1_C(X_t)\, dt = \int 1_C(y)\, \mu(dy) = \int 1_C(y) m(y)\, dy = 0.$$

By Proposition 8.7, $|C| = 0$.

Now

$$\int \varphi_\varepsilon(x - y) a_{ij}(y)\, \mu(dy) = \int \varphi_\varepsilon(x - y) a_{ij}(y) m(y)\, dy$$
$$\rightarrow a_{ij}(x) m(x)$$

for almost every x, since φ_ε is an approximation to the identity (see [PTA, Theorem IV.1.6]), a_{ij} is bounded, and $m \in L^{d/(d-1)}$. Similarly,

$$\int \varphi_\varepsilon(x - y)\, \mu(dy) = \int \varphi_\varepsilon(x - y) m(y)\, dy \rightarrow m(x)$$

for almost every x. Since $m > 0$ almost everywhere, the ratio, which is $a_{ij}^\varepsilon(x)$, converges to $a_{ij}(x)$ almost everywhere. $\qquad \square$

9. Notes

The results of Sections 1 and 2 are from Stroock and Varadhan [2]. The examples in Proposition 3.1 are due to Krylov [2], whereas Proposition 3.2 is due to Pucci [1]. Theorem 3.3 is a probabilistic formulation of a result of Gilbarg and Serrin [1].

For Section 4 we followed Gilbarg and Trudinger [1]. Section 5 is standard material. The approximation results in Section 6 are from Krylov [3].

To obtain the Harnack inequality in Section 7, we adapted Krylov and Safonov [2] and the account in Caffarelli [1]. Theorem 8.9 is due to Evans [1].

VI
MARTINGALE PROBLEMS

In this chapter we continue our discussion of nondivergence form operators. We introduced the martingale problem in Chapter V; now we investigate existence and uniqueness for the martingale problem for an operator $\mathcal{L}$.

Section 1 discusses existence. We will see that there exists a solution if the coefficients of $\mathcal{L}$ are continuous or if the diffusion coefficients are bounded and uniformly elliptic.

A solution to the martingale problem need not necessarily satisfy the strong Markov property. We see in Section 2 that regular conditional probabilities can act as a replacement. If the diffusion coefficients are uniformly elliptic, one can find strong Markov families of solutions.

The question of uniqueness of solutions is often quite difficult. Section 3 gives some techniques that allow us to reduce the problem to a simpler one.

Section 4 proves uniqueness in three cases, all under the assumption of uniform ellipticity. The first is when the diffusion coefficients are continuous, the second is when they are continuous except possibly at one point, and the third is when the dimension of the state space is two.

When uniqueness holds, there are some interesting consequences. These are discussed in Section 5.

A brief account of the submartingale problem and diffusions with reflection constitute Section 6.

1. Existence

In this section we discuss the existence of solutions to the martingale problem for an elliptic operator in nondivergence form. Let $\mathcal{L}$ be the elliptic operator in nondivergence form defined by

$$\mathcal{L}f(x) = \frac{1}{2} \sum_{i,j=1}^{d} a_{ij}(x)\partial_{ij}f(x) + \sum_{i=1}^{d} b_i(x)\partial_i f(x), \qquad f \in C^2. \tag{1.1}$$

We assume throughout that the a_{ij} and b_i are bounded and measurable. Since the coefficient of $\partial_{ij}f(x)$ is $(a_{ij}(x) + a_{ji}(x))/2$, there is no loss of generality in assuming that $a_{ij} = a_{ji}$. We let

$$\mathcal{N}(\Lambda_1, \Lambda_2) = \{\mathcal{L} : \sup_{i \leq d} \|b_i\|_\infty \leq \Lambda_2 \text{ and} \tag{1.2}$$

$$\Lambda_1 |y|^2 \leq \sum_{i,j=1}^{d} y_i y_j a_{ij}(x) \leq \Lambda_1^{-1} |y|^2 \text{ for all } x, y \in \mathbb{R}^d \}.$$

If $\mathcal{L} \in \mathcal{N}(A, B)$ for some $A > 0$, then we say $\mathcal{L}$ is uniformly elliptic.

A probability measure $\mathbb{P}$ is a *solution to the martingale problem for $\mathcal{L}$ started at x* if

$$\mathbb{P}(X_0 = x) = 1 \tag{1.3}$$

and

$$f(X_t) - f(X_0) - \int_0^t \mathcal{L}f(X_s)\, ds \tag{1.4}$$

is a local martingale under $\mathbb{P}$ whenever f is in $C^2(\mathbb{R}^d)$.

We begin by showing that continuity of the coefficients of $\mathcal{L}$ is a sufficient condition for the existence of a solution to the martingale problem.

(1.1) Theorem. *Suppose the a_{ij} and b_i are bounded and continuous and $x \in \mathbb{R}^d$. Then there exists a solution to the martingale problem for $\mathcal{L}$ started at x.*

Proof. Let a_{ij}^n and b_i^n be uniformly bounded C^2 functions on $\mathbb{R}^d$ that converge to a_{ij} and b_i uniformly on compacts. Let

$$\mathcal{L}_n f(x) = \frac{1}{2} \sum_{i,j=1}^{d} a_{ij}^n(x)\partial_{ij}f(x) + \sum_{i=1}^{d} b_i^n(x)\partial_i f(x), \tag{1.5}$$

let σ_n be a Lipschitz square root of a^n, and let X^n be the solution to

$$dX_t^n = \sigma^n(X_t^n)\, dW_t + b^n(X_t^n)\, dt, \qquad X_0^n = x,$$

where W_t is a d-dimensional Brownian motion. Let $\mathbb{P}_n$ be the law of X^n. Our desired $\mathbb{P}$ will be a limit point of the sequence $\{\mathbb{P}_n\}$.

Each $\mathbb{P}^n$ is a probability measure on $\Omega = C([0, \infty))$. A collection of continuous functions on a compact set has compact closure if they are uniformly bounded at one point and they are equicontinuous. This and Theorem V.2.4 imply easily that the $\mathbb{P}_n$ are tight.

Let $\mathbb{P}_{n_k}$ be a subsequence that converges weakly and call the limit $\mathbb{P}$. We must show that $\mathbb{P}$ is a solution to the martingale problem. If g is a continuous function on $\mathbb{R}^d$ with compact support, $g(X_0)$ is a continuous function on Ω, so

$$g(x) = \mathbb{E}_{n_k} g(X_0) \to \mathbb{E}\, g(X_0).$$

Since this is true for all such g, we must have $\mathbb{P}(X_0 = x) = 1$.

Next let $f \in C^2(\mathbb{R}^d)$ be bounded with bounded first and second partial derivatives. To show

$$\mathbb{E}\left[f(X_t) - f(X_s) - \int_s^t \mathcal{L}f(X_r)\, dr; A \right] = 0$$

whenever $A \in \mathcal{F}_s$, it suffices to show

$$\mathbb{E}\left[\left\{ f(X_t) - f(X_s) - \int_s^t \mathcal{L}f(X_r)\, dr \right\} \prod_{i=1}^m g_i(X_{r_i}) \right] = 0 \qquad (1.6)$$

whenever $m \geq 1$, $0 \leq r_1 \leq \cdots \leq r_m \leq s$, and the g_i are continuous functions with compact support on $\mathbb{R}^d$. Setting

$$Y(\omega) = \left\{ f(X_t) - f(X_s) - \int_s^t \mathcal{L}f(X_r)\, dr \right\} \prod_{i=1}^m g_i(X_{r_i}),$$

Y is a continuous bounded function on Ω, so $\mathbb{E}\,Y = \lim_{k \to \infty} \mathbb{E}_{n_k} Y$. Since $\mathbb{P}_{n_k}$ is a solution to the martingale problem for $\mathcal{L}_{n_k}$,

$$\mathbb{E}_{n_k}\left[\left\{ f(X_t) - f(X_s) - \int_s^t \mathcal{L}_{n_k} f(X_r)\, dr \right\} \prod_{i=1}^m g_i(X_{r_i}) \right] = 0.$$

Since the g_i are bounded, it suffices to show

$$\mathbb{E}_{n_k}\left[\int_s^t |(\mathcal{L}f - \mathcal{L}_{n_k} f)(X_r)|\, dr \right] \to 0 \qquad (1.7)$$

as $k \to \infty$.

Let $\varepsilon > 0$. Choose M large so that $\mathbb{P}_{n_k}(\sup_{r \leq t} |X_r| \geq M) \leq \varepsilon$. This can be done uniformly in k by Proposition V.2.3. Now choose k large so that $|a_{ij}(y) - a_{ij}^{n_k}(y)| < \varepsilon$ if $|y| \leq M$ and $i, j = 1, \ldots, d$, and similarly $|b_i(y) - b_i^{n_k}(y)| < \varepsilon$ if $|y| \leq M$ and $i = 1, \ldots, d$. Since $f \in C^2$ and the a_{ij}^n and b_i^n are uniformly bounded, there exist c_1 and c_2 such that

$$\sup_n \|\mathcal{L}_n f\|_\infty \leq c_1, \qquad \|\mathcal{L}f\|_\infty \leq c_1,$$

and

$$\sup_{|y| \leq M} |\mathcal{L}_{n_k} f(y) - \mathcal{L}f(y)| \leq c_2 \varepsilon.$$

Then

$$\mathbb{E}_{n_k} \int_s^t |(\mathcal{L}f - \mathcal{L}_{n_k}f)(X_r)| \, dr$$

$$\leq c_2(t-s)\varepsilon + 2(t-s)c_1 \mathbb{P}_{n_k} (\sup_{r \leq t} |X_r| \geq M)$$

$$\leq c_3 \varepsilon,$$

which proves (1.7).

Finally, suppose $f \in C^2$ but is not necessarily bounded. Let f_M be a C^2 function that is bounded with bounded first and second partial derivatives and that equals f on $B(0, M)$. If $T_M = \inf\{t : |X_t| \geq M\}$, the above argument applied to f_M shows that $f_M(X_t) - f_M(X_0) - \int_0^t \mathcal{L}f_M(X_s) \, ds$ is a martingale, and hence so is $f(X_{t \wedge T_M}) - f(X_0) - \int_0^{t \wedge T_M} \mathcal{L}f(X_s) \, ds$. Since X_t is continuous, $T_M \to \infty$ a.s., and therefore $f(X_t) - f(X_0) - \int_0^t \mathcal{L}f(X_s) \, ds$ is a local martingale. $\square$

If the operator $\mathcal{L}$ is uniformly elliptic, we can allow the b_i to be bounded without requiring any other smoothness. If $\mathcal{L}$ is given by (1.1), let $\mathcal{L}'$ be defined by

$$\mathcal{L}'f(x) = \frac{1}{2} \sum_{i,j=1}^{d} a_{ij}(x)\partial_{ij} f(x). \tag{1.8}$$

(1.2) Theorem. *Suppose $\mathcal{L} \in \mathcal{N}(\Lambda_1, \Lambda_2)$. If there exists a solution to the martingale problem for $\mathcal{L}'$ started at x, then there exists a solution to the martingale problem for $\mathcal{L}$ started at x.*

Proof. Let $\mathbb{P}'$ be a solution to the martingale problem for $\mathcal{L}'$ started at x. Let $\sigma(x)$ be a positive definite square root of $a(x)$. Then under $\mathbb{P}'$ (cf. Theorem II.5.1), X_t^i is a martingale and $d\langle X^i, X^j \rangle_t = a_{ij}(X_t) \, dt$. Letting $W_t = \int_0^t \sigma^{-1}(X_s) \, dX_s$, we see as in the proof of Theorem II.5.1 that W_t is a d-dimensional Brownian motion with quadratic variation $\langle W^i, W^j \rangle_t = \delta_{ij} t$. Hence under $\mathbb{P}'$ the process W_t is a Brownian motion and

$$dX_t = \sigma(X_t) \, dW_t.$$

Define a new probability measure $\mathbb{P}$ by setting the restriction of $d\mathbb{P}/d\mathbb{P}'$ to $\mathcal{F}_t$ equal to

$$M_t = \exp \left(\int_0^t (b\sigma^{-1})(X_s) \, dW_s + \frac{1}{2} \int_0^t |(b\sigma^{-1})(X_s)|^2 \, ds \right). \tag{1.9}$$

Under $\mathbb{P}'$, M_t is a martingale. By the Girsanov theorem (Section I.1), under $\mathbb{P}$ each component of

$$X_t - \left\langle \int_0^{\cdot} (b\sigma^{-1})(X_s)\,dW_s, X \right\rangle_t = X_t - \int_0^t b(X_s)\,ds$$

is a martingale and the quadratic variation of X remains the same. If

$$\widetilde{W}_t = \int_0^t \sigma^{-1}(X_s)\,d\left(X_s - \int_0^s b(X_r)\,dr\right),$$

then under $\mathbb{P}$, $\widetilde{W}_t$ is a martingale with $\langle \widetilde{W}^i, \widetilde{W}^j \rangle_t = \delta_{ij}t$, and hence $\widetilde{W}$ is a Brownian motion. Thus

$$dX_t = \sigma(X_t)\,d\widetilde{W}_t + b(X_t)\,dt.$$

By Theorem V.1.1, $\mathbb{P}$ is therefore a solution to the martingale problem for $\mathcal{L}$. $\qquad\square$

As a consequence of Theorems 1.1 and 1.2, there exists a solution to the martingale problem if $\mathcal{L} \in \mathcal{N}(\Lambda_1, \Lambda_2)$ for some $\Lambda_1, \Lambda_2 > 0$ and the a_{ij} are continuous.

Even if the a_{ij} are not continuous, a solution to the martingale problem will exist if uniform ellipticity holds.

(1.3) Theorem. *Suppose $\mathcal{L} \in \mathcal{N}(\Lambda_1, \Lambda_2)$, where the a_{ij} and b_i are measurable. If $x \in \mathbb{R}^d$, there exists a solution to the martingale problem for $\mathcal{L}$ started at x.*

Proof. By Theorem 1.2, we may assume the b_i are identically 0. Let a_{ij}^n be C^2 coefficients so that if $\mathcal{L}_n$ is defined by (1.5), then $\mathcal{L}_n \in \mathcal{N}(\Lambda_1, 0)$. Suppose also that the $a_{ij}^n(x)$ converge to $a_{ij}(x)$ almost everywhere. Let $\mathbb{P}_n$ be the solution to the martingale problem for $\mathcal{L}_n$ started at x, constructed as in the proof of Theorem 1.1. As in Theorem 1.1, the $\mathbb{P}_n$ are tight and a subsequence $\mathbb{P}_{n_k}$ converges weakly, say to $\mathbb{P}$.

As in the last paragraph of the proof of Theorem 1.1, it suffices to suppose $f \in C^2$ is bounded with bounded first and second partial derivatives. Let g_i be continuous functions on $\mathbb{R}^d$ with compact support. Write G for $\prod_{i=1}^m g_i(X_{r_i})$. Since

$$\left| \mathbb{E}_{n_k}\left[G \int_s^t \mathcal{L}_{n_k} f(X_r)\,dr \right] - \mathbb{E}\left[G \int_s^t \mathcal{L} f(X_r)\,dr \right] \right|$$

$$\leq \mathbb{E}_{n_k}\left[|G| \int_s^t |\mathcal{L}_{n_k} f - \mathcal{L} f)(X_r)|\,dr \right]$$

$$+ \left| \mathbb{E}_{n_k}\left[G \int_s^t \mathcal{L} f(X_r)\,dr \right] - \mathbb{E}\left[G \int_s^t \mathcal{L} f(X_r)\,dr \right] \right|,$$

we will have our result if (i) (1.7) holds and (ii)

$$\left| \mathbb{E}_{n_k}\left[G \int_s^t \mathcal{L} f(X_r)\,dr \right] - \mathbb{E}\left[G \int_s^t \mathcal{L} f(X_r)\,dr \right] \right| \to 0 \qquad (1.10)$$

for bounded $f \in C^2$. We first show (i). Let $\varepsilon > 0$. As in the proof of Theorem 1.1, there exist c_1 and M such that

$$\sup_n \mathbb{P}_n(\sup_{r \leq t} |X_r| \geq M) < \varepsilon$$

and

$$\sup_n \|\mathcal{L}_n f\|_\infty \leq c_1, \qquad \|\mathcal{L}f\|_\infty \leq c_1.$$

By Theorem V.6.2,

$$\mathbb{E}_{n_k}\left[\int_s^t |(\mathcal{L}f - \mathcal{L}_{n_k} f)(X_r)|\, dr \right]$$

$$\leq 2(t-s)c_1 \mathbb{P}_{n_k}(\sup_{r \leq t} |X_r| \geq M)$$

$$+ e^t \mathbb{E}_{n_k}\left[\int_0^\infty e^{-r} 1_{B(0,M)}(X_r)|(\mathcal{L}f - \mathcal{L}_{n_k} f)(X_r)|\, dr \right]$$

$$\leq c_2 \varepsilon + c_3 e^t \|(\mathcal{L}f - \mathcal{L}_{n_k} f)1_{B(0,M)}\|_d,$$

where c_3 is independent of n_k. The uniform boundedness of the a_{ij}^n and convergence of a_{ij}^n to a_{ij} almost everywhere imply that the right-hand side will be less than $c_4 \varepsilon$ if k is large, which proves (1.7).

To prove (ii), let $\varepsilon > 0$, let $M > 0$ be large so that $\mathbb{P}_{n_k}(\sup_{r \leq t} |X_r| \geq M) < \varepsilon$ and $\mathbb{P}(\sup_{r \leq t} |X_r| \geq M) < \varepsilon$, and let F be a continuous function on $\mathbb{R}^d$ with support in $B(0, M)$ such that

$$\int_{B(0,M)} |(\mathcal{L}f - F)(x)|^d\, dx < \varepsilon^d.$$

Then by Theorem V.6.2,

$$\left| \mathbb{E}_{n_k}\left[G \int_s^t (1_{B(0,M)}\mathcal{L}f - F)(X_r)\, dr \right] \right|$$

$$\leq e^t \prod_{i=1}^m \|g_i\|_\infty \mathbb{E}_{n_k} \int_0^\infty e^{-r}|(1_{B(0,M)}\mathcal{L}f - F)(X_r)|\, dr$$

$$\leq c_5 \varepsilon.$$

On the other hand,

$$\left| \mathbb{E}_{n_k}|G| \int_s^t (1_{B(0,M)^c}\mathcal{L}f)(X_r)\, dr \right| \leq c_6 \varepsilon$$

by our choice of M and the fact that $\mathcal{L}f$ is bounded. Similar equations hold with $\mathbb{E}_{n_k}$ replaced by $\mathbb{E}$. Since F is continuous,

$$\mathbb{E}_{n_k}\left[\int_s^t F(X_r)\, dr \prod g_i(X_{r_i}) \right] \to \mathbb{E}\left[\int_s^t F(X_r)\, dr \prod g_i(X_{r_i}) \right]$$

as $k \to \infty$. (1.10) follows. $\qquad \square$

Virtually the same proof shows

(1.4) Corollary. *Suppose $\mathcal{L} \in \mathcal{N}(A, B)$ and $x_n \to x$. If $\mathbb{P}_n$ is a solution to the martingale problem for $\mathcal{L}$ started at x_n, then any subsequential limit point of the $\mathbb{P}_n$ will be a solution to the martingale problem for $\mathcal{L}$ started at x.*

Proof. We follow the proof of Theorem 1.3, but we no longer have (1.7) to show and need only show (1.10). This is proved exactly as in the proof of Theorem 1.3. $\qquad\qquad\square$

2. The strong Markov property

We are not assuming that our solutions are part of a strong Markov family. As a substitute we have the following. Let $\mathbb{P}$ be a solution to the martingale problem for $\mathcal{L}$ started at x and let S be a finite stopping time. Define a probability measure $\mathbb{P}_S$ on $\Omega = C([0, \infty))$ by

$$\mathbb{P}_S(A) = \mathbb{P}(A \circ \theta_S). \tag{2.1}$$

Here θ_S is the shift operator that shifts the path by S. Recall the definition of regular conditional probability from Section I.5. Let $\mathbb{Q}_S(\omega, d\omega')$ be a regular conditional probability for $\mathbb{P}_S[\,\cdot\, \mid \mathcal{F}_S]$.

(2.1) Proposition. *With probability one, $Q_S(\omega, \cdot)$ is a solution to the martingale problem for $\mathcal{L}$ started at $X_S(\omega)$.*

Proof. If $A(\omega) = \{\omega' : X_0(\omega') = X_S(\omega)\}$, we first show that $\mathbb{Q}_S(\omega, A(\omega)) = 1$ for almost every ω. To do this, it suffices to show that

$$\mathbb{P}(B) = \mathbb{E}_{\mathbb{P}}[\mathbb{Q}_S(\omega, A(\omega)); B]$$

whenever $B \in \mathcal{F}_S$. The right-hand side, by the definition of $\mathbb{Q}_S$, is equal to

$$\mathbb{E}_{\mathbf{P}}[\mathbb{P}_S(A \mid \mathcal{F}_S); B] = \mathbb{E}_{\mathbb{P}}[\mathbb{P}(X_S = X_0 \circ \theta_S \mid \mathcal{F}_S); B]$$
$$= \mathbb{P}(X_S = X_S; B) = \mathbb{P}(B).$$

Next, if $f \in C^2$ and is bounded with bounded first and second partial derivatives, we need to show that

$$M_t = f(X_t) - f(X_0) - \int_0^t \mathcal{L}f(X_r)\, dr$$

is a martingale under $\mathbb{Q}_S$ for almost every ω. Let $u > t$. Since $M_t \circ \theta_S = M_{t+S} - M_S$ is a martingale with respect to $\mathcal{F}_{S+t}$, then

$$\mathbb{E}_{\mathbf{P}}[M_u \circ \theta_S; B \circ \theta_S \cap A] = \mathbb{E}_{\mathbb{P}}[M_t \circ \theta_S; B \circ \theta_S \cap A]$$

whenever $B \in \mathcal{F}_t$ and $A \in \mathcal{F}_S$. This is the same as saying

$$\mathbb{E}_{\mathbb{P}}[(M_u 1_B) \circ \theta_S; A] = \mathbb{E}_{\mathbb{P}}[(M_t 1_B) \circ \theta_S; A]. \tag{2.2}$$

Since (2.2) holds for all $A \in \mathcal{F}_S$, by the definition of $\mathbb{Q}_S$,

$$\mathbb{E}_{\mathbb{Q}_S}[M_u; B] = \mathbb{E}_{\mathbb{Q}_S}[M_t; B]$$

whenever $B \in \mathcal{F}_t$, which is what we needed to show.

Finally, if $f \in C^2$, then M_t is a local martingale under Q_S by the same argument as in the last paragraph of the proof of Theorem 1.1. $\square$

Essentially the same proof shows that

(2.2) Corollary. *Let $\mathbb{Q}'_S$ be a regular conditional probability for $\mathbb{P}_S[\cdot \mid X_S]$. Then with probability one, $\mathbb{Q}'_S$ is a solution to the martingale problem for $\mathcal{L}$ started at $X_S(\omega)$.*

If $\mathcal{L} \in \mathcal{N}(\Lambda_1, \Lambda_2)$, we can in fact show that there exists a family of solutions to the martingale problem that is a strong Markov family. We take $\Lambda_2 = 0$ for simplicity.

(2.3) Theorem. *Let $\Lambda > 0$ and suppose $\mathcal{L}_n \in \mathcal{N}(\Lambda, 0)$ with the $a_{ij}^n \in C^2$ and converging almost everywhere to the a_{ij}. Suppose $(\mathbb{P}_n^x, X_t)$ is a strong Markov family of solutions to the martingale problem for $\mathcal{L}_n$. Then there exists a subsequence n_k and a strong Markov family of solutions $(\mathbb{P}^x, X_t)$ to the martingale problem for $\mathcal{L}$ such that $\mathbb{P}_{n_k}^x$ converges weakly to $\mathbb{P}^x$ for all x.*

Note that part of the assertion is that the subsequence n_k does not depend on x.

Proof. Let $\{g_i\}$ be a countable dense subset of $C(\mathbb{R}^d)$, the continuous bounded functions on $\mathbb{R}^d$, and let $\{\lambda_j\}$ be a countable dense subset of $(0, \infty)$. Let

$$G_n^\lambda g(x) = \mathbb{E}_n^x \int_0^\infty e^{-\lambda t} g(X_t)\, dt.$$

Note that $\|G_n^\lambda g\|_\infty \leq \|g\|_\infty / \lambda$. By the equicontinuity of $G_n^\lambda g$ in n for each g (see Theorem V.8.1) and a diagonalization argument, we can find a subsequence n_k such that $G_{n_k}^{\lambda_j} g_i$ converges boundedly and uniformly on compacts. Since

$$\|G_n^{\lambda_j} g - G_n^{\lambda_j} h\|_\infty \leq \frac{1}{\lambda_j} \|g - h\|_\infty$$

it follows that $G_{n_k}^{\lambda_j} g$ converges uniformly on compacts for all $g \in C(\mathbb{R})$. Since

$$\|G_n^\lambda g - G_n^\mu g\|_\infty \leq \frac{c_1}{\lambda - \mu} \|g\|_\infty$$

(see Corollary IV.2.6), it follows that $G^\lambda_{n_k} g$ converges uniformly on compacts for all bounded $g \in C(\mathbb{R}^d)$ and all $\lambda \in (0, \infty)$. Call the limit $G^\lambda g$.

Suppose $x_n \to x$. By the tightness estimate Theorem V.2.4, $\mathbb{P}^{x_{n_k}}_{n_k}$ is a tight sequence. Let $\mathbb{P}$ be any subsequential limit point. By Corollary 1.4, $\mathbb{P}$ is a solution to the martingale problem for $\mathcal{L}$ started at x. If n' is a subsequence of n_k such that $\mathbb{P}^{x'_{n'}}_{n'}$ converges weakly to $\mathbb{P}$, by the equicontinuity of $G^\lambda_n g$,

$$\mathbb{E} \int_0^\infty e^{-\lambda t} g(X_t)\, dt = \lim_{n' \to \infty} \mathbb{E}^{x_{n'}}_{n'} \int_0^\infty e^{-\lambda t} g(X_t)\, dt$$
$$= \lim_{n' \to \infty} G^\lambda_{n'} g(x_{n'}) = G^\lambda g(x).$$

This holds for all bounded and continuous g; hence we see that if $\mathbb{P}_1$ and $\mathbb{P}_2$ are any two subsequential limit points of $\mathbb{P}^{x_{n_k}}_{n_k}$, their one-dimensional distributions agree by the uniqueness of the Laplace transform and the continuity of $g(X_t)$.

We next show that the two-dimensional distributions of any two subsequential limit points must agree. If g and h are bounded and continuous and $\mu > \lambda$,

$$\mathbb{E} \int_0^\infty \int_0^\infty e^{-\lambda t} e^{-\mu s} g(X_t) h(X_{t+s})\, ds\, dt$$
$$= \lim_{n' \to \infty} \mathbb{E}^{x_{n'}}_{n'} \int_0^\infty \int_0^\infty e^{-\lambda t} e^{-\mu s} g(X_t) h(X_{t+s})\, ds\, dt$$
$$= \lim_{n' \to \infty} \mathbb{E}^{x_{n'}}_{n'} \int_0^\infty e^{-\lambda t} g(X_t) \mathbb{E}^{X_t}_{n'} \int_0^\infty e^{-\mu s} h(X_s)\, ds\, dt$$
$$= \lim_{n' \to \infty} \mathbb{E}^{x_{n'}}_{n_k} \int_0^\infty e^{-\lambda t} g(X_t) G^\mu_{n'} h(X_t)\, dt$$
$$= \lim_{n' \to \infty} G^\lambda_{n'} \big(g(G^\mu_{n'} h) \big)(x_{n'}).$$

By the equicontinuity of the $G^\mu_n h$ and the fact that $G^\mu_{n_k} h$ converges boundedly and uniformly on compacts to $G^\mu h$, the right-hand side converges to $G^\lambda(g(G^\mu h))(x)$. By the uniqueness of the Laplace transform, we see that any two subsequential limit points have the same two-dimensional distributions.

Repeating the argument, we see that any two subsequential limit points have the same finite dimensional distributions. Since X_t is continuous, this implies that $\mathbb{P}_1 = \mathbb{P}_2$. We have thus shown that if $x_n \to x$, then $\mathbb{P}^{x_{n_k}}_{n_k}$ converges weakly to a probability measure; we call the limit $\mathbb{P}^x$. By the proof of Theorem 1.3, we know that $\mathbb{P}^x$ is a solution to the martingale problem for $\mathcal{L}$ started at x.

We now want to show that $(\mathbb{P}^x, X_t)$ forms a strong Markov family of solutions. We will do this by first showing that $\mathbb{E}^x_{n_k} f(X_t)$ converges uniformly on compacts to $\mathbb{E}^x f(X_t)$ if f is bounded and continuous. We have pointwise convergence of $\mathbb{E}^x_{n_k} f(X_t)$ for each x since we have weak convergence of $\mathbb{P}^x_{n_k}$ to $\mathbb{P}^x$.

We claim that the maps $x \mapsto \mathbb{E}^{x}_{n} f(X_t)$ are equicontinuous on compacts. If not, there exists $\varepsilon > 0$, $R > 0$, a subsequence n_m, and $x_m, y_m \in B(0, R)$ such that $|x_m - y_m| \to 0$ but

$$|\mathbb{E}^{x_m}_{n_m} f(X_t) - \mathbb{E}^{y_m}_{n_m} f(X_t)| > \varepsilon. \tag{2.3}$$

By compactness, there exists a further subsequence such that $\mathbb{P}^{x_{m_j}}_{n_{m_j}}$ converges weakly and also $x_{m_j} \to x \in \overline{B(0, R)}$; it follows that $y_{m_j} \to x$ also. By what we have already proved, $\mathbb{P}^{x_{m_j}}_{n_{m_j}}$ converges weakly to $\mathbb{P}^x$; hence $\mathbb{E}^{x_{m_j}}_{n_{m_j}} f(X_t)$ converges to $\mathbb{E}^x f(X_t)$ and the same with x_{m_j} replaced by y_{m_j}, a contradiction to (2.3). We thus have that the maps $x \mapsto \mathbb{E}^{x}_{n} f(X_t)$ are equicontinuous.

This implies that the convergence of $\mathbb{E}^{x}_{n_k} f(X_t)$ is uniform on compacts. In particular, the limit $\mathbb{E}^x f(X_t)$ is a continuous function of x. By [PTA, Section I.3], the map $x \mapsto \mathbb{E}^x f(X_t)$ being continuous when f is continuous implies that $(\mathbb{P}^x, X_t)$ is a strong Markov family of solutions. $\qquad \square$

3. Some useful techniques

In this section we want to provide a number of results that make proofs of uniqueness for the martingale problem easier. First, we show that if the diffusion coefficients are uniformly elliptic, then the drift coefficients do not matter. Second, we show that it is enough to look at λ-resolvents. Third, we prove that uniqueness of the martingale problem is a local property. Fourth, we see that it suffices to look at strong Markov solutions, and fifth, it is enough to look at 0-potentials in bounded domains. Finally, we examine time changes.

Let us show that for uniformly elliptic operators we may assume the drift coefficients are 0.

(3.1) Theorem. *Suppose $\mathcal{L}'$ is defined by (1.8) and suppose there is uniqueness for the martingale problem for $\mathcal{L}'$ started at x. If $\mathcal{L} \in \mathcal{N}(\Lambda_1, \Lambda_2)$, then there is uniqueness for the martingale problem for $\mathcal{L}$ started at x.*

Proof. Let $\mathbb{P}_1, \mathbb{P}_2$ be two solutions to the martingale problem for $\mathcal{L}$ started at x. From the definition of martingale problem, $\langle X^i, X^j \rangle_t = \int_0^t a_{ij}(X_s)\, ds$. Define $\mathbb{Q}_i$ on $\mathcal{F}_t$, $i = 1, 2$, by

$$d\mathbb{Q}_i/d\mathbb{P}_i = \exp\left(- \int_0^t (ba^{-1})(X_s)\, dX_s - \frac{1}{2} \int_0^t (ba^{-1}b^T)(X_s)\, ds \right),$$

where b^T denotes the transpose of b. A simple calculation shows that the quadratic variation of $\int_0^t (ba^{-1})(X_s)\, dX_s$ is $\int_0^t (ba^{-1}b^T)(X_s)\, ds$, so $d\mathbb{Q}_i/d\mathbb{P}_i$ is of the right form for use in the Girsanov theorem. If $f \in C^2$ and

$$M_t = f(X_t) - f(X_0) - \int_0^t \mathcal{L}f(X_s)\,ds, \tag{3.1}$$

then M_t is a local martingale under $\mathbb{P}_i$. By Itô's formula, the martingale part of M_t is the same as the martingale part of $\int_0^t \nabla f(X_s)\cdot dX_s$. We calculate

$$\left\langle \int_0^\cdot ba^{-1}(X_s)\,dX_s, M \right\rangle_t = \int_0^t \sum_{i,j=1}^d (ba^{-1})_j(X_s)\partial_i f(X_s)\,d\langle X^i, X^j\rangle_s$$

$$= \int_0^t \sum_{i=1}^d b_i(X_s)\partial_i f(X_s)\,ds.$$

Hence by the Girsanov theorem, under $\mathbb{Q}_i$ the process

$$M_t - \left(-\int_0^t b(X_s)\cdot \nabla f(X_s)\,ds \right) = f(X_t) - f(X_0) - \int_0^t \mathcal{L}' f(X_s)\,ds$$

is a local martingale. Clearly $\mathbb{Q}_i(X_0 = x) = 1$, so $\mathbb{Q}_i$ is a solution to the martingale problem for $\mathcal{L}'$ started at x. By the uniqueness assumption, $\mathbb{Q}_1 = \mathbb{Q}_2$. So if $A \in \mathcal{F}_t$,

$$\mathbb{P}_i(A) = \int_A \exp\left(\int_0^t (ba^{-1})(X_s)\,dX_s + \frac{1}{2}\int_0^t (ba^{-1}b^T)(X_s)\,ds \right) d\mathbb{Q}_i,$$

which implies $\mathbb{P}_1(A) = \mathbb{P}_2(A)$. $\qquad\qquad\square$

To prove uniqueness it turns out that it is sufficient to look at quantities which are essentially λ-potentials (that is, λ-resolvents). It will be convenient to introduce the notation

$$\mathcal{M}(\mathcal{L}, x) = \{\mathbb{P} : \mathbb{P} \text{ is a solution to the} \tag{3.2}$$
$$\text{martingale problem for } \mathcal{L} \text{ started at } x\}.$$

(3.2) Theorem. *Suppose for all $x \in \mathbb{R}^d$, $\lambda > 0$, and $f \in C^2(\mathbb{R}^d)$,*

$$\mathbb{E}_1 \int_0^\infty e^{-\lambda t} f(X_t)\,dt = \mathbb{E}_2 \int_0^\infty e^{-\lambda t} f(X_t)\,dt$$

whenever $\mathbb{P}_1, \mathbb{P}_2 \in \mathcal{M}(\mathcal{L}, x)$. Then for each $x \in \mathbb{R}^d$ the martingale problem for $\mathcal{L}$ has a unique solution.

Proof. By the uniqueness of the Laplace transform and the continuity of f and X_t, our hypothesis implies that $\mathbb{E}_1 f(X_t) = \mathbb{E}_2 f(X_t)$ for all $t > 0$ and $f \in C^2$ if $x \in \mathbb{R}^d$ and $\mathbb{P}_1, \mathbb{P}_2 \in \mathcal{M}(\mathcal{L}, x)$. A limit argument shows that equality holds for all bounded f. In other words, the one-dimensional distributions of X_t under $\mathbb{P}_1$ and $\mathbb{P}_2$ are the same.

We next look at the two-dimensional distributions. Suppose f, g are bounded and $0 < s < t$. For $i = 1, 2$, let $\mathbb{P}_{i,s}(A) = \mathbb{P}_i(A \circ \theta_s)$, and let $\mathbb{Q}_i$

be a regular conditional probability for $\mathbb{E}_{i,s}(\cdot \mid X_s)$. By Corollary 2.2, $\mathbb{Q}_i$ is a solution to the martingale problem for $\mathcal{L}$ started at X_s. By the first paragraph of this proof,

$$\mathbb{E}_{\mathbb{Q}_1} g(X_{t-s}) = \mathbb{E}_{\mathbb{Q}_2} g(X_{t-s}), \qquad \text{a.s.}$$

Since $\mathbb{Q}_1(A)$ is measurable with respect to the σ-field generated by the single random variable X_s for each A, then $\mathbb{E}_{\mathbb{Q}_1} g(X_{t-s})$ is also measurable with respect to the σ-field generated by X_s. So $\mathbb{E}_{\mathbb{Q}_1} g(X_{t-s}) = \varphi(X_s)$ for some function φ. Then

$$\begin{aligned}
\mathbb{E}_1 f(X_s) g(X_t) &= \mathbb{E}_1[f(X_s)\mathbb{E}_1(g(X_t) \mid X_s)] \\
&= \mathbb{E}_1 f(X_s)\mathbb{E}_{\mathbb{Q}_1}(g(X_{t-s})) = \mathbb{E}_1 f(X_s)\varphi(X_s).
\end{aligned}$$

By the uniqueness of the one-dimensional distributions, the right-hand side is equal to $\mathbb{E}_2 f(X_s)\varphi(X_s)$, which, similarly to the above, is equal to $\mathbb{E}_2 f(X_s) g(X_t)$. Hence the two-dimensional distributions of X_t under $\mathbb{P}_1$ and $\mathbb{P}_2$ are the same.

An induction argument shows that the finite dimensional distributions of X_t under $\mathbb{P}_1$ and $\mathbb{P}_2$ are the same. Since X_t has continuous paths, we deduce $\mathbb{P}_1 = \mathbb{P}_2$. $\qquad\square$

We now want to show that questions of uniqueness for martingale problems for elliptic operators are local questions. We start by giving a "piecing-together" lemma.

(3.3) Lemma. *Suppose $\mathcal{L}_1, \mathcal{L}_2$ are two elliptic operators with bounded coefficients. Let $S = \inf\{t : |X_t - x| \geq r\}$ and let $\mathbb{P}_1, \mathbb{P}_2$ be solutions to the martingale problems for $\mathcal{L}_1, \mathcal{L}_2$, respectively, started at x. Let $\mathbb{Q}_2$ be a regular conditional probability for $\mathbb{E}_{\mathbb{P}_{2S}}[\cdot \mid \mathcal{F}_S]$, where $\mathbb{P}_{2S}(A) = \mathbb{P}_2(A \circ \theta_S)$. Define $\overline{\mathbb{P}}$ by*

$$\overline{\mathbb{P}}(B \circ \theta_S \cap A) = \mathbb{E}_{\mathbb{P}_1}[\mathbb{Q}_2(B); A], \qquad A \in \mathcal{F}_S, B \in \mathcal{F}_\infty.$$

If the coefficients of $\mathcal{L}_1$ and $\mathcal{L}_2$ agree on $B(x, r)$, then $\overline{\mathbb{P}}$ is a solution to the martingale problem for $\mathcal{L}$ started at x.

$\overline{\mathbb{P}}$ represents the process behaving according to $\mathbb{P}_1$ up to time S and according to $\mathbb{P}_2$ after time S.

Proof. It is clear that the restriction of $\overline{\mathbb{P}}$ to $\mathcal{F}_S$ is equal to the restriction of $\mathbb{P}_1$ to $\mathcal{F}_S$. Hence

$$\overline{\mathbb{P}}(X_0 = x) = \mathbb{P}_1(X_0 = x) = 1.$$

If $f \in C^2$,

$$\begin{aligned}
M_t &= f(X_{t\wedge S}) - f(X_0) - \int_0^{t\wedge S} \mathcal{L}_1 f(X_s)\, ds \\
&= f(X_{t\wedge S}) - f(X_0) - \int_0^{t\wedge S} \mathcal{L}_2 f(X_s)\, ds
\end{aligned}$$

is a martingale under $\mathbb{P}_1$. Since for each t these random variables are $\mathcal{F}_S$ measurable, M_t is also a martingale under $\overline{\mathbb{P}}$. It remains to show that $N_t = f(X_{S+t}) - f(X_S) - \int_S^{S+t} \mathcal{L}_2 f(X_s) \, ds$ is a martingale under $\overline{\mathbb{P}}$. This follows from Proposition 2.1 and the definition of $\overline{\mathbb{P}}$. $\qquad\square$

(3.4) Theorem. *Suppose $\mathcal{L} \in \mathcal{N}(\Lambda_1, \Lambda_2)$. Suppose for each $x \in \mathbb{R}^d$ there exist $r_x > 0$ and $\mathcal{K}(x) \in \mathcal{N}(\Lambda_1, \Lambda_2)$ such that the coefficients of $\mathcal{K}(x)$ agree with those of $\mathcal{L}$ in $B(x, r_x)$ and the solution to the martingale problem for $\mathcal{K}(x)$ is unique for every starting point. Then the martingale problem for $\mathcal{L}$ started at any point has a unique solution.*

Proof. Fix x_0 and suppose $\mathbb{P}_1$ and $\mathbb{P}_2$ are two solutions to the martingale problem for $\mathcal{L}$ started at x_0. Suppose x_1 is such that $x_0 \in B(x_1, r_{x_1}/4)$. Let $S = \inf\{t : |X_t - x_1| > r_{x_1}/2\}$. Write $\mathbb{P}^{\mathcal{K}}$ for the solution to the martingale problem for $\mathcal{K}(x_1)$ started at x_0. Let $\mathbb{Q}_S^{\mathcal{K}}$ be the regular conditional probability defined as in (2.1). For $i = 1, 2$, define

$$\overline{\mathbb{P}}_i(B \circ \theta_S \cap A) = \mathbb{E}_i[\mathbb{Q}_S^{\mathcal{K}}(B); A], \qquad i = 1, 2, \quad A \in \mathcal{F}_S, B \in \mathcal{F}_\infty. \tag{3.3}$$

Since the coefficients of $\mathcal{L}$ and $\mathcal{K}(x_1)$ agree on $B(x_1, r_{x_1})$, by Lemma 3.3 applied to $\mathbb{P}_i$ and $\mathbb{P}^{\mathcal{K}}$, $\overline{\mathbb{P}}_i$ is a solution to the martingale problem for $\mathcal{K}(x_1)$ started at x_0. By the uniqueness assumption, they must both be equal to $\mathbb{P}^{\mathcal{K}}$. Hence the restriction of $\mathbb{P}_1$ and $\mathbb{P}_2$ to $\mathcal{F}_S$ must be the same, namely, the same as the restriction of $\mathbb{P}^{\mathcal{K}}$ to $\mathcal{F}_S$. We have thus shown that any two solutions to the martingale problem for $\mathcal{L}$ started at a point x_0 agree on $\mathcal{F}_S$ if $x_0 \in B(x_1, r_{x_1}/4)$ and $S = \inf\{t : |X_t - x_1| > r_{x_i}/2\}$.

Let $N > 0$. $\overline{B(x_0, N)}$ is compact and hence there exist finitely many points $x_1, \dots, x_m$ such that $\{B(x_i, r_{x_i}/4)\}$ is a cover for $\overline{B(x_0, N)}$. Let us define a measurable mapping $\psi : \overline{B(x_0, N)} \to \{1, \dots, m\}$ by letting $\psi(x)$ be the smallest index for which $x \in B(x_{\psi(x)}, r_{\psi(x)}/4)$. Let $S_0 = 0$ and $S_{i+1} = \inf\{t > S_i : X_t \notin B(\psi(X_{S_i}), r_{\psi(X(S_i))}/2)\}$. The S_i are thus stopping times describing when X_t has moved far enough to exit its current ball.

We now show that any two solutions $\mathbb{P}_1$ and $\mathbb{P}_2$ for the martingale problem for $\mathcal{L}$ started at x_0 agree on $\mathcal{F}_{S_i \wedge \tau(B(x_0, N))}$ for each i. We already have done the case $i = 1$ in the first paragraph of this proof.

Let $\mathbb{Q}_{i,S_1}$ be a regular conditional probability defined as in (2.1). If $A \in \mathcal{F}_{S_1}$ and $B \in (\mathcal{F}_\infty \circ \theta_{S_1}) \cap \mathcal{F}_{S_2}$, then

$$\mathbb{P}_i(A \cap B) = \mathbb{E}_i[\mathbb{Q}_{i,S_1}(B); A], \qquad i = 1, 2.$$

By Proposition 2.1, $\mathbb{Q}_{i,S_1}$ is a solution to the martingale problem for $\mathcal{L}$ started at X_{S_1}, so by what we have shown in the first paragraph $\mathbb{Q}_{1,S_1} = \mathbb{Q}_{2,S_1}$ on $(\mathcal{F}_\infty \circ \theta_{S_1}) \cap \mathcal{F}_{S_2}$. Since $\mathbb{Q}_{i,S_1}(B)$ is $\mathcal{F}_{S_1}$ measurable and $\mathbb{P}_1 = \mathbb{P}_2$ on $\mathcal{F}_{S_1}$, this shows $\mathbb{P}_1(A \cap B) = \mathbb{P}_2(A \cap B)$. The random variable $\int_0^{S_2} e^{-\lambda r} f(X_r) \, dr$ can be written

$$\int_0^{S_1} e^{-\lambda r} f(X_r) \, dr + e^{-\lambda S_1} \left(\int_0^{S_1} e^{-\lambda r} f(X_r) \, dr \circ \theta_{S_1} \right).$$

Hence $\mathbb{E}_1 \int_0^{S_2} e^{-\lambda r} f(X_r)\,dr = \mathbb{E}_2 \int_0^{S_2} e^{-\lambda r} f(X_r)\,dr$ whenever f is bounded and continuous and $\lambda > 0$. As in Theorem 3.2, this implies $\mathbb{P}_1 = \mathbb{P}_2$ on $\mathcal{F}_{S_2}$.

Using an induction argument, $\mathbb{P}_1 = \mathbb{P}_2$ on $\mathcal{F}_{S_i \wedge \tau(B(x_0, N))}$ for each i. Note that

$$r = \min_{1 \leq i \leq m} r_{x_i} > 0.$$

Since $S_{i+1} - S_i$ is greater than the time for X_t to move more than $r/4$, $S_i \uparrow \tau_{B(0,N)}$ by the continuity of the paths of X_t. Therefore $\mathbb{P}_1 = \mathbb{P}_2$ on $\mathcal{F}_{\tau(B(x_0, N))}$. Since N is arbitrary, this shows that $\mathbb{P}_1 = \mathbb{P}_2$. $\qquad\square$

It is often more convenient to work with strong Markov families. Recall the definition of $\mathcal{M}(\mathcal{L}, x)$ from (3.2).

(3.5) Theorem. *Let $\mathcal{L} \in \mathcal{N}(\Lambda, 0)$. Suppose there exists a strong Markov family $(\mathbb{P}_1^x, X_t)$ such that for each $x \in \mathbb{R}^d$, $\mathbb{P}_1^x$ is a solution to the martingale problem for $\mathcal{L}$ started at x. Suppose whenever $(\mathbb{P}_2^x, X_t)$ is another strong Markov family for which $\mathbb{P}_2^x \in \mathcal{M}(\mathcal{L}, x)$ for each x, we have $\mathbb{P}_1^x = \mathbb{P}_2^x$ for all x. Then for each x the solution to the martingale problem for $\mathcal{L}$ started at x is unique.*

In other words, if we have uniqueness within the class of strong Markov families, then we have uniqueness.

Proof. Let f be bounded and continuous, $\lambda > 0$, and $x \in \mathbb{R}^d$. Let $\mathbb{P}$ be any solution to the martingale problem for $\mathcal{L}$ started at x. By Theorem V.8.2, there exists a sequence a_{ij}^n converging to a_{ij} almost everywhere as $n \to \infty$ such that the coefficients of the a_{ij}^n are C^2, $\mathcal{L}^n \in \mathcal{N}(\Lambda, 0)$, and if $\widetilde{\mathbb{P}}_n^x$ is a solution to the martingale problem for $\mathcal{L}^n$ started at x,

$$\widetilde{\mathbb{E}}_n^x \int_0^\infty e^{-\lambda t} f(X_t)\,dt \to \mathbb{E} \int_0^\infty e^{-\lambda t} f(X_t)\,dt. \tag{3.4}$$

By Theorem 2.3, there exists a subsequence n_k such that $\widetilde{\mathbb{P}}_{n_k}^x$ converges weakly for all x, and if we call the limit $\widetilde{\mathbb{P}}^x$, then $(\widetilde{\mathbb{P}}^x, X_t)$ is a strong Markov family of solutions. By our hypothesis, $\widetilde{\mathbb{P}}^x = \mathbb{P}_1^x$. Using the weak convergence of $\widetilde{\mathbb{P}}_{n_k}^x$ to $\mathbb{P}_1^x$,

$$\widetilde{\mathbb{E}}_{n_k}^x \int_0^\infty e^{-\lambda t} f(X_t)\,dt \to \mathbb{E}_1^x \int_0^\infty e^{-\lambda t} f(X_t)\,dt.$$

Combining with (3.4), $\int_0^\infty e^{-\lambda t} f(X_t)\,dt$ has the same expectation under $\mathbb{P}$ and $\mathbb{P}_1^x$. Our result now follows by Theorem 3.2. $\qquad\square$

Besides equality of λ-potentials, it is enough to show equality of 0-potentials on bounded domains.

(3.6) Theorem. *Let $\mathcal{L} \in \mathcal{N}(\Lambda, 0)$. Suppose there exists a strong Markov family $(\mathbb{P}_1^x, X_t)$ such that for each $x \in \mathbb{R}^d$, $\mathbb{P}_1^x$ is a solution to the martingale problem* *for $\mathcal{L}$ started at x. Suppose whenever $(\mathbb{P}_2^x, X_t)$ is another strong Markov family for which $\mathbb{P}_2^x \in \mathcal{M}(\mathcal{L}, x)$ for each x, we have: for all f bounded and continuous and $M > 0$,*

$$\mathbb{E}_1^x \int_0^{\tau_{B(0,M)}} f(X_t)\, dt = \mathbb{E}_2^x \int_0^{\tau_{B(0,M)}} f(X_t)\, dt.$$

Then there is uniqueness for the martingale problem for $\mathcal{L}$ at each starting point x.

Proof. Let

$$G_i^\lambda f(x) = \mathbb{E}_i^x \int_0^{\tau_{B(0,M)}} e^{-\lambda t} f(X_t)\, dt, \qquad i = 1, 2.$$

By Proposition V.2.3 and scaling, there exists c_1 such that

$$\|G_i^\lambda\|_\infty \leq \left(\sup_x \mathbb{E}_i^x \tau_{B(0,M)} \right) \|f\|_\infty \leq c_1 \|f\|_\infty,$$

with c_1 independent of λ. Since $(\mathbb{P}_i^x, X_t)$ is a strong Markov family, $(\mathbb{P}_i^x, X_{t \wedge \tau(B(0,M))})$ has the Markov property. By Corollary IV.2.6, we have

$$G_i^\lambda G_i^\mu = \frac{G_i^\lambda - G_i^\mu}{\mu - \lambda}$$

and

$$G_i^\lambda f = G_i^\mu f + (\lambda - \mu)(G_i^\mu)^2 f + (\lambda - \mu)^2 (G_i^\mu)^3 f + \cdots$$

as long as $|\lambda - \mu| \leq 1/2c_1$. Our hypothesis is that $G_1^0 = G_2^0$. So for $\lambda < 1/2c_1$, we have $G_1^\lambda f = G_2^\lambda f$. By the uniqueness of the Laplace transform,

$$\mathbb{E}_1^x f(X_{t \wedge \tau(B(0,M))}) = \mathbb{E}_2^x f(X_{t \wedge \tau(B(0,M))})$$

for almost every t. By continuity, this equality must hold for all t. Since $(\mathbb{P}_i^x, X_{t \wedge \tau(B(0,M))})$ is a strong Markov family, this suffices to show that the finite dimensional distributions of $X_{t \wedge \tau(B(0,M))}$ under $\mathbb{P}_1^x$ and $\mathbb{P}_2^x$ are the same. Since M is arbitrary, this shows equality. Now apply Theorem 3.5. $\square$

Finally, we show that time changing a process preserves uniqueness.

(3.7) Theorem. *Let $\mathcal{L} \in \mathcal{N}(\Lambda_1, \Lambda_2)$. Suppose there exists $r : \mathbb{R}^d \to (0, \infty)$ and c_1 such that $c_1^{-1} \leq r(x) \leq c_1$ for all x. If there exists a unique solution to the martingale problem for $\mathcal{L}$ started at x, then there exists a unique solution to the martingale problem for $r\mathcal{L}$ started at x.*

Here $r\mathcal{L}$ is defined by $(r\mathcal{L})f(x) = r(x)\mathcal{L}f(x)$.

Proof. For simplicity, let us do the case where the drift coefficients are zero. If $\mathbb{P}_1$ and $\mathbb{P}_2$ are two solutions to the martingale problem for $r\mathcal{L}$ started at x, and σ is a positive definite square root of a, then under both $\mathbb{P}_1$ and $\mathbb{P}_2$, each component of X_t is a local martingale, and $d\langle X^i, X^j\rangle_t = r(X_t)a_{ij}(X_t)\,dt$. Let $A_t = \int_0^t r(X_s)\,ds$, $B_t = \inf\{u : A_u > t\}$, and $Y_t = X_{B_t}$. If $f \in C^2$, then

$$\int_0^{B_t} r(X_s)\mathcal{L}f(X_s)\,ds = \int_0^{B_t} r(X_s)\mathcal{L}f(X_s)\frac{ds}{dA_s}\,dA_s$$

$$= \int_0^t \mathcal{L}f(X_{B_u})\,du.$$

Hence

$$f(Y_t) - f(Y_0) - \int_0^t \mathcal{L}f(Y_u)\,du$$

$$= f(X_{B_t}) - f(X_0) - \int_0^{B_t} r(X_s)\mathcal{L}f(X_s)\,ds$$

is a local martingale. So the law of Y_t under $\mathbb{P}_i$ is a solution to the martingale problem for $\mathcal{L}$ started at x, $i = 1, 2$. By the uniqueness hypothesis, it follows that the law of Y_t under $\mathbb{P}_1$ and the law of Y_t under $\mathbb{P}_2$ are the same. A calculation similar to the one above shows that if $C_t = \int_0^t r^{-1}(Y_s)\,ds$ and $D_t = \inf\{u : C_u > t\}$, then $X_t = Y_{D_t}$. Moreover, D_t is measurable with respect to the σ-fields generated by Y_t. Hence the law of X_t is the same under $\mathbb{P}_1$ and $\mathbb{P}_2$, which means $\mathbb{P}_1 = \mathbb{P}_2$. $\square$

4. Some uniqueness results

We present some of the cases for which uniqueness of the martingale problem is known. We assume $\mathcal{L} \in \mathcal{N}(\Lambda_1, \Lambda_2)$ for some $\Lambda_1 > 0$, and by virtue of Theorem 3.1, we may take $\Lambda_2 = 0$ without loss of generality.

(4.1) Theorem. *Suppose $d \geq 3$. There exists ε_d (depending only on the dimension d) with the following property: if*

$$\sup_{i,j} \sup_x |a_{ij}(x) - \delta_{ij}| < \varepsilon_d,$$

then there exists a unique solution to the martingale problem for $\mathcal{L}$ started at any $x \in \mathbb{R}^d$.

Proof. Let $\mathbb{P}_1$, $\mathbb{P}_2$ be any two solutions to the martingale problem for $\mathcal{L}$ started at x. Define $G_i^\lambda f(x) = \mathbb{E}_i^x \int_0^\infty e^{-\lambda t} f(X_t)\,dt$. If $f \in C^2$ is bounded with bounded first and second partial derivatives, then by Itô's formula,

$$f(X_t) = f(X_0) + \text{ martingale } + \int_0^t \mathcal{L}f(X_s)\,ds.$$

Multiplying by $e^{-\lambda t}$, taking the expectation with respect to $\mathbb{P}_i$, and integrating over t from 0 to ∞,

$$\mathbb{E}_i \int_0^\infty e^{-\lambda t} f(X_t)\, dt = \frac{1}{\lambda} f(x) + \mathbb{E}_i \int_0^\infty e^{-\lambda t} \int_0^t \mathcal{L}f(X_s)\, ds\, dt \qquad (4.1)$$

$$= \frac{1}{\lambda} f(x) + \mathbb{E}_i \int_0^\infty \mathcal{L}f(X_s) \int_s^\infty e^{-\lambda t}\, dt\, ds$$

$$= \frac{1}{\lambda} f(x) + \frac{1}{\lambda} \mathbb{E}_i \int_0^\infty e^{-\lambda s} \mathcal{L}f(X_s)\, ds.$$

Set

$$u^\lambda(z) = \int_0^\infty e^{-\lambda t} \left((2\pi t)^{-d/2} e^{-z^2/2t} \right) dt,$$

the λ-potential density of Brownian motion. Let $U^\lambda f(x) = \int f(y) u^\lambda(x-y)\, dy$, the λ-potential of f with respect to Brownian motion. Then set

$$\mathcal{B} = \frac{1}{2} \sum_{i,j=1}^d (a_{ij}(x) - \delta_{ij}) \partial_{ij} f(x). \qquad (4.2)$$

If $f = U^\lambda g$ for $g \in C^2$ with compact support, then by Corollary IV.2.6,

$$U^\lambda g = U^0(g - \lambda U^\lambda g).$$

By [PTA, Proposition II.3.3],

$$\frac{\Delta U^\lambda g}{2} = \lambda U^\lambda g - g = \lambda f - g.$$

Since $\mathcal{L}f = (1/2)\Delta f + \mathcal{B}f$, we have from (4.1) that

$$G_i^\lambda f = \lambda^{-1} f(x) + \lambda^{-1} G_i^\lambda \left(\frac{\Delta U^\lambda g}{2} + \mathcal{B}f \right)(x)$$

$$= \lambda^{-1} f(x) + \lambda^{-1} G_i^\lambda (\lambda f - g) + \lambda^{-1} G_i^\lambda \mathcal{B}f,$$

or

$$G_i^\lambda g = f(x) + G_i^\lambda \mathcal{B}f(x).$$

Hence

$$G_i^\lambda g = U^\lambda g(x) + G_i^\lambda \mathcal{B}U^\lambda g(x), \qquad i = 1, 2. \qquad (4.3)$$

(We remark that if we were to iterate (4.3), that is, substitute for G_i^λ on the right-hand side, we would be led to

$$G_i^\lambda g = U^\lambda g + U^\lambda \mathcal{B}U^\lambda g + \cdots,$$

which indicates that (4.3) is essentially variation of parameters in disguise.) We return to the proof. Let

$$\rho = \sup_{\|g\|_d \leq 1} |G_1^\lambda g - G_2^\lambda g|.$$

By Theorem V.6.2, $\rho < \infty$. Taking the difference of (4.3) with $i = 1$ and $i = 2$, we have

$$G_i^\lambda g - G_2^\lambda g = (G_1^\lambda - G_2^\lambda)(\mathcal{B}U^\lambda g). \qquad (4.4)$$

The right-hand side is bounded by $\rho\|\mathcal{B}U^\lambda g\|_d$. By (III.8.1),

$$\|\mathcal{B}U^\lambda g\|_d \le \varepsilon_d \sum_{i,j=1}^{d} \|\partial_{ij}U^\lambda g\|_d \le \varepsilon_d c_1 d^2 \|g\|_d \le (1/2)\|g\|_d$$

if we take $\varepsilon_d < 1/2c_1 d^2$. Hence

$$|G_1^\lambda g - G_2^\lambda g| \le (\rho/2)\|g\|_d.$$

If we now take the supremum of the left-hand side over $g \in C^2$ with $\|g\|_d \le 1$, we obtain $\rho \le \rho/2$. Since we observed that $\rho < \infty$, this means that $\rho = 0$, or $G_1^\lambda g = G_2^\lambda g$ if $g \in L^d$. In particular, this holds if g is continuous with compact support. By a limit argument, this holds for all continuous bounded g. This is true for every starting point $x \in \mathbb{R}^d$, so by Theorem 3.2, $\mathbb{P}_1 = \mathbb{P}_2$.   □

(4.2) Corollary. *Let C be a positive definite matrix. There exists ε_d such that if*

$$\sup_{i,j} \sup_x |a_{ij}(x) - C_{ij}| < \varepsilon_d,$$

then there exists a unique solution to the martingale problem for $\mathcal{L}$ started at any $x \in \mathbb{R}^d$.

Proof. Let $\sigma(x)$ be a positive definite square root of $a(x)$ and $C^{1/2}$ a positive definite square root of C. By Theorem 1.1, to establish uniqueness it suffices to establish weak uniqueness of the stochastic differential equation $dX_t = \sigma(X_t)\,dW_t$. If X_t is a solution to this stochastic differential equation, it is easy to see that $Y_t = C^{-1/2}X_t$ is a solution to $dY_t = (\sigma C^{-1/2})(Y_t)\,dW_t$ and conversely. By Theorem 1.1 again, weak uniqueness for the latter stochastic differential equation will follow if we have weak uniqueness for the martingale problem for $\mathcal{L}^C$, where the coefficients of $\mathcal{L}^C$ are $C^{-1}a_{ij}$. The assumption $|a_{ij}(x) - C_{ij}| < \varepsilon_d$ implies $|C^{-1}a_{ij}(x) - \delta_{ij}| < c_1\varepsilon_d$, where c_1 depends on C. The result follows by Theorem 4.1 by taking ε_d sufficiently small.   □

We now can prove the important result due to Stroock and Varadhan.

(4.3) Theorem. *If $\mathcal{L} \in \mathcal{N}(\Lambda_1, \Lambda_2)$ and the a_{ij} are continuous, then the martingale problem for $\mathcal{L}$ started at x has a unique solution.*

Proof. By Theorem 3.1, we may suppose that $\Lambda_2 = 0$. If $x \in \mathbb{R}^d$, let $C = a(x)$ and then choose r_x such that if $y \in B(x, 2r_x)$, then $|a_{ij}(y) - a_{ij}(x)| < \varepsilon_d$ for $i,j = 1,\ldots,d$, where ε_d is given by Corollary 4.2. Let $a_{ij}^x(y)$ be continuous functions that agree with $a_{ij}(y)$ on $B(x, r_x)$ and such that if

$$\mathcal{K}^x f(z) = \sum_{i,j=1}^{d} a_{ij}^x(z)\partial_{ij}f(z),$$

then $\mathcal{K}^x \in \mathcal{N}(\Lambda_1, 0)$, and

$$\sup_{i,j} \sup_{y} |a_{ij}^x(y) - a_{ij}(x)| < \varepsilon_d.$$

By Corollary 4.2, we have uniqueness of the martingale problem for $\mathcal{K}^x$ starting at any point in $\mathbb{R}^d$. Moreover, the coefficients of $\mathcal{K}^x$ agree with those of $\mathcal{L}$ inside $B(x, r_x)$. The conclusion now follows by Theorem 3.4. $\square$

Next we prove uniqueness for the case where the diffusion coefficients are continuous except at a single point.

(4.4) Theorem. *Suppose $\mathcal{L} \in \mathcal{N}(\Lambda_1, \Lambda_2)$ and the a_{ij} are continuous except at $x = 0$. Then for each x there exists a unique solution to the martingale problem for $\mathcal{L}$ started at x.*

Proof. Again assume without loss of generality that $\Lambda_2 = 0$. We will use Theorem 3.6. So we may suppose $(\mathbb{P}_1^x, X_t)$ and $(\mathbb{P}_2^x, X_t)$ are two strong Markov families, $M \geq 0$, and we must show $G_1 = G_2$, where

$$G_i f(x) = \mathbb{E}_i^x \int_0^{\tau(B(0,M))} f(X_s) \, ds.$$

If we set $\widehat{a}_{ij}$ to be continuous and equal to a_{ij} outside of a neighborhood $B(0, \delta)$ of 0, and we define $\widehat{\mathcal{L}}$ by (1.1) with a_{ij} replaced by $\widehat{a}_{ij}$, then we have uniqueness of the martingale problem for $\widehat{\mathcal{L}}$ by Theorem 4.3. By the proof of Theorem 3.4, $\mathbb{P}_1^x = \mathbb{P}_2^x$ on $\mathcal{F}_{T(B(0,\delta))}$ if $|x| > \delta$. Since δ is arbitrary, $\mathbb{P}_1^x = \mathbb{P}_2^x$ on $\mathcal{F}_{T_0}$, where we have written T_0 for $T_{\{0\}}$, the hitting time to 0.

By the strong Markov property,

$$G_i f(x) = \mathbb{E}_i^x \int_0^{\tau(B(0,M)) \wedge T_0} f(X_s) \, ds$$
$$+ \mathbb{E}_i^x \left[\mathbb{E}_i^0 \int_0^{\tau(B(0,M))} f(X_s) \, ds; T_0 < \tau_{B(0,M)} \right].$$

By what we have just shown, the first term does not depend on i. So to show uniqueness we must show that $\mathbb{E}_i^0 \int_0^{\tau(B(0,M))} f(X_s) \, ds$ does not depend on i either; since $(T_0 < \tau_{B(0,M)})$ is in $\mathcal{F}_{T_0}$, then $\mathbb{P}_1(T_0 < \tau_{B(0,M)}) = \mathbb{P}_2(T_0 < \tau_{B(0,M)})$, and the result will follow by Theorem 3.6.

By a limit argument, it suffices to consider nonnegative f that are 0 in $B(0, \delta)$ for some δ. Let $\varepsilon < \delta$. Then

$$\mathbb{E}_i^0 \int_0^{\tau(B(0,M))} f(X_s)\,ds$$

$$= \mathbb{E}_i^0 \int_0^{\tau(B(0,\varepsilon))} f(X_s)\,ds + \mathbb{E}_i^0 G_i f(X_{\tau(B(0,\varepsilon))})$$

$$= \mathbb{E}_i^0 \int_0^{\tau(B(0,\varepsilon))} f(X_s)\,ds$$

$$+ \mathbb{E}_i^0 \mathbb{E}^{X(\tau(B(0,\varepsilon)))}\left[\int_0^{\tau(B(0,M))\wedge T_0} f(X_s)\,ds \right]$$

$$+ \mathbb{E}_i^0 \mathbb{E}_i^{X(\tau(B(0,\varepsilon)))}\left[\mathbb{E}_i^{X(T_0)} \int_0^{\tau(B(0,M))} f(X_s)\,ds; T_0 < \tau_{B(0,M)} \right].$$

Set $\overline{G}f(x) = \mathbb{E}^x \int_0^{\tau(B(0,M))\wedge T_0} f(X_s)\,ds$; this, as we have seen, does not depend on i. Therefore

$$G_i f(0) = \mathbb{E}_i^0 \int_0^{\tau(B(0,\varepsilon))} f(X_s)\,ds + \mathbb{E}_i^0 \overline{G}f(X_{\tau(B(0,\varepsilon))})$$

$$+ \mathbb{E}_i^0 \mathbb{E}_i^{X(\tau(B(0,\varepsilon)))}[G_i f(0); T_0 < \tau_{B(0,M)}].$$

Since f is 0 inside $B(0,\delta)$, the first term on the right is 0. Let $h(x) = \mathbb{P}^x(T_0 > \tau(B(0,M)))$; again this expression does not depend on i. By the support theorem, $h(x)$ is positive if $x \neq 0$. We then have

$$G_i f(0) = \mathbb{E}_i^0 \overline{G}f(X_{\tau(B(0,\varepsilon))}) + G_i f(0) - \mathbb{E}_i^0 h(X_{\tau(B(0,\varepsilon))}) G_i f(0),$$

and hence

$$G_i f(0) = \frac{\mathbb{E}_i^0 \overline{G}f(X_{\tau(B(0,\varepsilon))})}{\mathbb{E}_i^0 h(X_{\tau(B(0,\varepsilon))})}. \tag{4.5}$$

Neither $\overline{G}f$ nor h depends on i. We will show that if x_n is any sequence with $|x_n| \to 0$, then

$$\lim_{n\to\infty} \overline{G}f(x_n)/h(x_n) \tag{4.6}$$

exists. This will imply $\lim_{x\to 0} \overline{G}f(x)/h(x)$ exists. Letting $\varepsilon \to 0$ and using (4.5), we then conclude that $G_i f(0)$ does not depend on i, and the theorem will follow.

Let $r < \delta$. By looking at $C(\overline{G}f/h) + D$ for suitable C and D, we may assume that the infimum of $C(\overline{G}f/h) + D$ on $\partial B(0,r)$ is 0, the supremum is 1, and, moreover, that there exists $x_0 \in \partial B(0,r/2)$ such that $(C(\overline{G}f/h) + D)(x_0) \geq 1/2$. Observe that $\overline{G}f(X(t \wedge \tau(B(0,\delta)) \wedge T_0))$ is a martingale and the same is true if $\overline{G}f$ is replaced by h. Hence $C(\overline{G}f/h) + D = (C\overline{G}f + Dh)/h$ is still the ratio of functions, each of whose composition with $X_{t\wedge\tau(B(0,\delta))\wedge T_0}$ is a martingale. By our choice of C and D, $C\overline{G}f + Dh \geq 0$ on $\partial B(0,r)$, and hence by optional stopping is greater than or equal to 0 in $B(0,r)$. Similarly, $C\overline{G}f + Dh \leq Dh$ in $B(0,r)$. By the Harnack inequality, Corollary V.7.7, there exist c_1 and c_2 such that if $x \in \partial B(0,r/2)$, then

$$h(x) \leq c_1 h(x_0), \qquad (C\overline{G}f + Dh)(x) \geq c_2(C\overline{G}f + Dh)(x_0).$$

By scaling, c_1 and c_2 are independent of r. So

$$(C(\overline{G}f/h) + D)(x) \geq (c_2/c_1)(C(\overline{G}f/h) + D)(x_0) \geq c_2/2c_1,$$

or

$$\underset{\partial B(0,r/2)}{\text{Osc}}\,(C\overline{G}f/h + D) \leq 1 - c_2/2c_1 < 1.$$

(Recall $\text{Osc}_A\, g = \sup_A g - \inf_A g$.) It follows that

$$\underset{\partial B(0,r/2)}{\text{Osc}}\,(\overline{G}f/h) \leq \rho \underset{\partial B(0,r)}{\text{Osc}}\,(\overline{G}f/h)$$

for $\rho = 1 - c_2/2c_1$, which is independent of r. As in the proof of Theorem V.7.5, this implies that the limit in (4.6) exists. $\qquad\square$

Observe that the value of $a_{ij}(0)$ plays no role in the proof of Theorem 4.4. This is a special case of the fact that changing the values of the a_{ij} on a set of measure 0 makes no difference to the martingale problem for uniformly elliptic operators. To be more precise, suppose $\mathcal{L}_1, \mathcal{L}_2 \in \mathcal{N}(\Lambda, 0)$ and $\{x : \mathcal{L}_1 f(x) \neq \mathcal{L}_2 f(x) \text{ for some bounded } f \in C^2\}$ has zero Lebesgue measure. If $\mathbb{P}$ is a solution to the martingale problem for $\mathcal{L}_1$, then by Theorem V.6.2,

$$\mathbb{E} \int_0^\infty e^{-\lambda t} |\mathcal{L}_1 f(X_t) - \mathcal{L}_2 f(X_t)|\, dt = 0.$$

So for bounded $f \in C^2$

$$f(X_t) - f(X_0) - \int_0^t \mathcal{L}_1 f(X_s)\, ds = f(X_t) - f(X_0) - \int_0^t \mathcal{L}_2 f(X_s)\, ds$$

almost surely, from which it follows that $\mathbb{P}$ is also a solution to the martingale problem for $\mathcal{L}_2$.

The final case we wish to consider is the case when $\mathcal{L} \in \mathcal{N}(\Lambda, 0)$, no smoothness assumptions are made on the a_{ij}, and the dimension d is 2.

(4.5) Theorem. *Suppose $d = 2$, $\mathcal{L} \in \mathcal{N}(\Lambda, 0)$. The martingale problem for $\mathcal{L}$ started at any x is unique.*

Proof. By the time change result Theorem 3.7, we may assume that trace $a(x) = 2$. We follow the proof of Theorem 4.1. We will be done once we show

$$\|\mathcal{B}U^\lambda g\|_2 \leq \rho \|g\|_2 \tag{4.7}$$

for some $\rho < 1$, where $\mathcal{B}$ is defined by (4.2). We have

$$\mathcal{B}U^\lambda g(x) = \frac{1}{2}(a_{11}(x) - 1)\partial_{11} U^\lambda g(x) + \frac{1}{2}(a_{22}(x) - 1)\partial_{22} U^\lambda g(x)$$
$$+ a_{12}(x)\partial_{12} U^\lambda g(x).$$

Since $a_{11}(x) + a_{22}(x) = 2$,

$$|\mathcal{B}U^\lambda g(x)|$$
$$\leq \frac{1}{2}\left\{|a_{11}(x) - 1|\,|(\partial_{11}U^\lambda g - \partial_{22}U^\lambda g)(x)| + |a_{12}(x)|\,|2\partial_{12}U^\lambda g|\right\}$$
$$\leq \frac{1}{2}\left(|a_{11}(x) - 1|^2 + |a_{12}(x)|^2\right)^{1/2}$$
$$\times \left((\partial_{11}U^\lambda g(x) - \partial_{22}U^\lambda g(x))^2 + 4(\partial_{12}U^\lambda g(x))^2\right)^{1/2}.$$

By Parseval's identity,

$$\int_{\mathbb{R}^2} \left([\partial_{11}U^\lambda g(x) - \partial_{22}U^\lambda g(x)]^2 + 4[\partial_{12}U^\lambda g(x)]^2\right) dx$$
$$= c_1 \int_{\mathbb{R}^2} \left(|[\partial_{11}U^\lambda g - \partial_{22}U^\lambda g]\widehat{\ }(\xi)|^2 + 4|[\partial_{12}U^\lambda g]\widehat{\ }(\xi)|^2\right) d\xi,$$

where $[\,\cdot\,]\widehat{\ }$ denotes Fourier transform. The right-hand side is equal to

$$c_1 \int \left|\frac{-\xi_1^2\widehat{g}(\xi)}{\lambda + |\xi|^2/2} - \frac{-\xi_2^2\widehat{g}(\xi)}{\lambda + |\xi|^2/2}\right|^2 + 4\left|\frac{-\xi_1\xi_2\widehat{g}(\xi)}{\lambda + |\xi|^2/2}\right|^2 d\xi$$
$$= c_1 \int \frac{(\xi_1^2 + \xi_2^2)^2}{(\lambda + |\xi|^2/2)^2}|\widehat{g}(\xi)|^2\, d\xi \leq 4c_1 \int |\widehat{g}(\xi)|^2\, d\xi = 4\|g\|_2^2.$$

Since trace $a = 2$,

$$(a_{11} - 1)^2 + a_{12}(x)^2 = -\det(I - a(x)).$$

If U is an orthogonal matrix such that $D = U^{-1}a(x)U$ is a diagonal matrix, then trace $D = $ trace $a(x) = 2$ and $\det(I - a(x)) = \det(I - D)$. Since $\mathcal{L}$ is uniformly elliptic, $-\det(I - D) \leq c_2 < 1$ for a constant c_2 independent of x. Combining,

$$\|\mathcal{B}U^\lambda g\|_2 \leq \frac{1}{2}c_2^{1/2}(2\|g\|_2) = c_2^{1/2}\|g\|_2,$$

which proves (4.7) with $\rho = c_2^{1/2}$. $\qquad\qquad\qquad\qquad\qquad\qquad\quad\square$

There are some other cases where uniqueness is known. If the a_{ij} are continuous except on a set that is small in a certain sense, we have uniqueness (Krylov [4], Safonov [1]). If $\mathcal{L} \in \mathcal{N}(\Lambda_1, \Lambda_2)$ and $\mathbb{R}^d$ can be divided into the union of finitely many disjoint polyhedra such that the a_{ij} are constant on the interior of each polyhedron, then uniqueness holds (Bass and Pardoux [1]). Gao [1] showed uniqueness when $\mathbb{R}^d$ is divided into two by a hyperplane and the a_{ij} are uniformly continuous in each half space, but not necessarily on the boundary. On the other hand, Nadirashvili [1] has constructed an example in $\mathbb{R}^3$ of an elliptic operator in $\mathcal{N}(\Lambda, 0)$ for some $\Lambda > 0$ for which uniqueness does not hold.

5. Consequences of uniqueness

We mention some conclusions that one can draw when uniqueness holds.

(5.1) Theorem. *Suppose there exists a unique solution $\mathbb{P}^x$ to the martingale problem for $\mathcal{L}$ started at x for each $x \in \mathbb{R}^d$. Then $(\mathbb{P}^x, X_t)$ forms a strong Markov family.*

Proof. This is a consequence of Theorem 1.1 and Theorem I.5.1. □

Uniqueness implies some convergence results.

(5.2) Theorem. *Suppose $\mathcal{L}_n \in \mathcal{N}(\Lambda_1, \Lambda_2)$ and the diffusion coefficients a_{ij}^n converge to a_{ij} almost everywhere, and similarly for the drift coefficients b_i^n. Suppose $x_n \to x$, $\mathbb{P}$ is the unique solution to the martingale problem for $\mathcal{L}$ started at x, and for each n, $\mathbb{P}_n$ is a solution to the martingale problem for $\mathcal{L}_n$ started at x_n. Then $\mathbb{P}_n$ converges weakly to $\mathbb{P}$.*

Proof. By Theorem V.2.4, the probability measures $\mathbb{P}_n$ are tight. By Theorem 1.3 and its proof, any subsequential limit point is a solution to the martingale problem for $\mathcal{L}$ started at x. By the uniqueness hypothesis, any subsequential limit point must be equal to $\mathbb{P}$; this implies that the whole sequence converges to $\mathbb{P}$. □

A more interesting application is to Markov chains converging to a diffusion. See Stroock and Varadhan [2].

(5.3) Theorem. *If there exists a unique solution $\mathbb{P}^x$ to the martingale problem for all x and f is continuous on $\partial B(0,1)$, then $u(x) = \mathbb{E}^x f(X_{\tau(B(0,1))})$ is a continuous function on $\overline{B}(0,1)$.*

Proof. This follows by Theorem III.3.4. □

If $\mathcal{L}$ is an elliptic operator in $\mathcal{N}(\Lambda_1, \Lambda_2)$, u is said to be a *good solution* for the equation $(\lambda - \mathcal{L})u = f$ if whenever $\mathcal{L}_n$ is a sequence in $\mathcal{N}(\Lambda_1, \Lambda_2)$ whose drift and diffusion coefficients are smooth and converge almost everywhere to those of $\mathcal{L}$, then the solution to $(\lambda - \mathcal{L}_n)u_n = f$ converges to u uniformly on compacts. When a good solution exists, *weak uniqueness* is said to hold for the equation $(\lambda - \mathcal{L})u = f$. The point here is that the solution u is stable under slight perturbations of the coefficients of $\mathcal{L}$.

(5.4) Theorem. *Suppose $\mathcal{L} \in \mathcal{N}(\Lambda_1, \Lambda_2)$. The martingale problem for $\mathcal{L}$ started at x has a unique solution for every $x \in \mathbb{R}^d$ if and only if weak uniqueness holds for the equation $(\lambda - \mathcal{L})u = f$ for all f continuous and bounded.*

Proof. Let us suppose we have uniqueness for the martingale problem. Let f be bounded and continuous. By Theorem 5.2, if $x \in \mathbb{R}^d$, $\mathbb{P}_n^x$ converges weakly to $\mathbb{P}^x$, where $\mathbb{P}_n^x$ is the solution to the martingale problem for $\mathcal{L}_n$ started at x. The solution to $(\lambda - \mathcal{L}_n)u_n = f$ is given by $u_n(x) = \mathbb{E}_n^x \int_0^\infty e^{-\lambda t} f(X_t)\, dt$. By weak convergence, this converges to $u(x) = \mathbb{E}^x \int_0^\infty e^{-\lambda t} f(X_t)\, dt$. We thus have pointwise convergence of u_n to u. By Theorem V.8.1, the u_n are equicontinuous, which implies that the convergence is uniform on compacts.

Conversely, suppose we have weak uniqueness. By Theorem 3.2, to show uniqueness of the martingale problem, we must show that if f is bounded and continuous, then the value of $\mathbb{E} \int_0^\infty e^{-\lambda t} f(X_t)\, dt$ must be the same no matter which solution $\mathbb{P}$ of the martingale problem started at x is used. Let u be a good solution to $(\lambda - \mathcal{L})u = f$. By Theorems V.8.2 and V.8.8, there exist $\mathcal{L}_n \in \mathcal{N}(\Lambda_1, \Lambda_2)$ with smooth coefficients such that if $\mathbb{P}_n$ is the solution to the martingale problem for $\mathcal{L}_n$ started at x, then

$$u_n(x) = \mathbb{E}_n \int_0^\infty e^{-\lambda t} f(X_t)\, dt \to \mathbb{E} \int_0^\infty e^{-\lambda t} f(X_t)\, dt.$$

Since u is a good solution, $u_n(x) \to u(x)$. Hence $\mathbb{E} \int_0^\infty e^{-\lambda t} f(X_t)\, dt = u(x)$, and the value of $\mathbb{E} \int_0^\infty e^{-\lambda t} f(X_t)\, dt$ is determined uniquely. $\qquad\square$

6. Submartingale problems

Suppose X_t is a solution to

$$dX_t = \sigma(X_t)\, dW_t + b(X_t)\, dt + v(X_t)\, dL_t, \qquad X_t \in \overline{D}, \quad X_0 = x_0, \qquad (6.1)$$

where L_t is a continuous nondecreasing process that increases only when $X_t \in \partial D$ and v is a vector field on ∂D. If f is C^2 in D and C^1 in $\overline{D}$, Itô's formula says that

$$f(X_t) = f(X_0) + \text{ martingale } + \int_0^t \mathcal{L}f(X_s)\, ds \qquad (6.2)$$

$$+ \int_0^t (\nabla f \cdot v)(X_s)\, dL_s,$$

where $\mathcal{L}$ is defined by (1.1). If in addition,

$$\nabla f \cdot v \geq 0 \text{ on } \partial D, \qquad (6.3)$$

then the process $f(x_t) - f(X_0) - \int_0^t \mathcal{L}f(X_s)\, ds$ is a submartingale since $\int_0^t (\nabla f \cdot v)(X_s)\, dL_s$ is nondecreasing.

A probability measure $\mathbb{P}$ on $C([0,\infty), \overline{D})$ is called a solution to the *submartingale problem* corresponding to $(\mathcal{L}, v)$ started at x_0 if $\mathbb{P}(X_0 = x_0) = 1$ and

$$f(X_t) - f(X_0) - \int_0^t \mathcal{L}f(X_s)\,ds$$

is a submartingale whenever f is C^2 on D, C^1 on $\overline{D}$, and $\nabla f \cdot v \geq 0$ on ∂D. Suppose D is a C^2 domain, the a_{ij} and b_i are bounded, the a_{ij} are continuous, $\mathcal{L}$ is strictly elliptic, v is C^1, and $\inf_{x \in \partial D} \nu(x) \cdot v(x) > 0$, where $\nu(x)$ is the inward pointing unit normal vector at $x \in \partial D$. Stroock and Varadhan [1] have shown that in this case there is a unique solution to the submartingale problem for $(\mathcal{L}, v)$ starting at each $x_0 \in \overline{D}$.

7. Notes

The martingale problem was originally formulated by Stroock and Varadhan; see Stroock and Varadhan [2]. Most of Sections 1 through 4 follows Stroock and Varadhan [2]. Theorem 4.4 is due to Cerutti, Escauriaza, and Fabes [1]. For more on the subject of Section 6 see Stroock and Varadhan [1].

VII
DIVERGENCE FORM OPERATORS

In this chapter we consider elliptic operators in divergence form. Probabilistic techniques play a much smaller role here than in Chapters 5 and 6. However, one can still say a great deal about the processes associated to operators in divergence form.

Section 1 consists mostly of definitions. Section 2 discusses a number of classical analytic inequalities.

One of the major results in this subject is the Harnack inequality of Moser. This is proved in Section 3.

Sections 4, 5, and 6 are devoted to obtaining upper and lower bounds on the transition densities of processes associated to operators in divergence form. Section 4 obtains an upper bound by a method due to Nash. Better bounds can be obtained for the off-diagonal terms by a method of Davies; this is in Section 5. Section 6 contains the lower bounds.

Section 7 contains some extensions of the results in Sections 4 through 6, primarily the Hölder continuity of the transition densities and bounds for Green functions.

Section 8 discusses some path properties for the associated processes.

1. Preliminaries

Elliptic operators in *divergence form* are operators $\mathcal{L}$ defined on C^2 functions by

$$\mathcal{L}f(x) = \frac{1}{2} \sum_{i,j=1}^{d} \partial_i(a_{ij}\partial_j f)(x), \tag{1.1}$$

where the a_{ij} are measurable functions of x and $a_{ij}(x) = a_{ji}(x)$ for all pairs i, j and all x. Let $\mathcal{D}(\Lambda)$ be the set of operators in divergence form such that for all x and all $y = (y_1, \ldots, y_d)$,

$$\Lambda |y|^2 \leq \sum_{i,j=1}^{d} a_{ij}(x) y_i y_j \leq \Lambda^{-1} |y|^2. \tag{1.2}$$

Throughout this chapter we assume the operator $\mathcal{L}$ is *uniformly elliptic*, that is, $\mathcal{L} \in \mathcal{D}(\Lambda)$ for some $\Lambda > 0$.

If the a_{ij} are not differentiable, an interpretation has to be given to $\mathcal{L}f$; see (1.6). For most of this chapter we will assume the a_{ij} are smooth. With this assumption,

$$\mathcal{L}f(x) = \frac{1}{2} \sum_{i,j=1}^{d} a_{ij}(x) \partial_{ij} f(x) + \frac{1}{2} \sum_{j=1}^{d} \Big(\sum_{i=1}^{d} \partial_i a_{ij}(x) \Big) \partial_j f(x), \tag{1.3}$$

and so $\mathcal{L}$ is equivalent to an operator in nondivergence form (cf. Chapter V) with $b_j(x) = (1/2) \sum_{i=1}^{d} \partial_i a_{ij}(x)$. However, all of our estimates for $\mathcal{L} \in \mathcal{D}(\Lambda)$ will depend only on Λ and not on any smoothness of the a_{ij}. So by a limit procedure, our results and estimates will be valid for operators $\mathcal{L}$ where the a_{ij} are only bounded and strictly elliptic. See Sections 7 and 8 for a bit more information on these more general a_{ij}.

We refer to the conclusion of the following proposition as *scaling*.

(1.1) Proposition. *Let $\mathcal{L} \in \mathcal{D}(\Lambda)$ and let $(\mathbb{P}^x, X_t)$ be the associated process (in the sense of Section I.2). If $r > 0$, $a_{ij}^r(x) = a_{ij}(x/r)$, and $\mathcal{L}^r f(x) = \sum_{i,j=1}^{d} \partial_i(a_{ij}^r \partial_j f)(x)$, then $\mathcal{L}^r \in \mathcal{D}(\Lambda)$ and $(\mathbb{P}^{x/r}, rX_{t/r^2})$ is the process associated to $\mathcal{L}^r$.*

Proof. Using (1.3), this is proved entirely analogously to Proposition V.2.2. See also the proof of Proposition I.8.6. $\qquad\square$

An important example of operators in divergence form is given by the Laplace-Beltrami operators on Riemannian manifolds. Such an operator is the infinitesimal generator of a Brownian motion on the manifold. After a time change (cf. Theorem VI.3.7), the Laplace-Beltrami operator in local coordinates is an operator in divergence form, where the a_{ij} matrix is the inverse of the matrix g_{ij} that determines the Riemannian metric.

Recall the divergence theorem. Suppose D is a nice region, F is a smooth vector field, $\nu(x)$ is the outward pointing normal vector at $x \in \partial D$, and σ is surface measure on ∂D. The divergence theorem then says that

$$\int_{\partial D} F \cdot \nu(y)\, \sigma(dy) = \int_D \operatorname{div} F(x)\, dx. \tag{1.4}$$

(1.2) Proposition. *Let g be a C^∞ function with compact support and f a bounded C^∞ function. Then*

$$\int_{\mathbb{R}^d} g(x)\mathcal{L}f(x)\,dx = -\frac{1}{2}\int_{\mathbb{R}^d}\Big(\sum_{i,j=1}^d \partial_i g(x)a_{ij}(x)\partial_j f(x)\Big)\,dx.$$

The integrand on the right could be written $\nabla g \cdot a\nabla f$.

Proof. We apply the divergence theorem. Let D be a ball large enough to contain the support of g and let $F(x)$ be the vector field whose ith component is

$$\frac{g(x)}{2}\sum_{j=1}^d a_{ij}(x)\partial_j f(x).$$

Since g is 0 on ∂D, then $F \cdot \nu = 0$ on ∂D, and also,

$$\operatorname{div} F(x) = \frac{1}{2}\sum_{i=1}^d \partial_i\Big(g(x)\sum_{j=1}^d a_{ij}(x)\partial_j f(x)\Big)$$

$$= \frac{1}{2}\sum_{i,j=1}^d \partial_i g(x)a_{ij}(x)\partial_j f(x) + g(x)\mathcal{L}f(x).$$

We now substitute into (1.4). $\qquad\qquad\qquad\qquad\qquad\qquad\qquad\square$

Applying Proposition 1.2 twice, if f and g are smooth with compact support,

$$\int g(x)\mathcal{L}f(x)\,dx = \int f(x)\mathcal{L}g(x)\,dx. \tag{1.5}$$

This equation says that $\mathcal{L}$ is *self-adjoint* with respect to Lebesgue measure.

Note that Proposition 1.2 allows us to give an interpretation to $\mathcal{L}f = 0$ even when the a_{ij} are not differentiable. We say f is a solution to $\mathcal{L}f = 0$ if f is differentiable in some sense, e.g., $f \in W^{1,p}$ for some p and

$$\int \sum_{i,j=1}^d \partial_i f(x)a_{ij}(x)\partial_j g(x)\,dx = 0 \tag{1.6}$$

whenever g is in C^∞ with compact support. Here $W^{1,p}$ is the closure of $C^2 \cap L^\infty$ with respect to the norm

$$\|f\|_{W^{1,p}} = \|f\|_p + \sum_{i=1}^d \|\partial_i f\|_p.$$

See Stein [1] for further information about the space $W^{1,p}$.

The expression

$$\int \frac{1}{2} \sum_{i,j=1}^{d} a_{ij}(x) \partial_i f(x) \partial_j g(x)\, dx = \frac{1}{2} \int \nabla f(x) \cdot a(x) \nabla g(x)\, dx$$

is an example of what is known as a *Dirichlet form*. If we denote it by $\mathcal{E}(f,g)$, then Proposition 1.2 says that

$$\int g \mathcal{L} f\, dx = -\mathcal{E}(f,g)$$

for g with compact support. In the case of Brownian motion, the Dirichlet form is

$$\mathcal{E}_{BM}(f,g) = \frac{1}{2} \int \nabla f(x) \cdot \nabla g(x)\, dx.$$

Part of defining a Dirichlet form is specifying the domain. For example, the Dirichlet form for Brownian motion in $\mathbb{R}^d$ has domain $\{f \in L^2 : \mathcal{E}_{BM}(f,f) < \infty\}$. The Dirichlet form for reflecting Brownian motion in a domain $D \subseteq \mathbb{R}^d$ operates on $\{f \in L^2(D) : \int_D |\nabla f(x)|^2\, dx < \infty\}$, whereas the Dirichlet form for Brownian motion killed on exiting a set D has domain $\{f \in L^2(D) : \int_D |\nabla f(x)|^2\, dx < \infty, f = 0 \text{ on } \partial D\}$.

Note that the uniform ellipticity of $\mathcal{L}$ implies that

$$\Lambda \mathcal{E}_{BM}(f,f) \le \mathcal{E}_{\mathcal{L}}(f,f) \le \Lambda^{-1} \mathcal{E}_{BM}(f,f). \tag{1.7}$$

An active area of research is the construction of Markov processes corresponding to a given Dirichlet form and seeing how properties of the Dirichlet form are reflected in properties of the process; see Fukushima, Oshima, and Takeda [1].

2. Inequalities

We will make use of several classical inequalities. The first is the Sobolev inequality.

(2.1) Theorem. *Suppose $d > 2$. There exists c_1 such that if $f \in C^2$ and $\nabla f \in L^2$, then*

$$\left(\int_{\mathbb{R}^d} |f(x)|^{2d/(d-2)}\, dx \right)^{(d-2)/2d} \le c_1 \left(\int_{\mathbb{R}^d} |\nabla f(x)|^2\, dx \right)^{1/2}.$$

There are many different proofs of this. See Stein [1] or [PTA, Theorem IV.3.10]. An elegant proof using isoperimetric inequalities can be found in Maz'ja [1]. An elementary proof can be found in Nirenberg [1].

A variant of the Sobolev inequality is the following for bounded domains.

(2.2) Corollary. *Suppose $d > 2$. Let Q be the unit cube. Suppose f is C^2 on Q and $\nabla f \in L^2(Q)$. There exists c_1 such that*

$$\left(\int_Q |f|^{2d/(d-2)} \right)^{(d-2)/d} \leq c_1 \left[\int_Q |\nabla f|^2 + \int_Q |f|^2 \right].$$

Proof. Let Q^* be the cube with the same center as Q but side length twice as long. By reflecting over the boundaries of Q, we can extend f to Q^* so that $\int_{Q^*} |f|^p \leq c_2 \int_Q |f|^p$ for $p = 2d/(d-2)$ and also $\int_{Q^*} |f|^2 \leq c_2 \int_Q |f|^2$ and $\int_{Q^*} |\nabla f|^2 \leq c_2 \int_Q |\nabla f|^2$, where c_2 is a constant not depending on f. Let φ be a C^∞ function taking values in $[0, 1]$ with support in Q^* and so that $\varphi = 1$ on Q. Applying Theorem 1.1 to φf,

$$\left(\int_Q |f|^p \right)^{2/p} \leq \left(\int |\varphi f|^p \right)^{2/p} \leq c_1 \int |\nabla(\varphi f)|^2,$$

where $p = 2d/(d-2)$. Since

$$|\nabla(\varphi f)|^2 \leq 2|\nabla \varphi|^2 |f|^2 + 2|\varphi|^2 |\nabla f|^2,$$

and φ and $\nabla \varphi$ are bounded by constants independent of f and have support in Q^*, the result follows. $\qquad\square$

Another closely related inequality is the Nash inequality.

(2.3) Theorem. *Suppose $d \geq 2$. There exists c_1 such that if $f \in C^2$, $f \in L^1 \cap L^2$, and $\nabla f \in L^2$, then*

$$\left(\int |f|^2 \right)^{1+2/d} \leq c_1 \left(\int |\nabla f|^2 \right) \left(\int |f| \right)^{4/d}.$$

Proof. If $\widehat{f}(\xi) = \int e^{ix \cdot \xi} f(x)\, dx$ is the Fourier transform of f, then the Fourier transform of $\partial_j f$ is $i\xi_j \widehat{f}(\xi)$. Recall $|\widehat{f}(\xi)| \leq \int |f|$. By the Plancherel theorem, $\int |f|^2 = c_2 \int |\widehat{f}(\xi)|^2\, d\xi$ and $\int |\nabla f|^2 = c_2 \int |\xi|^2 |\widehat{f}(\xi)|^2\, d\xi$. We have

$$\int |f|^2 = c_2 \int |\widehat{f}(\xi)|^2\, d\xi \leq c_2 \int_{|\xi| \leq R} |\widehat{f}|^2 + c_2 \int_{|\xi| > R} \frac{|\xi|^2}{R^2} |\widehat{f}|^2$$

$$\leq c_3 R^n \left(\int |f| \right)^2 + c_4 R^{-2} \int |\nabla f|^2.$$

We now choose R to minimize the right-hand side. $\qquad\square$

When $d \geq 3$, we can also derive Theorem 2.3 from Theorem 2.1. Let

$$p = \frac{d+2}{d-2}, \quad q = \frac{d+2}{4}, \quad a = \frac{2d}{d+2}, \quad \text{and} \quad b = \frac{4}{d+2}.$$

Since $a + b = 2$, $|f|^2 = |f|^a |f|^b$, and then Hölder's inequality with the given p and q tells us that

$$\int |f|^2 \leq \left(\int |f|^{ap} \right)^{1/p} \left(\int |f|^{bq} \right)^{1/q}.$$

Since $bq = 1$ and $ap = 2d/(d-2)$, an application of Theorem 2.1 gives Theorem 2.3.

The Poincaré inequality states the following.

(2.4) Theorem. *Suppose Q is a unit cube of side length h and f is C^2 on Q with $\nabla f \in L^2(Q)$. There exists c_1 not depending on f such that*

$$\int_Q |f(x) - f_Q|^2 \, dx \leq c_1 h^2 \int_Q |\nabla f(x)|^2 \, dx,$$

where $f_Q = |Q|^{-1} \int_Q f(x) \, dx$.

Proof. By a translation of the coordinate axes, we may suppose Q is centered at the origin. Since $\nabla(f - f_Q) = \nabla f$, by subtracting a constant from f we may suppose without loss of generality that $f_Q = 0$. Let us also suppose for now that $h = 1$.

If $m = (m_1, \ldots, m_d)$, let C_m denote the Fourier coefficient of $e^{2\pi i m \cdot x}$, that is,

$$C_m = \int_Q e^{-2\pi i m \cdot x} f(x) \, dx.$$

Since $\int_Q f = 0$, then $C_0 = 0$. The mth Fourier coefficient of $\partial_j f$ is $2\pi i m_j C_m$. By the Parseval identity and the fact that $C_0 = 0$,

$$\int_Q |\nabla f|^2 = \sum_m (2\pi)^2 |m|^2 |C_m|^2 \tag{2.1}$$

$$\geq c_2 \sum_m |C_m|^2 = c_2 \int_Q |f|^2.$$

We eliminate the supposition that $h = 1$ by a scaling argument, namely, we apply (2.1) to $f(x) = g(xh)$ for x in the unit cube, and then replace g by f. $\qquad\square$

Finally, we will need the John-Nirenberg inequality. We continue to use the notation

$$f_Q = |Q|^{-1} \int_Q f. \tag{2.2}$$

(2.5) Theorem. *Suppose Q_0 is a cube, $f \in L^1(Q_0)$, and for all cubes $Q \subseteq Q_0$,*

$$\frac{1}{|Q|} \int_Q |f(x) - f_Q| \leq 1. \tag{2.3}$$

Then there exist c_1 and c_2 independent of f such that

$$\int_{Q_0} e^{c_1 f(x)}\, dx \le c_2.$$

An f satisfying (2.3) is said to be in BMO, the space of functions of bounded mean oscillation. A proof of Theorem 2.5 may be found in [PTA, Proposition IV.7.6] or Garnett [1].

3. Moser's Harnack inequality

Let $Q(h)$ denote the cube centered at the origin with side length h. Moser's Harnack inequality (Theorem 3.5) says that if $\mathcal{L} \in \mathcal{D}(\Lambda)$, there exists c_1 depending only on Λ such that if $\mathcal{L}u = 0$ and $u \ge 0$ in $Q(4)$, then

$$\sup_{Q(1)} u \le c_1 \inf_{Q(1)} u.$$

We begin proving this important fact by establishing a sort of converse to Poincaré's inequality for powers of u. Recall that u is $\mathcal{L}$-harmonic in $Q(r)$ if u is C^2 on $Q(r)$ and $\mathcal{L}u = 0$ on $Q(r)$.

(3.1) Proposition. *Suppose $r > 1$ and u is nonnegative and $\mathcal{L}$-harmonic in $Q(r)$. There exists c_1 depending only on the ellipticity bound Λ such that if $v = u^p$ for $p \in \mathbb{R}$, then*

$$\int_{Q(1)} |\nabla v|^2 \le c_1 \left(\frac{2p}{2p-1}\right)^2 \frac{1}{(r-1)^2} \int_{Q(r)} |v|^2.$$

Proof. The result is trivial if $p = 1/2$. The result is also trivial if $p = 0$, for then v is identically 1 and $\nabla v = 0$. So we suppose p is some value other than 0 or 1/2. Let φ be a smooth function taking values in $[0,1]$ with support in $Q(r)$ such that $\varphi = 1$ on $Q(1)$ and $|\nabla\varphi| \le c_2/(r-1)$. Let $w = u^{2p-1}\varphi^2$. Since u is $\mathcal{L}$-harmonic and $w = 0$ outside of $Q(r)$, Proposition 1.2 tells us that

$$0 = 2\int_{Q(r)} w\mathcal{L}u = -\int_{Q(r)} \nabla w \cdot a\nabla u$$

$$= -(2p-1)\int_{Q(r)} u^{2p-2}\varphi^2 \nabla u \cdot a\nabla u - 2\int_{Q(r)} u^{2p-1}\varphi\nabla\varphi \cdot a\nabla u.$$

We then have, using (1.2) and the Cauchy-Schwarz inequality,

$$\int_{Q(r)} |\nabla v|^2 \varphi^2 = \int_{Q(r)} p^2 u^{2p-2} |\nabla u|^2 \varphi^2$$

$$\leq \Lambda p^2 \int_{Q(r)} u^{2p-2} \varphi^2 \nabla u \cdot a \nabla u$$

$$= c_2 \frac{2p^2}{|2p-1|} \int_{Q(r)} u^{2p-1} \varphi \nabla \varphi \cdot a \nabla u$$

$$= \frac{2c_2 p^2}{|2p-1|} \int_{Q(r)} u^p \varphi \nabla \varphi \cdot a u^{p-1} \nabla u$$

$$= \frac{2c_2 p}{|2p-1|} \int_{Q(r)} v \nabla \varphi \cdot a \varphi \nabla v$$

$$\leq \frac{2c_3 |p|}{|2p-1|} \Big(\int_{Q(r)} |\nabla v|^2 \varphi^2 \Big)^{1/2} \Big(\int_{Q(r)} v^2 |\nabla \varphi|^2 \Big)^{1/2}.$$

Dividing both sides by $(\int_{Q(r)} |\nabla v|^2 \varphi^2)^{1/2}$, we obtain

$$\int_{Q(1)} |\nabla v|^2 \leq \int_{Q(r)} |\nabla v|^2 \varphi^2$$

$$\leq c_3^2 \Big(\frac{2p}{2p-1} \Big)^2 \int_{Q(r)} v^2 |\nabla \varphi|^2$$

$$\leq c_3^2 \Big(\frac{2p}{2p-1} \Big)^2 \frac{1}{(r-1)^2} \int_{Q(r)} v^2. \qquad \square$$

Let us define

$$\Phi(p, h) = \Big(\int_{Q(h)} u^p \Big)^{1/p}.$$

(3.2) Proposition. *Suppose $d \geq 3$. If $u \geq 0$ in $Q(2)$ and $\mathcal{L}u = 0$ in $Q(2)$, then for all $q_0 > 0$ there exists c_1 (depending on q_0 but not u) such that*

$$\sup_{Q(1)} u \leq c_1 \Phi(q_0, 2).$$

Proof. Let $R = d/(d-2)$, $p > 0$, and $2 > r > 1$. By Corollary 2.2 and Proposition 3.1,

$$\Big(\int_{Q(1)} u^{2pR} \Big)^{1/R} \leq c_2 \Big[\int_{Q(1)} |\nabla(u^p)|^2 + \int_{Q(1)} |u^p|^2 \Big]$$

$$\leq c_3 \Big[\frac{1}{(r-1)^2} \Big(\frac{2p}{2p-1} \Big)^2 \int_{Q(r)} u^{2p} + \int_{Q(1)} |u^p|^2 \Big]$$

$$\leq \frac{c_4}{(r-1)^2} \Big(\frac{2p}{2p-1} \Big)^2 \int_{Q(r)} u^{2p}.$$

Taking both sides to the $1/2p$ power and using scaling, if $r < s < 2r$,

$$\Phi(2Rp, r) \leq \left(\frac{c_4}{(s/r - 1)^2} \frac{(2p)^2}{(2p - 1)^2} \right)^{1/2p} \Phi(2p, s). \tag{3.1}$$

Suppose $p_0 = R^{-m-1/2}/2$, where m is the smallest positive integer such that $2p_0 < q_0$. Let $p_n = R^n p_0$, $r_n = 1 + 2^{-n}$. Then

$$\frac{r_n}{r_{n-1}} - 1 = \frac{2^{-n-1}}{1 + 2^{-(n-1)}} \geq 2^{-n}/2$$

and by our assumption on p_0,

$$\left(\frac{2p_n}{2p_n - 1} \right)^2 \leq c_5,$$

where c_5 depends only on R. Substituting in (3.1),

$$\Phi(2p_{n+1}, r_{n+1}) \leq (c_6 2^{2n})^{1/(2R^n p_0)} \Phi(2p_n, r_n).$$

By induction,

$$\Phi(2p_n, r_n) \leq c_6^\alpha 2^\beta \Phi(2p_0, 2),$$

where

$$\alpha = \sum_{j=0}^\infty \frac{1}{2R^j p_0} < \infty, \qquad \beta = \sum_{j=0}^\infty \frac{2j}{2R^j p_0} < \infty.$$

Therefore $\Phi(2p_n, r_n) \leq c_7 \Phi(2p_0, 2)$. By Hölder's inequality,

$$\Phi(2p_0, 2) \leq c_8 \Phi(q_0, 2).$$

The conclusion now follows from the fact that

$$\sup_{Q(1)} u \leq \limsup_{n \to \infty} \Phi(2p_n, r_n). \qquad \square$$

(3.3) Proposition. *Suppose u is bounded below by a positive constant on $Q(2)$ and $q_0 > 0$. Then there exists c_1 (depending only on q_0 but not u) such that*

$$\inf_{Q(1)} u \geq \left(\int_{Q(2)} u^{-q_0} \right)^{-1/q_0}.$$

Proof. The proof is almost identical to the above, working with

$$\Phi(-p, h) = \left(\int_{Q(h)} u^{-p} \right)^{-1/p} \tag{3.2}$$

instead of $\Phi(p, h)$. $\qquad \square$

To connect $\Phi(p, h)$ for $p > 0$ and $p < 0$, we look at $\log u$.

(3.4) Proposition. *Suppose u is positive and $\mathcal{L}$-harmonic in $Q(4)$. There exists c_1 independent of u such that if $w = \log u$, then*

$$\int_Q |\nabla w|^2 \leq c_1 h^{d-2}$$

for all cubes Q of side length h contained in $Q(2)$.

Proof. Let Q^* be the cube with the same center as Q but side length twice as long. Note $Q^* \subseteq Q(4)$. Let φ be C^∞ with values in $[0,1]$, equal to 1 on Q, supported in Q^*, and such that $\|\nabla\varphi\|_\infty \leq c_2/h$. Since $\nabla w = \nabla u/u$ and u is $\mathcal{L}$-harmonic in $Q(4)$,

$$\begin{aligned}
0 = 2\int \frac{\varphi^2}{u}\mathcal{L}u &= -\int \nabla(\varphi^2/u)\cdot a\nabla u \\
&= -\int \frac{2\varphi\nabla\varphi}{u}\cdot a\nabla u + \int \frac{\varphi^2}{u^2}\nabla u\cdot a\nabla u \\
&= -2\int \varphi\nabla\varphi\cdot a\nabla w + \int \varphi^2\nabla w\cdot a\nabla w.
\end{aligned}$$

So by the Cauchy-Schwarz inequality and (1.2),

$$\begin{aligned}
\int_{Q^*} \varphi^2|\nabla w|^2 &\leq c_3\int_{Q^*} \varphi^2\nabla w\cdot a\nabla w = c_4\int_{Q^*}\nabla\varphi\cdot a\varphi\nabla w \\
&\leq c_5\left(\int_{Q^*}|\nabla\varphi|^2\right)^{1/2}\left(\int_{Q^*}\varphi^2|\nabla w|^2\right)^{1/2}
\end{aligned}$$

Dividing by the second factor on the right, squaring, and using the bound on $|\nabla\varphi|$,

$$\int_Q |\nabla w|^2 \leq \int_{Q^*}\varphi^2|\nabla w|^2 \leq c_5^2|Q^*|(c_2/h)^2,$$

which implies our result. $\qquad\square$

Putting all the pieces together, we have Moser's Harnack inequality.

(3.5) Theorem. *There exists c_1 such that if u is $\mathcal{L}$-harmonic and nonnegative in $Q(4)$, then*

$$\sup_{Q(1)} u \leq c_1 \inf_{Q(1)} u.$$

Proof. By looking at $u + \varepsilon$ and letting $\varepsilon \to 0$, we may suppose u is bounded below in $Q(4)$. Set $w = \log u$. By Proposition 3.4 and Theorem 2.4, there exists c_3 such that if Q is a cube contained in $Q(2)$, then

$$\left(\frac{1}{|Q|}\int_Q |w - w_Q|\right)^2 \leq \frac{1}{|Q|}\int_Q |w - w_Q|^2 \leq c_2\frac{h^2}{|Q|}\int_Q |\nabla w|^2 \leq c_3.$$

By the John-Nirenberg inequality applied to $w/c_3^{1/2}$ and $-w/c_3^{1/2}$, there exist c_4 and q_0 such that

$$\int_{Q(2)} e^{q_0 w} \le c_4, \qquad \int_{Q(2)} e^{-q_0 w} \le c_4.$$

This can be rewritten as

$$\int_{Q(2)} u^{q_0} \int_{Q(2)} u^{-q_0} \le c_4^2,$$

or

$$\Phi(q_0, 2) \le c_4^{2/q_0} \Phi(-q_0, 2). \tag{3.3}$$

This and Propositions 3.2 and 3.3 show

$$\sup_{Q(1)} u \le c_5 \Phi(q_0, 2) \le c_6 \Phi(-q_0, 2) \le c_7 \inf_{Q(1)} u. \qquad \square$$

An easy corollary proved by repeated use of Theorem 3.5 to a suitable overlapping sequence of cubes (cf. Corollary V.7.7) is the following.

(3.6) Corollary. *Suppose $D_1 \subseteq \overline{D_1} \subseteq D_2$, where D_1 and D_2 are bounded connected domains in $\mathbb{R}^d$ and $d \ge 3$. There exists c_1 depending only on D_1 and D_2 such that if u is nonnegative and $\mathcal{L}$-harmonic in D_2, then*

$$\sup_{D_1} u \le c_1 \inf_{D_2} u.$$

Another corollary of the Moser Harnack inequality is that $\mathcal{L}$-harmonic functions must be Hölder continuous with a modulus of continuity independent of the smoothness of the a_{ij}.

(3.7) Theorem. *Suppose $d \ge 3$ and suppose u is $\mathcal{L}$-harmonic in $Q(2)$. There exist c_1 and α not depending on u such that if $x, y \in Q(1)$,*

$$|u(x) - u(y)| \le c_1 |x - y|^\alpha \sup_{Q(2)} |u|.$$

Proof. Fix x and let $r < 1$. Our result will follow (cf. [PTA, Proposition II.2.2]) if we show there exists $\rho < 1$ independent of r such that

$$\underset{B(x,r/2)}{\operatorname{Osc}}\, u \le \rho \underset{B(x,r)}{\operatorname{Osc}}\, u. \tag{3.4}$$

By looking at $Cu + D$ for suitable C and D, we may suppose that the infimum of $Cu + D$ on $B(x, r)$ is 0, the supremum is 1, and there exists $x_0 \in B(x, r/2)$ such that $(Cu + D)(x_0) \ge 1/2$. By Corollary 3.6 with $D_1 = B(x, r/2)$ and $D_2 = B(x, r)$, there exists c_2 such that

$$(Cu + D)(y) \ge c_2 (Cu + D)(x_0) \ge c_2/2, \qquad y \in B(x, r/2).$$

On the other hand, if $(\mathbb{P}^x, X_t)$ is the process associated with $\mathcal{L}$, then

$$(Cu + D)(y) = \mathbb{E}^y (Cu + D)(X_{\tau(B(x,r))}) \le 1$$

by optional stopping. Hence $\mathrm{Osc}_{B(x,r/2)}(Cu+D) \le 1-c_2/2$, and (3.4) follows.

$\square$

4. Upper bounds on heat kernels

We are now going to investigate bounds on the transition densities of X_t, where $(\mathbb{P}^x, X_t)$ is the process associated to an operator $\mathcal{L} \in \mathcal{D}(\Lambda)$. Let P_t be the operator defined by

$$P_t f(x) = \mathbb{E}^x f(X_t).$$

We shall see that there exists a symmetric function $p(t,x,y)$ such that

$$P_t f(x) = \int f(y) p(t,x,y)\, dy$$

and that $p(t,x,y)$ has upper and lower bounds similar to those of Brownian motion. Recall that $\partial_t u$ means $\partial u / \partial t$. Since $u(x,t) = \mathbb{E}^x f(X_t)$ is also a solution to the Cauchy problem $\partial_t u = \mathcal{L}u$ in $\mathbb{R}^d \times (0,\infty)$ with initial condition $u(x,0) = f(x)$ and, as we have seen in Chapter II, $u(x,t) = \int f(y) p(t,x,y)\, dy$, then $p(t,x,y)$ is also the fundamental solution to the Cauchy problem for $\mathcal{L}$. The equation $\partial_t u = \mathcal{L}u$ is a model for heat flow in a nonhomogeneous medium, which leads to the name *heat kernel* for $p(t,x,y)$.

It is possible to derive bounds on $p(t,x,y)$ from Moser's Harnack inequality via arguments using capacities and some probabilistic arguments; see Littman, Stampacchia, and Weinberger [1] and Barlow and Bass [1]. We present instead an approach due to Nash [1], Davies [1], and Fabes and Stroock [2] for the upper bound, which is quite elegant and useful.

First, we derive some properties of P_t. We continue to assume that the coefficients a_{ij} are smooth and that $\mathcal{L} \in \mathcal{D}(\Lambda)$ for some $\Lambda > 0$.

(4.1) Proposition. *If $f \in C^\infty$ is bounded and in L^1, then $P_t f$ is differentiable in t and*

$$\partial_t P_t f = P_t \mathcal{L} f = \mathcal{L} P_t f.$$

Proof. By Itô's formula,

$$P_{t+h} f(x) - P_t f(x) = \mathbb{E}^x \int_t^{t+h} \mathcal{L} f(X_s)\, ds,$$

so

$$\partial_t P_t f(x) = \mathbb{E}^x \mathcal{L} f(X_t) = P_t \mathcal{L} f(x)$$

by the continuity of $\mathcal{L} f$.

By Proposition I.9.1 and its proof, $P_t f$ is a smooth function of x. Applying Itô's formula to $P_t f$,

$$P_h(P_t f)(x) - P_t f(x) = \mathbb{E}^x P_t f(X_h) - \mathbb{E}^x P_t f(X_0)$$
$$= \mathbb{E}^x \int_0^h \mathcal{L}(P_t f)(X_s)\, ds.$$

However, $P_h(P_t f) = P_{t+h} f$ by the Markov property. Dividing by h, letting $h \to 0$, and using the continuity of $\mathcal{L}(P_t f)$,

$$\partial_t P_t f(x) = \mathbb{E}^x \mathcal{L}(P_t f)(X_0) = \mathcal{L} P_t f(x). \qquad \square$$

Next we show that P_t is a symmetric operator.

(4.2) Proposition. *If f and g are bounded and in L^1,*

$$\int f(x) P_t g(x)\, dx = \int g(x) P_t f(x)\, dx.$$

Proof. Let $\overline{f}, \overline{g} \in L^1 \cap C^2$ be bounded with bounded first and second partial derivatives. By (1.5),

$$\int \overline{f}(\mathcal{L}\overline{g}) = \int \overline{g}(\mathcal{L}\overline{f}).$$

Therefore

$$\int \overline{f}\big((\lambda - \mathcal{L})\overline{g}\big) = \int \overline{g}\big((\lambda - \mathcal{L})\overline{f}\big). \tag{4.1}$$

If f, g are bounded C^∞ functions and $\lambda > 0$, let $\overline{f} = G^\lambda f$, $\overline{g} = G^\lambda g$, where G^λ is defined by

$$G^\lambda f(x) = \mathbb{E}^x \int_0^\infty e^{-\lambda t} f(X_t)\, dt.$$

By Section III.6, $\overline{f}$ and $\overline{g}$ are smooth and $(\lambda - \mathcal{L})\overline{f} = f$, $(\lambda - \mathcal{L})\overline{g} = g$. By Jensen's inequality, $\overline{f}$ and $\overline{g}$ are in L^1. Substituting in (4.1),

$$\int (G^\lambda f)g = \int (G^\lambda g)f, \qquad \lambda > 0.$$

We have seen in Proposition 4.1 that $P_t f$ is differentiable in t, and hence continuous in t. Noting $G^\lambda f = \int_0^\infty e^{-\lambda t} P_t f\, dt$, the uniqueness of the Laplace transform tells us that

$$\int (P_t f)g = \int (P_t g)f, \qquad t > 0.$$

We now use a limit argument to extend this to the case where f and g are arbitrary bounded functions in L^1. $\qquad \square$

With these preliminaries out of the way, we can now present Nash's method, which yields an upper bound for the transition density.

(4.3) Theorem. *There exists a function $p(t, x, y)$ mapping $(0, \infty) \times \mathbb{R}^d \times \mathbb{R}^d$ to $[0, \infty)$ that is symmetric in x and y for almost every pair (x, y) (with respect to Lebesgue measure on $\mathbb{R}^d \times \mathbb{R}^d$) and such that $P_t f(x) = \int f(y) p(t, x, y) \, dy$ for all bounded functions f. There exists c_1 depending only on Λ such that*

$$p(t, x, y) \leq c_1 t^{-d/2}, \qquad t > 0, \quad x, y \in \mathbb{R}^d.$$

Proof. Let f be C^∞ with compact support with $\int f = 1$. We observe that

$$\int P_t f(x) \, dx = \int 1(P_t f) = \int (P_t 1) f = \int f = 1$$

because $P_t 1 = 1$.

Set

$$E(t) = \int (P_t f(x))^2 \, dx,$$

and note $E(0) = \int f(x)^2 \, dx < \infty$. By Proposition 4.1,

$$E'(t) = 2 \int P_t f(x) \partial_t (P_t f(x)) \, dx = 2 \int P_t f(x) \mathcal{L} P_t f(x) \, dx.$$

By Proposition 1.2, this is equal to

$$-\int \nabla(P_t f) \cdot a \nabla(P_t f)(x) \, dx \leq -\Lambda \int |\nabla(P_t f)(x)|^2 \, dx,$$

since $\mathcal{L} \in \mathcal{D}(\Lambda)$. By Theorem 2.3 (the Nash inequality), we have the right-hand side bounded above in turn by

$$-c_2 \left(\int (P_t f(x))^2 \right)^{1+2/d} \left(\int P_t f(x) \right)^{4/d} = -c_2 E(t)^{1+2/d}.$$

Therefore

$$E'(t) \leq -c_2 E(t)^{1+2/d}, \tag{4.2}$$

or

$$(E(t)^{-2/d})' \geq c_3.$$

(We are treating the differential inequality (4.2) by the same methods we would use if it were an equality and we had a first order separable differential equation.) An integration yields

$$E(t)^{-2/d} - E(0)^{-2/d} \geq c_3 t,$$

or

$$E(t)^{-2/d} \geq c_3 t.$$

We conclude from this that

$$E(t) \le c_4 t^{-d/2}.$$

Using the linearity of P_t, we thus have that

$$\|P_t f\|_2 \le c_4^{1/2} t^{-d/4} \|f\|_1 \tag{4.3}$$

for f smooth. A limit argument extends this to all $f \in L^1$. We now use a duality argument. If $g \in L^1$ and $f \in L^2$,

$$\int g(P_t f) = \int f(P_t g) \le \|f\|_2 \|P_t g\|_2 \le c_4^{1/2} t^{-d/4} \|g\|_1 \|f\|_2.$$

Taking the supremum over $g \in L^1$ with $\|g\|_1 \le 1$,

$$\|P_t f\|_\infty \le c_4^{1/2} t^{-d/4} \|f\|_2. \tag{4.4}$$

By the semigroup property, (4.4) applied to $P_{t/2}f$, and (4.3) applied to f,

$$\|P_t f\|_\infty = \|P_{t/2}(P_{t/2}f)\|_\infty \le c_4^{1/2} (t/2)^{-d/4} \|P_{t/2}f\|_2$$
$$\le c_4 (t/2)^{-d/2} \|f\|_1.$$

This says

$$|P_t f(x)| \le c_5 t^{-d/2} \int |f(y)|\, dy. \tag{4.5}$$

Applying this to $f = 1_B$, B a Borel set, we see that $\mathbb{P}^x(X_t \in dy)$ is absolutely continuous with respect to Lebesgue measure and the density, which we shall call $p(t, x, y)$, is nonnegative and bounded by $c_5 t^{-d/2}$ for almost all pairs (x, y). The symmetry (except for a null set of pairs) follows easily by Proposition 4.2. $\qquad\square$

5. Off-diagonal upper bounds

A more sophisticated variant of the argument of the previous section, due to Davies, allows us to obtain a better estimate on $p(t, x, y)$ when $|x - y|$ is large relative to $t^{1/2}$. We will use the notation

$$\mathcal{E}(f, g) = \frac{1}{2} \int \nabla f \cdot a \nabla g = \frac{1}{2} \int \sum_{i,j=1}^d \partial_i f(x) a_{ij}(x) \partial_j g(x)\, dx.$$

Let $R : \mathbb{R}^d \to \mathbb{R}$ be a smooth function.

(5.1) Proposition. *If $f \in C^1$, then*

$$\mathcal{E}(e^R f^{2p-1}, e^{-R} f) \ge \frac{1}{p} \mathcal{E}(f^p, f^p) - \frac{p}{2} \|\nabla R \cdot a \nabla R\|_\infty \int |f|^{2p}.$$

Proof. We calculate:

$$\nabla(e^R f^{2p-1}) = (2p-1)e^R f^{2p-2}\nabla f + e^R f^{2p-1}\nabla R$$

and

$$\nabla(e^{-R}f) = e^{-R}\nabla f - e^{-R}f\nabla R.$$

So

$$
\begin{aligned}
\nabla(e^R f^{2p-1}) &\cdot a\nabla(e^{-R}f) \\
&= (2p-1)f^{2p-2}\nabla f \cdot a\nabla f - (2p-2)f^{2p-1}\nabla f \cdot a\nabla R \\
&\quad - f^{2p}\nabla R \cdot a\nabla R \\
&= pf^{2p-2}\nabla f \cdot a\nabla f - pf^{2p}\nabla R \cdot a\nabla R \\
&\quad + (p-1)\Big[f^{2p-2}\nabla f \cdot a\nabla f - 2f^{2p-1}\nabla f \cdot a\nabla R \\
&\qquad + f^{2p}\nabla R \cdot a\nabla R\Big].
\end{aligned}
$$

The expression in brackets is

$$
\begin{aligned}
(f^{p-1}\nabla f) &\cdot a(f^{p-1}\nabla f) - 2(f^{p-1}\nabla f) \cdot a(f^p \nabla R) + (f^p \nabla R) \cdot a(f^p \nabla R) \\
&= (f^{p-1}\nabla f - f^p \nabla R) \cdot a(f^{p-1}\nabla f - f^p \nabla R) \geq 0.
\end{aligned}
$$

Since $\nabla(f^p) = pf^{p-1}\nabla f$, so that $\nabla(f^p) \cdot a\nabla(f^p) = p^2 f^{2p-2}\nabla f \cdot a\nabla f$, we have

$$
\begin{aligned}
\nabla(e^R f^{2p-1}) \cdot a\nabla(e^{-R}f) &\geq pf^{2p-2}\nabla f \cdot a\nabla f - pf^{2p}\nabla R \cdot a\nabla R \\
&\geq \frac{1}{p}\nabla(f^p) \cdot a\nabla(f^p) - p\|\nabla R \cdot a\nabla R\|_\infty |f|^{2p}.
\end{aligned}
$$

Integrating over $\mathbb{R}^d$ and multiplying by $1/2$ completes the derivation. $\square$

We use Proposition 5.1 to obtain a differential inequality for the L^p norms of a semigroup related to P_t. Let $f \geq 0$ be smooth, let R be smooth, and define

$$P_t^R f(x) = e^{R(x)} P_t(e^{-R}f)(x).$$

Note that

$$\partial_t P_t^R f(x) = e^R \partial_t P_t(e^{-R}f) = e^R \mathcal{L}(P_t(e^{-R}f)).$$

Let $p \geq 1$, set $f_t = P_t^R f$, and $E(t) = (\int f_t^{2p})^{1/2p}$. Let $\Gamma = \|\nabla R \cdot a\nabla R\|_\infty$.

(5.2) Proposition. *There exists c_1 independent of f, R, and p such that*

$$E'(t) \leq -\frac{c_1}{p}\frac{E(t)^{1+4p/d}}{\|f_t\|_p^{4p/d}} + \frac{p}{2}\Gamma E(t). \tag{5.1}$$

Proof. Let $D(t) = \int f_t(x)^{2p}\, dx$. Then by Proposition 5.1,

$$D'(t) = 2p \int f_t^{2p-1} \partial_t f_t = 2p \int f_t^{2p-1} e^R \mathcal{L}(P_t(e^{-R}f))$$

$$= -p \int \nabla(e^R f_t^{2p-1}) \cdot a\nabla(P_t(e^{-R}f))$$

$$= -2p\mathcal{E}(e^R f_t^{2p-1}, e^{-R}f_t)$$

$$\leq -2\mathcal{E}(f_t^p, f_t^p) + p^2 \Gamma \int f_t^{2p}.$$

Since $E(t) = D(t)^{1/2p}$,

$$E'(t) = \frac{1}{2p} D(t)^{1/2p-1} D'(t)$$

$$\leq D(t)^{1/2p-1}\left[-\frac{1}{p}\mathcal{E}(f_t^p, f_t^p) + \frac{p}{2}\Gamma D(t)\right].$$

Since $\mathcal{L} \in \mathcal{D}(\Lambda)$,

$$\mathcal{E}(f_t^p, f_t^p) = \int \nabla(f_t^p) \cdot a\nabla f_t^p \geq \Lambda \int |\nabla f_t^p|^2.$$

So using Theorem 2.3 (the Nash inequality),

$$E'(t) \leq D(t)^{1/2p-1}\left[-\frac{c_2}{p}\left(\int f_t^{2p}\right)^{1+2/d}\left(\int f_t^p\right)^{-4/d} + \frac{p}{2}\Gamma D(t)\right]$$

$$\leq -\frac{c_2}{p} D(t)^{1/2p+2/d}\|f_t\|_p^{-4p/d} + \frac{p}{2}\Gamma D(t)^{1/2p}.$$

Since $D(t) = E(t)^{2p}$, the conclusion follows. $\qquad\square$

(5.3) Corollary.

$$\|f_t\|_2 \leq e^{\Gamma t/2}\|f\|_2.$$

Proof. Let $p = 1$. We have $f_0 = e^R P_0(e^{-R}f) = f$, so $E(0) = \|f\|_2$. From Proposition 5.2, $E'(t) \leq \Gamma E(t)/2$. Then

$$(\log E(t))' = E'(t)/E(t) \leq \Gamma/2,$$

or

$$\log E(t) - \log E(0) \leq \Gamma t/2.$$

We now solve for $E(t)$. $\qquad\square$

To handle the differential inequality obtained in Proposition 5.2, we proceed in the same manner as we would if we had equality: we multiply by an integrating factor to obtain a separable ordinary differential equation.

(5.4) Proposition. *Suppose w is a nonnegative, nondecreasing continuous function on $[0, \infty)$. Suppose $p \geq 2$ and $a, b, \Gamma > 0$. There exists c_1 depending only on a and b such that if*

$$u'(t) \leq -\frac{a}{p}\frac{t^{p-2}u(t)^{1+bp}}{w(t)^{bp}} + \frac{p}{2}\Gamma u(t),$$

then

$$u(t) \leq t^{-(1-1/p)/b}(c_1 p^2)^{1/bp}e^{\Gamma t/2p}w(t).$$

Proof. Let $v(t) = u(t)e^{-p\Gamma t/2}$, so that

$$v'(t) = \left[u'(t) - \frac{p\Gamma}{2}u(t)\right]e^{-p\Gamma t/2}.$$

Hence

$$v'(t) \leq -\frac{a}{p}\frac{t^{p-2}v^{1+bp}e^{p\Gamma t(1+bp)/2}}{w^{bp}}e^{-p\Gamma t/2},$$

or

$$(v^{-bp})' \geq \frac{abt^{p-2}e^{bp^2\Gamma t/2}}{w^{bp}}.$$

Therefore

$$v^{-bp} \geq ab\int_0^t \frac{s^{p-2}e^{bp^2\Gamma s/2}}{w(s)^{bp}}\,ds$$

$$\geq \frac{ab}{w(t)^{bp}}\int_0^t s^{p-2}e^{bp^2\Gamma s/2}\,ds.$$

Using a change of variables, the integral on the right is

$$\left(\frac{2t}{bp^2\Gamma}\right)^{p-1}\int_0^{bp^2\Gamma/2} e^{st}s^{p-2}\,ds$$

$$\geq \left(\frac{2t}{bp^2\Gamma}\right)^{p-1}e^{t(1-1/p^2)bp^2\Gamma/2}\int_{(1-1/p^2)bp^2\Gamma/2}^{bp^2\Gamma/2} s^{p-2}\,ds$$

$$= t^{p-1}e^{t(bp^2\Gamma-b\Gamma)/2}\frac{1-(1-1/p^2)^{p-1}}{p-1}.$$

Since

$$\frac{1-(1-1/p^2)^{p-1}}{p-1} \geq c_2/p^2$$

for $p \in [2,\infty)$, then

$$v^{-bp} \geq \frac{c_3}{w^{bp}}\frac{t^{p-1}}{p^2}e^{bp^2\Gamma t/2}e^{-b\Gamma t/2},$$

and our estimate follows easily. □

Our estimate of the upper bound is completed by

(5.5) Theorem. *There exist c_1 and c_2 depending only on Λ such that if $t > 0$, then*

$$p(t,x,y) \leq c_1 t^{-d/2}e^{-|x-y|^2/c_2 t}$$

for almost every pair (x,y) (with respect to Lebesgue measure on $\mathbb{R}^d \times \mathbb{R}^d$).

Proof. For $p \geq 1$, let

$$u_p(t) = \|f_t\|_{2p},$$

and if $p \geq 2$,

$$w_p(t) = \sup_{0 \leq s \leq t} s^{d(p-2)/4p} u_{p/2}(s).$$

By Corollary 5.3,

$$w_2(t) \leq e^{t\Gamma/2}\|f\|_2.$$

By Proposition 5.2,

$$u_p'(t) \leq -\frac{c_3}{p}\frac{u_p(t)^{1+4p/d}}{u_{p/2}(t)^{4p/d}} + \frac{p}{2}\Gamma u_p(t)$$

$$\leq -\frac{c_3}{p}\frac{u_p(t)^{1+4p/d}\,t^{p-2}}{w_p(t)^{4p/d}} + \frac{p}{2}\Gamma u_p(t);$$

hence by Proposition 5.4 (with $a = c_3$ and $b = 4/d$) and the definition of $w_{2p}(t)$,

$$w_{2p}(t) \leq (c_4 p^2)^{d/4p} e^{\Gamma t/2p} w_p(t).$$

Using induction and the fact that

$$\prod_{n=1}^{\infty}(c_4 2^{2n})^{d/4\cdot 2^n} < \infty, \qquad \sum_{n=1}^{\infty}\frac{\Gamma t}{2^n} \leq \Gamma t,$$

we have

$$w_{2^n}(t) \leq c_5 e^{\Gamma t}\|f\|_2.$$

Therefore

$$\|f_t\|_\infty \leq \limsup_{n\to\infty}\|f_t\|_{2^n} \leq c_6 t^{-d/4} e^{\Gamma t}\|f\|_2.$$

We have thus shown that P_t^R maps L^2 into L^∞ with norm bounded by $c_6 t^{-d/4}e^{\Gamma t}$. Clearly the same bound holds for P_t^{-R}. Note that

$$\int f(P_t^R g) = \int f e^R P_t(e^{-R}g) = \int g e^{-R} P_t(e^R f) = \int g(P_t^{-R}f).$$

By the duality argument of Theorem 4.3,

$$\|P_t^R f\|_\infty \leq c_6 t^{-d/4} e^{\Gamma t}\|f\|_2.$$

Therefore, since

$$P_t^R f = e^R P_t(e^{-R}f) = e^R P_{t/2}\big(e^{-R}e^R P_{t/2}(e^{-R}f)\big) = P_{t/2}^R P_{t/2}^R f,$$

$$\|P_t^R f\|_\infty \leq c_6^2 (t/2)^{-d/2} e^{2\Gamma(t/2)}\|f\|_1.$$

This shows

$$e^{R(x)}p(t,x,y)e^{-R(y)} \leq c_7 t^{-d/2} e^{\Gamma t}$$

for almost every pair (x,y). Fix t_0 and then fix (x_0, y_0) not in the null set. Let $R(x)$ be a smooth bounded function equal to $x \cdot (x_0 - y_0)/2\Lambda^{-1}t_0$ for

$|x| \leq N$ and $\|\nabla R\|_\infty \leq |x_0 - y_0|/2\Lambda^{-1}t_0$, where N is a large real number. Then

$$p(t, x, y) \leq c_7 t^{-d/2} \exp\left(\frac{x_0 - y_0}{2\Lambda^{-1}t_0} \cdot (y - x) + \Lambda^{-1}\left(\frac{|y_0 - x_0|}{2\Lambda^{-1}t_0}\right)^2 t\right). \qquad (5.2)$$

Applying (5.2) with $x = x_0$ and $y = y_0$ and $t = t_0$ yields the desired result.

$\square$

Set $Gf(x) = \mathbb{E}^x \int_0^\infty f(X_s)\, ds$.

(5.6) Proposition. *Suppose $d \geq 3$. There exist $\alpha > 0$ and $c_1 > 0$ such that*

(a) $$|Gf(x)| \leq c_1(\|f\|_\infty + \|f\|_1)$$

(b) $$|Gf(x) - Gf(y)| \leq c_1(\|f\|_\infty + \|Gf\|_\infty)|x - y|^\alpha.$$

Proof. For $f \geq 0$,

$$Gf(x) = \int_0^\infty P_s f(x)\, ds = \int_0^\infty \int f(y) p(s, x, y)\, dy\, ds$$
$$\leq \int f(y) \int_0^\infty p(s, x, y)\, dy\, ds \leq c_2 \int f(y)|x - y|^{2-d}\, dy,$$

using Theorem 5.5 (cf. [PTA, Proposition II.3.1]). For bounded f, we have the estimate

$$|Gf(x)| \leq \mathbb{E}^x \int_0^\infty |(1_{B(x,1)}f)(X_s)|\, ds + \mathbb{E}^x \int_0^\infty |(1_{B(x,1)^c}f)(X_s)|\, ds$$
$$\leq \|f\|_\infty \int_0^\infty P_s 1_{B(x,1)}(x)\, ds + \int_0^\infty P_s(1_{B(x,1)^c}|f|)(x)\, ds.$$

Since

$$\int 1_{B(x,1)}(y)|x - y|^{2-d}\, dy \leq c_3$$

and

$$\int |f(y)|1_{B(x,1)^c}(y)|x - y|^{2-d}\, dy \leq c_4 \int |f(y)|\, dy,$$

(a) follows.

Now fix x_0 and suppose $x, y \in B(x_0, r)$, where r will be chosen in a moment.

$$Gf(x) = \mathbb{E}^x \int_0^{\tau(B(x_0,r))} f(X_s)\, ds + \mathbb{E}^x \mathbb{E}^{X_{\tau(B(x_0,r))}} \int_0^\infty f(X_s)\, ds \qquad (5.3)$$
$$= \mathbb{E}^x \int_0^{\tau(B(x_0,r))} f(X_s)\, ds + \mathbb{E}^x Gf(X_{\tau(B(x_0,r))}).$$

We have a similar expression for $Gf(y)$. The first term on the right of (5.3) is bounded in absolute value by $\|f\|_\infty \mathbb{E}^x \tau_{B(x_0,r)}$ and

$$\mathbb{E}^x \tau_{B(x_0,r)} \le \mathbb{E}^x \int_0^\infty 1_{B(x_0,r)}(X_s)\, ds$$

$$\le c_5 \int_{B(x_0,r)} |x-y|^{2-d}\, dy \le c_6 r^2.$$

On the other hand, $h(x) = \mathbb{E}^x Gf(X_{\tau(B(x_0,r))})$ is $\mathcal{L}$-harmonic in $B(x_0, r/2)$ by Proposition III.5.1. By Theorem 3.7 and scaling, there exists $\beta > 0$ such that

$$|h(x) - h(y)| \le c_7(|x-y|/r)^\beta \|h\|_\infty \le c_7(|x-y|/r)^\beta \|Gf\|_\infty$$

if $x, y \in B(x_0, r/2)$. Combining,

$$|Gf(x) - Gf(y)| \le c_8 \big(\|f\|_\infty + \|Gf\|_\infty\big)\Big(r^2 + \frac{|x-y|^\beta}{r^\beta}\Big).$$

If we take $r = |x-y|^{\beta/(\beta+2)}$, we obtain (b). $\qquad\square$

Let N be fixed, $D = B(x_0, N)$, and $\overline{G}f(x) = \mathbb{E}^x \int_0^{\tau_D} f(X_s)\, ds$. A similar proof shows

(5.7) Corollary. *There exist* $c_1, c_2 > 0$ *and* $\alpha > 0$ *such that*

(a)
$$|\overline{G}f(x)| \le c_1 \|f\|_\infty$$

(b)
$$|\overline{G}f(x) - \overline{G}f(y)| \le c_2 \big(\|f\|_\infty + \|\overline{G}f\|_\infty\big)|x-y|^\alpha.$$

c_1 *may depend on* N, *but* c_2 *does not.*

Proof. The only difference with the proof of Proposition 5.6 is the observation that $\|f\|_{L^1(D)} \le |D|\,\|f\|_\infty$. $\qquad\square$

6. Lower bounds

In this section we show there exist c_1 and c_2 such that

$$p(t, x, y) \ge c_1 t^{-d/2} e^{-|x-y|^2/c_2 t}.$$

We assume throughout that $d \ge 3$. This restriction will be removed in Section 7.

Let $x_0 \in \mathbb{R}^d$, $N > 0$, and let $D = B(x_0, N)$. We consider the process X_t killed on exiting D. Observe that

$$\mathbb{P}^x(X_t \in B, \tau_D > t) \le \mathbb{P}^x(X_t \in B) = \int_B p(t, x, y)\, dy. \tag{6.1}$$

This shows that $\mathbb{P}^x(X_t \in dy, \tau_D > t)$ has a density bounded by $p(t, x, y)$. We denote the density by $p_D(t, x, y)$, or when it is clear which domain D is meant, by $\bar{p}(t, x, y)$.

(6.1) Proposition. $\bar{p}(t, x, y) = \bar{p}(t, y, x)$ *for almost every pair* (x, y) *(with respect to Lebesgue measure on* $D \times D$*).*

Proof. We know that X_t has continuous paths and that $p(t, x, y)$ is symmetric for almost every pair. With this observation the proof is now the same as that for Brownian motion; see [PTA, Proposition II.4.1] or Port and Stone [1]. $\qquad\square$

Since $\bar{p}(t, x, y) \le p(t, x, y) \le c_1 t^{-d/2}$, we have

$$\int_D \int_D \bar{p}(t, x, y)^2 \, dx \, dy \le c_1^2 t^{-d} |D|^2, \tag{6.2}$$

so $\overline{P}_t f(x) = \int_D f(y) \bar{p}(t, x, y) \, dy$ is a bounded linear operator on $L^2(D)$.

A linear operator T in L^2 is said to be *completely continuous* if the closure of $\{T f_n\}$ is compact whenever $\{f_n\}$ is a bounded sequence in L^2. The identity I is an example of a bounded linear operator on L^2 that is not completely continuous, for if $\{f_n\}$ is an orthonormal set in $L^2(D)$, then

$$\|I f_n - I f_m\|_2^2 = \|f_n - f_m\|_2^2 = \|f_n\|_2^2 + \|f_m\|^2 - 2 \int f_n f_m = 2,$$

and no subsequence can converge. See Riesz and Sz.-Nagy [1] for further information on completely continuous operators.

(6.2) Proposition. $\overline{P}_t$ *is a completely continuous operator on* L^2.

Proof. An operator T of the form

$$T f(x) = \left(\int f(y) \varphi(y) \, dy \right) \varphi(x)$$

for $\varphi \in L^2$ is completely continuous. To see this, if f_n is a bounded sequence in L^2, then the numbers $a_n = \int f_n(y) \varphi(y) \, dy$ form a bounded sequence of reals, and some subsequence a_{n_j} converges, say to a. Hence $T f_{n_j} = a_{n_j} \varphi \to a\varphi$ in L^2. Since L^2 is a metric space, the fact that $\{T f_n\}$ has a convergent subsequence whenever $\{f_n\}$ is bounded implies that T is completely continuous. Such a T has a kernel $\varphi(x)\varphi(y)$, that is,

$$T f(x) = \int f(y) \big[\varphi(x)\varphi(y) \big] \, dy.$$

By the same argument, any operator with kernel $\sum_{i=1}^n b_i \varphi_i(x)\varphi_i(y)$ for some n and some finite sequence $\{\varphi_i\} \in L^2$ will be completely continuous. Here the $b_i \in \mathbb{R}$.

If T_i is completely continuous and $\|T_i - T\|_2 \to 0$ as $i \to \infty$, where $\|T_i - T\|_2 = \sup_{\|f\|_2 \leq 1} \|(T_i - T)f\|_2$, we show T is completely continuous. Suppose $\{f_n\}$ is a bounded sequence in L^2. For each i, $\{T_i f_n\}$ has compact closure, hence a subsequence converges. By a diagonalization procedure, we can find a subsequence n_j such that $T_i f_{n_j}$ converges in L^2 as $j \to \infty$ for each i. Since the triangle inequality tells us that

$$\|Tf_{n_j} - Tf_{n_k}\|_2 \leq \|T - T_i\|_2 \|f_{n_j}\|_2 + \|T - T_i\|_2 \|f_{n_k}\|_2$$
$$+ \|T_i f_{n_j} - T_i f_{n_k}\|_2,$$

then

$$\limsup_{j,k \to \infty} \|Tf_{n_j} - Tf_{n_k}\|_2 \leq 2\|T - T_i\|_2 \sup_n \|f_n\|_2.$$

This holds for all i, so $\limsup_{j,k \to \infty} \|Tf_{n_j} - Tf_{n_k}\|_2 = 0$. By the completeness of L^2, the sequence Tf_{n_j} converges in L^2, and hence the sequence has compact closure.

We now examine $\overline{P}_t$. From (6.2) and Proposition 6.1, we can approximate $\overline{p}(t,x,y)$ in $L^2(D \times D)$ by simple functions that are symmetric in x and y. A function $1_B(x,y)$ where $(x,y) \in B$ if $(y,x) \in B$ for $B \subseteq D \times D$ can be approximated in $L^2(D \times D)$ by functions of the form $\sum_{i=1}^n a_i 1_{C_i}(x) 1_{C_i}(y)$, $C_i \subseteq D$. Therefore $\overline{p}(t,x,y)$ can be approximated in $L^2(D \times D)$ by functions of the form

$$T(x,y) = \sum_{j=1}^m b_i \varphi_i(x) \varphi_i(y), \tag{6.3}$$

where each $\varphi_i \in L^2(D)$. If T is the operator on L^2 defined by $Tf(x) = \int f(y) T(x,y)\, dy$, then

$$\|\overline{P}_t f - Tf\|_2^2 = \int_D \left(\int_D f(y) \left[\overline{p}(t,x,y) - T(x,y) \right] dy \right)^2 dx$$
$$\leq \int_D \left[\int_D |f(y)|^2\, dy \right] \left[\int_D \left[\overline{p}(t,x,y) - T(x,y) \right]^2 dy \right] dx$$
$$= \|f\|_2^2 \int_D \int_D \left[\overline{p}(t,x,y) - T(x,y) \right]^2 dy\, dx.$$

Each operator T of the form (6.3) is completely continuous by what we have proved, and $\overline{P}_t$ can be approximated by such operators, and hence $\overline{P}_t$ is completely continuous. $\square$

(6.3) Theorem. *There exist a sequence of reals $0 < \lambda_1 \leq \lambda_2 \leq \cdots$ tending to ∞ and a complete orthonormal system φ_i for $L^2(D)$ with $\varphi_1 > 0$ almost everywhere such that for each $t > 0$*

$$\overline{p}(t,x,y) = \sum_{i=1}^{\infty} e^{-\lambda_i t} \varphi_i(x) \varphi_i(y) \tag{6.4}$$

for every pair $(x, y) \in D \times D$ *except for a null set with respect to Lebesgue measure on* $D \times D$ *(the null set may depend on t). The series in (6.4) converges absolutely with respect to* $L^\infty(D \times D)$.

Proof. We first show $\overline{P}_t$ has only nonzero eigenvalues. By Jensen's inequality, $\|\overline{P}_t f\|_2 \leq \|f\|_2$. Suppose $f \in L^2(D)$ and f is 0 outside of D. Given ε, let g be a continuous function with compact support such that $\|f - g\|_2 < \varepsilon$. Then

$$\limsup_{t \to 0} \|\overline{P}_t f - f\|_2 \leq \limsup_{t \to 0} \|\overline{P}_t f - \overline{P}_t g\|_2 + \limsup_{t \to 0} \|\overline{P}_t g - g\|_2$$
$$+ \|f - g\|_2$$
$$\leq 2\varepsilon,$$

since $\overline{P}_t g \to g$ for each x by the continuity of X_t. Therefore $\overline{P}_t f \to f$ in L^2 as $t \to 0$.

If $\overline{P}_t \varphi = 0$ for some φ, then

$$0 = \int \varphi (\overline{P}_t \varphi) = \int (\overline{P}_{t/2} \varphi)(\overline{P}_{t/2} \varphi)$$

by the semigroup property and the symmetry of $\overline{P}_t$. So $\overline{P}_{t/2} \varphi = 0$. By induction, $\overline{P}_{t/2^n} \varphi = 0$. Since $\overline{P}_{t/2^n} \varphi \to \varphi$ as $n \to \infty$, then $\varphi = 0$ almost everywhere. Hence $\overline{P}_t$ cannot have any zero eigenvalues.

We now proceed just as in [PTA, Theorem II.4.13], except we substitute the compactness in L^2 for the compactness in $C(D)$, where $C(D)$ denotes the continuous functions on D. In that proof a key fact was that $\overline{P}_t$ mapped bounded sequences in L^2 into a set whose closure in $C(D)$ was compact. Here we use the fact that $\overline{P}_t$ is completely continuous on L^2 and hence maps bounded sequences in L^2 into a set whose closure in $L^2(D)$ is compact. Substituting L^2 for $C(D)$ in the proof of [PTA, Theorem II.4.13], we obtain Theorem 6.3. $\qquad\square$

We are going to show that $\overline{p}(t, x, y)$ is jointly continuous in x and y and then later show that we can take the modulus to depend only on N and Λ.

(6.4) Proposition. *There exists a version of* $\overline{p}(t, x, y)$ *that is jointly continuous in* x *and* y.

Proof. Since we have that the convergence in (6.4) is absolute and takes places in $L^\infty(D \times D)$, to prove the proposition we need only show that φ_i has a continuous version. We have

$$|\overline{P}_t \varphi_i(x)| = \left| \int_D \overline{p}(t, x, y) \varphi_i(y) \, dy \right|$$
$$\leq \left(\int_D (\overline{p}(t, x, y))^2 \, dy \right)^{1/2} \left(\int_D \varphi_i(y)^2 \, dy \right)^{1/2} \leq c_1$$

since $\overline{p}(t, x, y) \leq p(t, x, y) \leq c_2 t^{-d/2}$ and $|D| < \infty$. From

$$\overline{P}_1\varphi_i = e^{-\lambda_i}\varphi_i, \qquad \text{a.e.}$$

we obtain

$$\|\varphi_i\|_\infty \le c_1 e^{\lambda_1} < \infty.$$

Integrating (6.4) over t from 0 to ∞,

$$\overline{G}\varphi_i = \lambda_i^{-1}\varphi_i, \qquad \text{a.e.,} \tag{6.5}$$

so by Corollary 5.7, $\lambda_i\overline{G}\varphi_i$ is a continuous version of φ_i. $\qquad\square$

Our estimates on $p(t,x,y)$ from Section 5 allow us to obtain a tightness estimate.

(6.5) Proposition. *There exist c_1 and c_2 such that*

$$\mathbb{P}^x(\sup_{s\le t}|X_s - x| \ge \lambda) \le c_1 e^{-c_2\lambda^2/t}.$$

Proof. Let $T = \inf\{t : |X_t - x| \ge \lambda\}$. We can write

$$\mathbb{P}^x(\sup_{s\le t}|X_s - x| \ge \lambda) \le \mathbb{P}^x(T < t, |X_t - x| < \lambda/2) \tag{6.6}$$

$$+ \mathbb{P}^x(|X_t - x| \ge \lambda/2).$$

The second term on the right is bounded by

$$\int_{B(x,\lambda/2)^c} p(t,x,y)\,dy \le \int_{|y-x|\ge\lambda/2} c_3 t^{-d/2} e^{-c_4|y-x|^2/t}\,dy \tag{6.7}$$

$$\le c_5 e^{-c_6\lambda^2/t}$$

by Theorem 5.5. Conditioning on $T \in ds$, the first term is bounded by

$$\int_0^t \mathbb{P}^x(T \in ds, |X_t - X_T| \ge \lambda/2) \tag{6.8}$$

$$= \int_0^t \mathbb{E}^x\left[\mathbb{P}^{X_T}(|X_{t-T} - X_0| \ge \lambda/2); T \in ds\right]$$

$$\le \int_0^t \mathbb{E}^x\left[\mathbb{P}^{X_s}(|X_{t-s} - X_0| \ge \lambda/2); T \in ds\right]$$

by the strong Markov property. By the same argument we used in deriving (6.7), there exist c_7 and c_8 such that

$$\sup_{r\le t}\mathbb{P}^z(|X_r - z| \ge \lambda/2) \le c_7 e^{-c_8\lambda^2/t}.$$

Substituting in (6.8) bounds the second term on the right in (6.6). $\qquad\square$

We can obtain a lower bound for $\overline{p}(t,x,x)$ on the diagonal as follows.

(6.6) Proposition. *There exist c_1 and c_2 depending only on Λ and N such that*

$$\bar{p}(t,x,x) \geq c_1 t^{-d/2}, \qquad t > 0, \quad x \in B(0, c_2 N).$$

Proof. If c_3 is sufficiently large, $\mathbb{P}^x(\sup_{s \leq t/2} |X_s - x| > c_3 t^{1/2}) \leq 1/2$ by Proposition 6.5. So

$$\int_{B(x, c_3 t^{1/2})} \bar{p}(t/2, x, y)\, dy \tag{6.9}$$

$$= \mathbb{P}^x(X_{t/2} \in B(x, c_3 t^{1/2}), \sup_{s \leq t/2} |X_s - x| < c_3 t^{1/2}) \geq 1/2.$$

We have by the semigroup property and the Cauchy-Schwartz inequality,

$$\left(\int_{B(x, c_3 t^{1/2})} \bar{p}(t/2, x, y)\, dy \right)^2 \leq |B(x, c_3 t^{1/2})| \int \bar{p}(t/2, x, y)^2\, dy$$

$$\leq c_4 t^{d/2} \bar{p}(t, x, x).$$

By (6.9),

$$\bar{p}(t, x, x) \geq c_4 t^{-d/2}/4,$$

which is the estimate we wanted. $\qquad\qquad\qquad\qquad\qquad\qquad\qquad\qquad \square$

The key estimate needed is a lower bound for $\bar{p}(t, x, y)$ when y is close to x.

(6.7) Proposition. *There exist c_1, c_2, and c_3 such that if $N = t^{1/2}$, $|y - x| \leq c_1 t^{1/2}$, and $x \in B(x_0, c_3 N)$, then $\bar{p}(t, x, y) \geq c_2 t^{-d/2}$.*

Proof. Fix t and x. Let

$$S(z) = \sum_{i=1}^{\infty} \lambda_i e^{-\lambda_i t} \varphi_i(x) \varphi_i(z), \qquad z \in D. \tag{6.10}$$

By the Cauchy-Schwarz inequality,

$$|S(z)| \leq \left(\sum_{i=1}^{\infty} \lambda_i e^{-\lambda_i t} \varphi_i(x)^2 \right)^{1/2} \left(\sum_{i=1}^{\infty} \lambda_i e^{-\lambda_i t} \varphi_i(z)^2 \right)^{1/2}.$$

If we let $a = \sup_{\lambda \geq 0} \lambda e^{-\lambda t/2}$, then $a < \infty$ and

$$\sum_{i=1}^{\infty} \lambda_i e^{-\lambda_i t} \varphi_i(z)^2 \leq a \sum_{i=1}^{\infty} e^{-\lambda_i t/2} \varphi_i(z)^2 \tag{6.11}$$

$$= a \bar{p}(t/2, z, z) \leq c_4 a t^{-d/2},$$

and similarly with z replaced with x. It follows that S is bounded.

Let us compute $\overline{G}S(y)$. By (6.5),

$$\overline{G}S(y) = \sum_{i=1}^{\infty} \lambda_i e^{-\lambda_i t} \varphi_i(x) \overline{G}\varphi_i(y) \tag{6.12}$$

$$= \sum_{i=1}^{\infty} \lambda_i e^{-\lambda_i t} \varphi_i(x) \lambda_i^{-1} \varphi_i(y)$$

$$= \overline{p}(t, x, y).$$

By Proposition 6.6, $\overline{p}(t, x, x) \geq c_5 t^{-d/2}$. Since both S and $\overline{G}S$ are bounded by constant multiples of $t^{-d/2}$, Corollary 5.7 and scaling tell us that there exists c_6 such that

$$|\overline{G}S(x) - \overline{G}S(y)| \leq c_5 t^{-d/2}/2, \qquad |x - y| \leq c_6 t^{1/2}, \quad x \in B(0, c_6 t^{1/2}).$$

Hence for such y, $\overline{G}S(y) \geq c_5 t^{-d/2}/2$. Combining with (6.12) completes the proof. $\qquad\qquad\square$

We now complete the proof of the lower bound by what is known as a *chaining argument*. Note $p(t, x, y) \geq \overline{p}(t, x, y)$.

(6.8) Theorem. *There exist c_1 and c_2 depending only on Λ such that*

$$p(t, x, y) \geq c_1 t^{-d/2} e^{-c_2 |x-y|^2/t}.$$

Proof. By Proposition 6.7 with $x_0 = x$, there exists c_3 such that if $|x - y| < c_3 t^{1/2}$, then $p(t, x, y) \geq c_4 t^{-d/2}$. Thus to prove the theorem, it suffices to consider the case $|x - y| \geq c_3 t^{1/2}$.

By Proposition 6.7 (with $x_0 = w$) and scaling, there exist c_4 and c_5 such that if $|z - w| \leq c_4 (t/n)^{1/2}$,

$$p(t/n, w, z) \geq p_{B(w,(t/n)^{1/2})}(t/n, w, z) \geq c_5 (t/n)^{-d/2}. \tag{6.13}$$

Let $R = |x - y|$ and let n be the smallest positive integer greater than $9R^2/c_4^2 t$. So $3R/n \leq c_4 (t/n)^{1/2}$. Let $v_0 = x$, $v_n = y$, and $v_1, \ldots, v_{n-1}$ be points equally spaced on the line segment connecting x and y. Let $B_i = B(v_i, R/n)$. If $w \in B_i$ and $z \in B_{i+1}$, then $|z - w| \leq 3R/n \leq c_4 (t/n)^{1/2}$, and so $p(t/n, w, z) \geq c_5 (t/n)^{-d/2}$ by (6.13).

By the semigroup property,

$$p(t, x, y)$$

$$= \int \cdots \int p(t/n, x, z_1) p(t/n, z_1, z_2) \cdots p(t/n, z_{n-1}, y) \, dz_1 \cdots dz_{n-1}$$

$$\geq \int_{B_{n-1}} \cdots \int_{B_1} p(t/n, x, z_1) p(t/n, z_1, z_2) \cdots p(t/n, z_{n-1}, y) \, dz_1 \cdots dz_{n-1}$$

$$\geq (c_6 (t/n)^{-d/2})^n \prod_{i=1}^{n-1} |B_i|.$$

Since $|B_i| \geq c_7(R/n)^d$ and $(R/n)(t/n)^{-1/2}$ is bounded below by a positive constant, then

$$p(t,x,y) \geq c_8 c_9^n (n/t)^{d/2} \geq c_8 t^{-d/2} \exp(-n \log c_9^{-1}).$$

If $n > 2$, then $n/2 \leq 9R^2/c_4^2 t$, so

$$p(t,x,y) \geq c_8 t^{-d/2} \exp(-18R^2 \log c_9^{-1}/c_4^2 t).$$

If $n \leq 2$, then $9R^2/c_4^2 t \leq 2$, and

$$p(t,x,y) \geq c_8 t^{-d/2} \exp(-2 \log c_9^{-1}).$$

The result follows with $c_1 = c_8(c_9^2 \wedge 1)$ and $c_2 = 18(\log(c_9^{-1}) \wedge 1)/c_4^2$. $\square$

7. Extensions

Until later on in this section we continue to suppose that $d \geq 3$ and that the a_{ij} are smooth.

(7.1) Theorem. *There exist c_1 and α such that*

$$|p(t,x,y) - p(t,x,y')| \leq c_1 t^{-d/2}|y - y'|^\alpha.$$

Proof. We showed in Proposition 6.7 that $\overline{p}(t,x,y) = \overline{G}S(y)$, where S was defined by (6.10); here $\overline{p}(t,x,y) = p_{B(x_0,N)}(t,x,y)$ and $\overline{G}$ are the transition densities and potential operator, respectively, for X_t killed on exiting $B(x_0,N)$ for some x_0 and N. By Corollary 5.7,

$$|\overline{G}S(y) - \overline{G}S(y')| \leq c_2|y - y'|^\alpha(\|S\|_\infty + \|\overline{G}S\|_\infty).$$

Moreover, both S and $\overline{G}S$ were bounded by $c_3 t^{-d/2}$, where c_3 was independent of x_0 and N. Hence

$$|\overline{p}(t,x,y) - \overline{p}(t,x,y')| \leq c_4 t^{-d/2}|y - y'|^\alpha, \tag{7.1}$$

c_4 independent of N. We let $N \to \infty$; using the fact that $p_{B(x_0,N)}(t,x,y)$ increases up to $p(t,x,y)$ as $N \to \infty$, we obtain (7.1) with $\overline{p}$ replaced by p.
   $\square$

By symmetry, we obtain an analogous result with the roles of x and y reversed. Joint continuity of $p(t,x,y)$ in (x,y) follows by the semigroup property.

The partial derivatives of $p(t,x,y)$ with respect to t exist and are continuous in x and y. We let $\partial_t^k p(t,x,y)$ denote $\partial^k p(t,x,y)/\partial t^k$.

(7.2) Proposition. *$p(t, x, y)$ has derivatives in t of all orders on $(0, \infty) \times \mathbb{R}^d \times \mathbb{R}^d$ and $\partial_t^k p(t, x, y)$ is Hölder continuous in x and y for all k.*

Proof. Fix N and recall the notation $\overline{p}(t, x, y)$ and the eigenvalue expansion (6.4). If we let

$$R_k(z) = -\sum_{i=1}^{\infty} (-\lambda_i)^{k+1} e^{-\lambda_i t} \varphi_i(x) \varphi_i(z),$$

then since $\sup_{\lambda \geq 0} \lambda^{k+1} e^{-\lambda t/2} < \infty$, we see just as in the proof in Proposition 6.7 that $\|R_k\|_\infty \leq c_1 < \infty$, where c_1 depends on t and k. Moreover,

$$\overline{G} R_k(x) = \sum_{i=1}^{\infty} (-\lambda_i)^k e^{-\lambda_i t} \varphi_i(x) \varphi_i(y), \tag{7.2}$$

and $\|\overline{G} R_k\|_\infty \leq c_2$ with c_2 depending on t and k but not N. Since

$$\overline{P}_s \varphi_i = e^{-\lambda_i s} \varphi_i,$$

then

$$\|\varphi_i\|_\infty = e^{\lambda_i s} \|\overline{P}_s \varphi_i\|_\infty$$
$$\leq e^{\lambda_i s} \left(\int_D \overline{p}(t, x, y)^2 \, dy \right)^{1/2} \left(\int_D \varphi_i(y)^2 \, dy \right)^{1/2}$$
$$\leq c_3 e^{\lambda_i s} s^{-d/2}, \qquad s > 0.$$

With this bound and (6.4), it is easy to use dominated convergence to see that $\overline{p}(t, x, y)$ is differentiable in t and the kth partial derivative is $\overline{G} R_k(x)$. We thus have by Corollary 5.7 that $\partial_t^k \overline{p}(t, x, y)$ is bounded and Hölder continuous for all k. Letting $N \to \infty$, a limit argument shows that the same is true for $\partial_t^k p(t, x, y)$ (cf. Barlow and Bass [1]).   □

We obtain the following Green function estimates.

(7.3) Proposition. *If $d \geq 3$, there exists a symmetric function $G(x, y)$ that is continuous in x on $\mathbb{R}^d - \{y\}$ such that*

$$\mathbb{E}^x \int_0^\infty 1_B(X_s) \, ds = \int_B G(x, y) \, dy,$$

for all Borel sets B. There exist c_1 and c_2 such that

$$c_1 |x - y|^{2-d} \leq G(x, y) \leq c_2 |x - y|^{2-d}.$$

The function $G(\cdot, y)$ is $\mathcal{L}$-harmonic on $\mathbb{R}^d - \{y\}$.

Proof. Let $G(x, y) = \int_0^\infty p(s, x, y) \, ds$. The bounds, continuity, and symmetry are immediate from integration and the corresponding facts about $p(s, x, y)$. By Fubini's theorem,

$$\mathbb{E}^x \int_0^\infty 1_B(X_s)\, ds = \int_0^\infty \mathbb{P}^x(X_s \in B)\, ds = \int_0^\infty \int_B p(s,x,y)\, dy\, ds$$

$$= \int_B \int_0^\infty p(s,x,y)\, ds\, dy = \int_B G(x,y)\, dy.$$

Fix y and let B' be a ball about y. By the strong Markov property, if $B \subseteq B'$,

$$G1_B(x) = \mathbb{E}^x \int_{T_{B'}}^\infty 1_B(X_s)\, ds = \mathbb{E}^x \mathbb{E}^{X_{T(B')}} \int_0^\infty 1_B(X_s)\, ds$$

$$= \mathbb{E}^x G1_B(X_{T_{B'}}).$$

So for $x \notin B'$, $G1_B(x)$ is $\mathcal{L}$-harmonic as a consequence of Theorem III.5.1. On the other hand, $G1_B(z) = \int_B G(z,y)\, dy$. Letting $B = B(y,\varepsilon)$, dividing by $|B(y,\varepsilon)|$, and letting $\varepsilon \to 0$, it follows that $G(x,y) = \mathbb{E}^x G(X_{T_{B'}})$. This proves that $G(x,y)$ is $\mathcal{L}$-harmonic for $x \notin B'$. Since B' is arbitrary, $G(x,y)$ is $\mathcal{L}$-harmonic in $\mathbb{R}^d - \{y\}$. $\qquad\square$

We can also obtain a lower bound for the Green function in a ball.

(7.4) Proposition. *Let $x_0 \in \mathbb{R}^d$, $N > 0$. Suppose $\overline{G}(x,y) = \int_0^\infty \overline{p}(s,x,y)\, ds$. If $r < 1$, there exists c_1 depending on r, Λ, and N such that $\overline{G}(x,y) \geq c_1 |x - y|^{2-d}$ if $x, y \in B(x_0, rN)$.*

Proof. Fix r. If $n > 4/(1-r)^2$, then $n^{-1/2} < (1-r)/2$, and hence $B(x, n^{-1/2}N)$ and $B(y, n^{-1/2}N)$ are contained in $B(x_0, N)$. If we take $n > 4(1-r)^2$ sufficiently large, the chaining argument in the proof of Theorem 6.8 shows that

$$\overline{p}(1,x,y) \geq c_2 > 0, \qquad x, y \in B(x_0, (1+r)N/2). \tag{7.3}$$

Let $\varepsilon < (1-r)/2$. If $B \subseteq B(y, \varepsilon N)$ is a small ball about y, then by the strong Markov property,

$$G1_B(x) = \overline{G}1_B(x) + \mathbb{E}^x G1_B(X_{\tau(B(0,N))}). \tag{7.4}$$

By Proposition 7.3, $G1_B(z) \leq c_3 |B|\, |z - y|^{2-d}$ if $z \in \partial B(0,N)$. On the other hand,

$$G1_B(x) \geq c_4 |x - y|^{2-d} |B|.$$

If ε is also less than $(c_4/2c_3)^{1/(d-2)}(1-r)$ and $w \in B(y, \varepsilon N)$, then

$$\sup_{z \in \partial B(x_0, N)} c_3 |z - y|^{2-d} \leq \inf_{w \in B(y, \varepsilon N)} c_4 |w - y|^{2-d}/2.$$

Substituting in (7.4),

$$\overline{G}1_B(w) \geq c_5 |w - y|^{2-d} |B|.$$

Letting B shrink to $\{y\}$ shows $\overline{G}(w,y) \geq c_5 |w - y|^{2-d}$ if $w \in B(y, \varepsilon N)$. This establishes the proposition if $x \in B(y, \varepsilon N)$. If $x \notin B(y, \varepsilon N)$, from the Markov property and (7.3) we have

$$\overline{G}(x,y) \geq \int_{B(y,\varepsilon N)} \overline{p}(1,x,w)\overline{G}(w,y)\,dw$$

$$\geq c_2 c_5 |B(y,\varepsilon N)| \varepsilon^{2-d} \geq c_6 |x-y|^{2-d},$$

where $c_6 = c_2 c_5 |B(y,\varepsilon N)|$ depends on r and N. $\qquad\square$

Up until now we have required on our proofs that $d \geq 3$. However, Theorems 5.5, 6.8, and 7.1 are still valid when $d = 2$.

(7.5) Theorem. *Suppose $d = 2$. Then the conclusions of Theorems 5.5, 6.8, and 7.1 hold.*

Proof. Let X_t be the process associated with $\mathcal{L} \in \mathcal{D}(\Lambda)$. Let Y_t be an independent one-dimensional Brownian motion and let $\widetilde{X}_t = (X_t, Y_t)$. It is easy to see that the operator $\widetilde{\mathcal{L}}$ associated with $\widetilde{X}_t$ is in $\mathcal{D}(\widetilde{\Lambda})$ for some $\widetilde{\Lambda} > 0$ and $\widetilde{a}_{ij}(x_1,x_2,x_3) = a_{ij}(x_1,x_2)$ if $i,j \leq 2$ and equals $(1/2)\delta_{ij}$ if one or both of i,j equals 3. We have Theorems 5.5, 6.8, and 7.1 holding for the transition densities $\widetilde{p}(t,x,y)$.

By the independence, if $x = (x_1,x_2)$ and $y = (y_1,y_2)$ are in $\mathbb{R}^2$ and $\widetilde{x} = (x_1,x_2,x_3)$, $\widetilde{y} = (y_1,y_2,y_3)$, then

$$\widetilde{p}(t,\widetilde{x},\widetilde{y}) = p(t,x,y)q(t,x_3,y_3),$$

where $q(t,x_3,y_3) = (2\pi t)^{-1/2} \exp(|x_3 - y_3|^2/2t)$ is the transition density for one-dimensional Brownian motion. Our estimates for $p(t,x,y)$ follow by dividing the estimates for $\widetilde{p}(t,\widetilde{x},\widetilde{y})$ by $q(t,x_3,y_3)$. The joint continuity follows similarly. $\qquad\square$

We have required that the a_{ij} be smooth functions of x. The conclusions of Theorems 5.5 and 6.8 and the estimates in this section are still valid when the a_{ij} are not smooth but $\mathcal{L} \in \mathcal{D}(\Lambda)$. See Moser [1] and Littman, Stampacchia, and Weinberger [1] for a discussion of what a solution to $\mathcal{L}u = 0$ means in this case.

8. Path properties

In this section we want to use our analytic estimates to derive some properties of the process X_t associated with an operator $\mathcal{L}$.

(8.1) Proposition. *Suppose $d \geq 2$, $\mathcal{L} \in \mathcal{D}(\Lambda)$. There exist c_1, c_2, c_3, c_4 such that*

$$c_1 e^{-c_2 \lambda^2/t} \leq \mathbb{P}^x(\sup_{s \leq t} |X_s - x| \geq \lambda) \leq c_3 e^{-c_4 \lambda^2/t}.$$

Proof. The upper bound for the case $d \geq 3$ was done in Proposition 6.2. The same argument takes care of the case $d = 2$ as well, using Theorem 7.5.

The lower bound is a consequence of the fact that

$$\mathbb{P}^x(\sup_{s\leq t}|X_s - x| \geq \lambda) \geq \mathbb{P}^x(|X_t - x| \geq \lambda) = \int_{B(x,\lambda)^c} p(t,x,y)\,dy$$

and Theorem 6.8. $\qquad\qquad\square$

(8.2) Proposition. *Suppose $d \geq 2$, $\mathcal{L} \in \mathcal{D}(\Lambda)$. There exist c_1, c_2, c_3, c_4 such that*

$$c_1 e^{-c_2 t/\lambda^2} \leq \mathbb{P}^x(\sup_{s\leq t}|X_s - x| \leq \lambda) \leq c_3 e^{-c_4 t/\lambda^2}.$$

Proof. Let us first look at the upper bound. By Theorem 5.5,

$$\sup_{y\in B(x,\lambda)} \mathbb{P}^y(X_{t_0} \in B(x,\lambda)) \leq \sup_{y\in B(x,\lambda)} \int_{B(x,\lambda)} p(t_0, y, z)\,dz$$
$$\leq c_5 t_0^{-d/2}|B(x,\lambda)| \leq c_6 \lambda^d t_0^{-d/2}.$$

If we take $t_0 = (2c_6)^{2/d}\lambda^2$, the right-hand side will be less than $1/2$. By the Markov property,

$$\mathbb{P}^x(\sup_{s\leq 2t_0} |X_s - x| \leq \lambda) \leq \mathbb{P}^x(X_{t_0} \in B(x,\lambda), X_{2t_0} \in B(x,\lambda))$$
$$\leq \mathbb{E}^x\left[\mathbb{P}^{X_{t_0}}(X_{t_0} \in B(x,\lambda); X_{t_0} \in B(x,\lambda)\right]$$
$$\leq (1/2)\mathbb{P}^x(X_{t_0} \in B(x,\lambda)) \leq 1/4,$$

and by induction,

$$\mathbb{P}^x(\sup_{s\leq nt_0} |X_s - x| \leq \lambda) \leq 2^{-n} = e^{-n\log 2}. \tag{8.1}$$

So if $t > t_0$, the upper bound follows from (8.1) by letting n be the largest integer less than $t/t_0 = (2c_6)^{-2/d}(t/\lambda^2)$. If $t \leq t_0$, the upper bound follows by taking c_3 large enough, since probabilities are bounded above by 1.

We now turn to the lower bound. We will show there exist a and b such that if $t_0 = a\lambda^2$ and

$$C = \{X_{t_0} \in B(x, \lambda/3), \sup_{s\leq t_0}|X_s - x| \leq 2\lambda/3\},$$

then

$$\inf_{y\in B(x,\lambda/3)} \mathbb{P}^y(C) \geq b. \tag{8.2}$$

Given (8.2), by the Markov property,

$$\mathbb{P}^x(\sup_{s \le 2t_0} |X_s - x| \le \lambda)$$

$$\ge \mathbb{P}^x\Big(X_{t_0} \in B(x, \lambda/3), \sup_{s \le t_0} |X_s - x| \le 2\lambda/3,$$

$$X_{2t_0} \in B(x, \lambda/3), \sup_{t_0 \le s \le 2t_0} |X_s - X_{t_0}| \le 2\lambda/3\Big)$$

$$= \mathbb{P}^x(C, C \circ \theta_{t_0}) = \mathbb{E}^x\Big[\mathbb{P}^{X_{t_0}}(C); C\Big]$$

$$\ge b\mathbb{P}^x(C) \ge b^2;$$

in the next to the last inequality we used the fact that on the set C we have $X_{t_0} \in B(x, \lambda/3)$. By induction,

$$\mathbb{P}^x(\sup_{s \le nt_0} |X_s - x| \le \lambda) \ge b^n,$$

and this suffices to prove the proposition.

We now prove (8.2). If $y \in B(x, \lambda/3)$, on the one hand, we have

$$\mathbb{P}^y(X_{t_0} \in B(x, \lambda/3)) \ge \int_{B(x,\lambda/3)} p(a\lambda^2, y, z)\, dz \tag{8.3}$$

$$\ge c_7(a\lambda^2)^{-d/2}|B(x, \lambda/3)| \ge c_8 a^{-d/2}$$

by Theorem 6.8. On the other hand, by Proposition 8.1,

$$\mathbb{P}^y(\sup_{s \le t_0} |X_s - x| > 2\lambda/3) \le \mathbb{P}^y(\sup_{s \le t_0} |X_s - y| > \lambda/3) \tag{8.4}$$

$$\le c_9 e^{-c_{10}\lambda^2/t_0} = c_9 e^{-c_{10}/a}.$$

We can now choose a small, depending only on c_8, c_9, c_{10}, such that $c_8 a^{-d/2} \ge 2c_9 e^{-c_{10}/a}$, and we deduce that $\mathbb{P}^y(C) \ge c_8 a^{-d/2}/2$, which proves (8.2). $\square$

From Proposition 8.2 we obtain

(8.3) Proposition. *There exist c_1 and c_2 such that*

$$c_1 r^2 \le \mathbb{E}^x \tau_{B(x,r)} \le c_2 r^2.$$

Proof. We have $\mathbb{P}^x(\tau_{B(x,r)} > t) = \mathbb{P}^x(\sup_{s \le t} |X_s - x| \le r)$. We now use Proposition 8.2 and integrate over t from 0 to ∞. $\square$

Higher moments of $\tau_{B(x,r)}$ under $\mathbb{P}^x$ can be estimated similarly.

(8.4) Theorem. (Support theorem) *Let $\psi : [0, t] \to \mathbb{R}^d$ be continuous and let $\varepsilon > 0$. There exists c_1 depending only on Λ, ε, t, and the modulus of continuity of ψ such that*

$$\mathbb{P}^{\psi(0)}(\sup_{s \le t} |X_s - \psi(s)| < \varepsilon) \ge c_1.$$

Proof. As in the proof of Theorem I.8.5, it suffices to consider the case where ψ is differentiable. Using (1.3) we see there exists a Brownian motion W_t such that

$$dX_t = \sigma(X_t)\,dW_t + b(X_t)\,dt,$$

where σ is a positive definite square root of a and

$$b_j(x) = (1/2)\sum_{i=1}^{d}\partial_i a_{ij}(x).$$

Let $M_t = \int_0^t \psi'(s)\sigma^{-1}(X_s)\,dW_s$. Note that $\langle M\rangle_t$ is bounded by $c_2 t$, where c_2 depends on Λ and the size of ψ' but not on the smoothness of the a_{ij}. Define a probability measure $\mathbb{Q}$ by setting its Radon-Nikodym derivative with respect to $\mathbb{P}^x$ to be equal to $\exp(M_t - \langle M\rangle_t/2)$. As in the proof of Theorem I.8.5, under $\mathbb{Q}$, $X_t - \psi(t)$ is a process associated to the operator $\mathcal{L}$. So $\mathbb{Q}(C) > c_3$ by Proposition 8.2, where $C = \{\sup_{s\le t}|X_s - \psi(s)| < \varepsilon\}$. Then

$$c_3 \le \mathbb{Q}(C) = \int_C e^{M_t - \langle M\rangle_t/2}\,d\mathbb{P}^x \le \left(\mathbb{E}^x_{\mathbb{P}}\left[e^{2M_t - \langle M\rangle_t}\right]\right)^{1/2}\left(\mathbb{P}^x(C)\right)^{1/2}$$

by the Cauchy-Schwarz inequality. Since

$$\mathbb{E}^x_{\mathbb{P}}e^{2M_t - \langle M\rangle_t} = \mathbb{E}^x_{\mathbb{P}}e^{2M_t - \langle 2M\rangle_t/2}e^{\langle M\rangle_t} < \infty,$$

we obtain $\mathbb{P}^x(C) \ge c_4$ with c_4 depending on the a_{ij} only through Λ and not the smoothness of the a_{ij}. $\qquad\square$

As with processes associated to operators in nondivergence form, a process associated to $\mathcal{L} \in \mathcal{D}(\Lambda)$ hits small sets.

(8.5) Proposition. *Suppose $B \subseteq B(x,1)$. There exists c_1 depending only on Λ and $|B|$ such that*

$$\mathbb{P}^x(T_B < \tau_{B(x,1)}) \ge c_1.$$

Proof. By looking at $B \cap B(x,r)$ for r sufficiently close to 1, we may suppose $B \subseteq B(x,r)$ for some $r < 1$ and still $|B| > 0$. We write τ for $\tau_{B(x,1)}$. By Proposition 7.4, if $\overline{G}(x,y)$ is the Green function for $B(x,1)$,

$$\mathbb{E}^x\int_0^\tau 1_B(X_s)\,ds = \int \overline{G}(x,y)1_B(y)\,dy \ge c_2,$$

c_2 depending only on $|B|$. By the strong Markov property and the fact that $\overline{G}1_B(z) \le \overline{G}1_{B(x,1)}(z) \le c_3$,

$$\mathbb{E}^x\int_0^\tau 1_B(X_s)\,ds = \mathbb{E}^x\left[\mathbb{E}^{X_{T_B}}\int_0^\tau 1_B(X_s)\,ds; T_B < \tau\right]$$
$$= \mathbb{E}^x[\overline{G}1_B(X_{T_B}); T_B < \tau] \le c_3\mathbb{P}^x(T_B < \tau).$$

Hence $\mathbb{P}^x(T_B < \tau) \ge c_2/c_3$. $\qquad\square$

Because the Green function for Brownian motion and for X_t are comparable, much of the theory of capacity for Brownian motion has analogues for X_t. We recall some definitions and facts; see [PTA, Section II.5] for the Brownian case. A point x is regular for B if $\mathbb{P}^x(T_B = 0) = 0$. If B is a bounded set, the capacitary measure is the unique measure μ_B supported on $\overline{B}$ such that $G\mu_B(x) = 1$ on the regular points of B. $G\mu_B(x) \to 0$ as $|x| \to \infty$. The capacity of B is $\mu_B(\overline{B})$.

We content ourselves with proving the following.

(8.6) Proposition. *Suppose $d \geq 3$. Suppose D is a closed set such that every point of ∂D is regular for D. Then $C(D)$, the capacity of D, is equal to*

$$I(D) = \inf\Big\{\frac{1}{2}\int \nabla f \cdot a\nabla f : 0 \leq f \leq 1, f = 1 \text{ on } D,$$

$$f(x) \to 0 \text{ as } |x| \to \infty\Big\}.$$

Proof. Let μ_D be the capacitary measure for D. Then $f_D = G\mu_D$ is a function with values in $[0,1]$, equal to one on D, and tending to 0 at ∞. If g is an integrable smooth function with compact support so that Gg is bounded, then $\mathcal{L}(Gg) = -g$ and by Proposition 1.2,

$$\frac{1}{2}\int \nabla(Gg) \cdot a\nabla(Gg) = \int g(Gg) \leq \|Gg\|_\infty \int |g|.$$

This, a limit argument, and Fatou's lemma show that $(1/2)\int \nabla(G\mu_D) \cdot a\nabla(G\mu_D)$ is finite and equal to

$$\int (G\mu_D)(x)\mu_D(dx) = \mu_D(\overline{D}) = C(D).$$

Therefore $I(D) \leq C(D)$.

On the other hand, $\mathcal{E}(f,g) = (1/2)\int \nabla f \cdot a\nabla g$ forms an inner product. If $f_D = G\mu_D$, then f_D is $\mathcal{L}$-harmonic in D^c and so $\mathcal{E}(f_D,g) = -\int g(\mathcal{L}f_D) = 0$ whenever g is 0 on $\overline{D}$ and tends to 0 at ∞. Therefore

$$\mathcal{E}(f_D + g, f_D + g) = \mathcal{E}(f_D, f_D) + 2\mathcal{E}(f_D, g) + \mathcal{E}(g, g) \geq \mathcal{E}(f_D, f_D).$$

If h is a function with values in $[0,1]$ that is 1 on D and tends to 0 at ∞, let $g = h - f_D$. Thus $\mathcal{E}(h,h) \geq \mathcal{E}(f_D, f_D)$, or f_D is the function minimizing the infimum in the definition of $I(D)$. Therefore $C(D) = I(D)$. $\square$

Finally, we consider briefly the case where the a_{ij} need not be smooth. Suppose $\mathcal{L} \in \mathcal{D}(\Lambda)$. Take a_{ij}^n tending to a_{ij} almost everywhere so that if $\mathcal{L}_n$ is defined in terms of the a_{ij}^n, then $\mathcal{L}_n \in \mathcal{D}(\Lambda)$ for all n. By the equicontinuity of $p_n(t, x, y)$, the functions

$$G_n^\lambda g(x) = \int \int_0^\infty e^{-\lambda t} p_n(t, x, y)\, dt\, g(y)\, dy$$

are equicontinuous whenever g is a continuous bounded function on $\mathbb{R}^d$ and $\lambda > 0$. As in Theorem VI.2.3, there exists a subsequence such that $G^\lambda_{n_j} g$ converges uniformly on compacts, say to $G^\lambda g(x)$, for all continuous and bounded g and all $\lambda > 0$. For each x, $\mathbb{P}^x_{n_j}$ is tight, and by the argument of Theorem VI.2.3, any two subsequential limit points must agree. So $\mathbb{P}^x_{n_j}$ converges weakly, say to $\mathbb{P}^x$. Using the equicontinuity of $p_n(t, x, y)$ and arguing as in Theorem VI.2.3, $(\mathbb{P}^x, X_t)$ is a strong Markov process. One can show that the Dirichlet form for this process is $\mathcal{E}(f, f) = (1/2) \int \nabla f \cdot a \nabla f$.

9. Notes

Section 2 contains classical material that can be found many places. See Stroock [2] for further discussion of the Nash inequality. In Section 3 we followed Moser [1]. A proof of the Harnack inequality that does not use the John-Nirenberg inequality can be found in Moser [2]. Upper and lower bounds for $p(t, x, y)$ were first proved by Aronson [1]. Sections 4 and 5 are based on Fabes and Stroock [2], whereas much of Sections 6, 7, and 8 are derived from the material in Barlow and Bass [1].

VIII
THE MALLIAVIN CALCULUS

The Malliavin calculus is a method originally developed for proving smoothness of $p(t, x, y)$ in the variable y, where $p(t, x, y)$ is the transition density of a process associated to an operator with smooth coefficients. The basic idea involves an integration by parts formula in an infinite-dimensional space.

There are two main approaches, one using the Girsanov transformation and the other using the Ornstein-Uhlenbeck operator. Both are interesting and both are useful.

We present the Girsanov approach first. The integration by parts formula is given in Section 1 and extended to solutions to SDEs in Section 2. This is then applied to derive a criterion for a process to have smooth densities in Section 3.

Section 4 considers a class of operators $\mathcal{L}$ given in terms of vector fields and their connection with Stratonovich integrals. Section 5 contains a proof of Hörmander's theorem on the smoothness of $p(t, x, y)$ in y.

The second approach, using the Ornstein-Uhlenbeck operator, is presented in Section 6. Another proof of the criterion for a process to have a smooth density is derived.

1. Integration by parts formula

In this chapter all our functions will be *smooth*, that is, each function is in C^∞ and it and all its derivatives are bounded.

Suppose $F : \mathbb{R}^d \to \mathbb{R}$ is a smooth function. The directional derivative of F at x in the direction h is given by

$$D_x F(h) = \nabla F(x) \cdot h, \qquad h \in \mathbb{R}^d.$$

This definition can be used even when h is not a unit vector. Observe that $D_x F$ is a linear functional on $\mathbb{R}^d$. By Taylor's theorem,

$$|F(x+h) - F(x) - D_x F(h)| = |F(x+h) - F(x) - \nabla F(x) \cdot h|$$
$$= o(|h|)$$

as $|h| \to 0$. Here $f(h) = o(|h|)$ means $f(h)/|h| \to 0$ as $|h| \to 0$.

The *Fréchet derivative* is an extension of the idea of directional derivative to Banach spaces. Let B be a Banach space. $F : B \to \mathbb{R}$ is *Fréchet differentiable* at $x \in B$ if there exists a linear functional $T = T_x$ on B such that

$$|F(x+h) - F(x) - T_x(h)| = o(\|h\|), \qquad h \in B,$$

as $\|h\| \to 0$. We write $D_x F(h)$ or $DF(h)$ for $T_x(h)$.

For each x, the map $D_x F$ is a linear functional from B to $\mathbb{R}$, hence an element of B^*, the dual space of B. Since B^* is another Banach space, we can talk about the Fréchet derivative of $D_x F$, which would be the second-order Fréchet derivative.

A key step in the Malliavin calculus is the following integration by parts formula. Let $C[0,1]$ denote the $\mathbb{R}^d$-valued continuous functions on $[0,1]$.

(1.1) Theorem. *Suppose F maps $C[0,1]$ to $\mathbb{R}$, F is bounded, and F has a bounded Fréchet derivative at each point of $C[0,1]$. Let W_t be d-dimensional Brownian motion. Suppose h_s is adapted and bounded and let $H_t = \int_0^t h_s\, ds$. Then*

$$\mathbb{E}\left[F(W) \int_0^1 h_s\, dW_s \right] = E[D_W F(H)].$$

For each ω, both W_t and H_t are in $C[0,1]$. The right-hand side represents the expectation of the Fréchet derivative at $W.(\omega)$ in the direction $H.(\omega)$.

Proof. Let

$$X_t^\varepsilon = W_t + \varepsilon \int_0^t h_s\, ds.$$

Let

$$M_t^\varepsilon = \exp\left(-\varepsilon \int_0^t h_s\, dW_s - \frac{\varepsilon^2}{2} \int_0^t |h_s|^2\, ds \right).$$

Let $\mathbb{P}_\varepsilon$ be defined by $d\mathbb{P}_\varepsilon/d\mathbb{P} = M_t^\varepsilon$ on $\mathcal{F}_t$. By Girsanov's theorem (see Section I.1), under $\mathbb{P}_\varepsilon$ the process

$$W_t - \left\langle W, -\varepsilon \int_0^{\cdot} h_s \, dW_s \right\rangle_t = W_t + \varepsilon \int_0^t h_s \, ds = X_t^{\varepsilon}$$

is a martingale with the same quadratic variation as W_t, namely t. By Lévy's theorem (Section I.1), under $\mathbb{P}_\varepsilon$ the process X_t is a Brownian motion. Therefore

$$\mathbb{E}_\varepsilon F(X^\varepsilon) = \mathbb{E} \, F(W). \tag{1.1}$$

On the other hand,

$$\mathbb{E}_\varepsilon F(X^\varepsilon) \tag{1.2}$$
$$= \mathbb{E} \left[F\left(W + \varepsilon \int_0^{\cdot} h_s \, ds \right) \exp \left(-\varepsilon \int_0^1 h_s \, dW_s + \frac{\varepsilon^2}{2} \int_0^1 |h_s|^2 \, ds \right) \right].$$

By (1.1), the right-hand side of (1.2) is independent of ε. We differentiate (1.2) with respect to ε and set $\varepsilon = 0$. The assumptions on h and F allow us to interchange the operations of differentiation and expectation by use of the dominated convergence theorem and we obtain

$$0 = -\mathbb{E} \left[F(W) \int_0^1 h_s \, dW_s \right] + \mathbb{E} \, D_W F(H). \qquad \square$$

2. Smooth functionals

We can make some easy extensions of Theorem 1.1. If $D_w F(H)$ is bounded and continuous as a function of H and

$$\mathbb{E} \int_0^1 |h_s|^2 \, ds < \infty, \tag{2.1}$$

then taking limits in Theorem 1.1 gives us Theorem 1.1 for h satisfying (2.1).

The direction in which we want to generalize, however, is to more general functionals F. We want to consider functionals such as $F(W) = f(X_1)$, where f is smooth and X_t is the solution to a SDE with respect to W.

Let us say that a functional F on $C[0,1]$ is L^p-*smooth* with derivative $D_W F(H) = DF(H)$ if there exist functionals F_n on $C[0,1]$ that are bounded, continuous, Fréchet differentiable with bounded and continuous Fréchet derivatives, and with $F_n(W) \to F(W)$ in L^p and $D_W F_n(H) \to D_W F(H)$ in L^p whenever $H_t = \int_0^t h_s \, ds$ and h satisfies (2.1).

(2.1) Theorem. *If $p \geq 2$ and F is L^p-smooth, then*

$$\mathbb{E}\left[F(W)\int_0^1 h_s\, dW_s\right] = \mathbb{E}\left[D_W F(H)\right].$$

Proof. We apply the generalization of Theorem 1.1 discussed above to F_n and let $n \to \infty$. The convergence of $\mathbb{E}\left[D_W F_n(H)\right]$ is a consequence of the L^p convergence of $D_W F_n(H)$ to $D_W F(H)$ in L^p, $p \geq 2$. Since $\mathbb{E}\,(\int_0^1 h\, dW)^2 = \mathbb{E}\int_0^1 |h_s|^2\, ds$, the convergence of $\mathbb{E}\left[F_n(W)\int h\, dW\right]$ follows from the L^p convergence of $F_n(W)$ to $F(W)$ in L^p with $p \geq 2$ and the Cauchy-Schwarz inequality. $\qquad\square$

Before we get to the main result of this section, we need the following lemma.

(2.2) Lemma. *Suppose σ_n are C^∞ functions satisfying $\sup_n |\sigma_n(x)| \leq c_1(1+|x|)$ and σ_n and its derivatives converge to σ and its derivatives, respectively, uniformly on compacts. Suppose $x_n \to x$. If $X_n(t)$ solves*

$$dX_n(t) = \sigma_n(X_n(t))\, dW_t, \qquad X_n(0) = x_n,$$

then $X_n(t)$ converges in L^p for $p \geq 2$ to the solution to $dX_t = \sigma(X_t)\, dW_t$, $X_0 = x_0$.

Proof. Suppose first that σ_n and σ are bounded and σ_n and its derivatives converge to σ and its derivatives, respectively, uniformly. We have

$$X_t - X_n(t) = (x_0 - x_n) + \int_0^t [\sigma(X_s) - \sigma_n(X_n(s))]\, dW_s$$

$$= (x_0 - x_n) + \int_0^t [\sigma(X_s) - \sigma(X_n(s))]\, dW_s$$

$$+ \int_0^t [\sigma(X_n(s)) - \sigma_n(X_n(s))]\, dW_s.$$

So by Doob's inequality,

$$\mathbb{E}\sup_{s \leq t} |X_s - X_n(s)|^p \leq c_2 \mathbb{E}\, |X_t - X_n(t)|^p$$

$$\leq c_3 |x_0 - x_n|^p + c_3 \mathbb{E}\left|\int_0^t [\sigma(X_s) - \sigma(X_n(s))]\, dW_s\right|^p$$

$$+ c_3 \mathbb{E}\left|\int_0^t [\sigma(X_n(s)) - \sigma_n(X_n(s))]\, dW_s\right|^p.$$

By the Burkholder-Davis-Gundy inequalities (see (I.1.5)), the right-hand side is less than

$$c_3|x_0 - x_n|^p + c_4 \mathbb{E}\left[\int_0^t |\sigma(X_s) - \sigma(X_n(s))|^2\, ds\right]^{p/2}$$

$$+ c_4 \mathbb{E}\left[\int_0^t |\sigma(X_n(s)) - \sigma_n(X_n(s))|^2\, ds\right]^{p/2}$$

$$\leq c_3|x_0 - x_n|^p + c_4\|\sigma'\|_\infty^p \mathbb{E}\left[\int_0^t |X_s - X_n(s)|^2\, ds\right]^{p/2}$$

$$+ c_4\|\sigma - \sigma_n\|_\infty^p t^{p/2}.$$

If $t \leq 1$, by Hölder's inequality, we have the bound

$$\mathbb{E}\sup_{s \leq t} |X_s - X_n(s)|^p \leq c_3|x_0 - x_n|^p + c_4\|\sigma - \sigma_n\|_\infty^p$$

$$+ c_5\|\sigma'\|_\infty^p \mathbb{E}\int_0^t |X_s - X_n(s)|^p\, ds.$$

We apply Gronwall's inequality (Lemma I.3.3) with

$$g_n(t) = \mathbb{E}\sup_{s \leq t} |X_s - X_n(s)|^p,$$

so

$$g_n(t) \leq (c_3|x_0 - x_n|^p + c_4\|\sigma - \sigma_n\|_\infty^p)\exp(c_5\|\sigma'\|_\infty^p t).$$

Letting $n \to \infty$ shows $g_n(t) \to 0$, which proves the lemma in this special case.

We now consider the general case. If

$$T_n(M) = \inf\{t : |X_n(t)| \geq M \text{ or } |X_t| \geq M\},$$

the above argument shows that

$$\|X_n(t \wedge T_n(M)) - X(t \wedge T_n(M))\|_p \to 0$$

as $n \to \infty$. Hence for each M,

$$\mathbb{E}\left[|X_n(t) - X_t|^p; T_n(M) > t\right] \to 0. \tag{2.2}$$

On the other hand, by the Cauchy-Schwarz inequality,

$$\mathbb{E}\left[|X_n(t) - X_t|^p; T_n(M) \leq t\right]$$

$$\leq c_6\left(\mathbb{E}|X_n(t)|^{2p} + \mathbb{E}|X_t|^{2p}\right)^{1/2}\left(\mathbb{P}(T_n(M) \leq t)\right)^{1/2}.$$

By Proposition I.7.4, $X_n(t)$ and X_t are in L^{2p} with a norm independent of n, whereas

$$\mathbb{P}(T_n(M) \leq t) \leq \mathbb{P}(\sup_{s \leq t}|X_n(s)| \geq M) + \mathbb{P}(\sup_{s \leq t}|X_s| \geq M)$$

$$\leq M^{-2p}\left(\mathbb{E}\sup_{s \leq t}|X_n(s)|^{2p} + \mathbb{E}\sup_{s \leq t}|X_s|^{2p}\right)$$

$$\leq c_7 M^{-2p}\left(\mathbb{E}|X_n(t)|^{2p} + \mathbb{E}|X_t|^{2p}\right),$$

using Doob's inequality. Therefore $\mathbb{E}\left[|X_n(t) - X_t|^p; T_n(M) \le t\right]$ can be made small uniformly in n by taking M large. This and (2.2) complete the proof.

$\square$

The main examples of L^p-smooth functionals will be $f(X_1)$, where f is smooth and X_t solves an SDE.

(2.3) Theorem. *Let W_t be d-dimensional Brownian motion. Suppose X_t is an m-dimensional process that is a solution to*

$$dX_t = \sigma(X_t)\,dW_t + b(X_t)\,dt, \qquad X_0 = x_0, \qquad (2.3)$$

where σ and b are C^∞ and

$$\sum_{i=1}^{m}\sum_{j=1}^{d}|\sigma_{ij}(x)| + \sum_{i=1}^{m}|b_i(x)| \le c_1(1+|x|).$$

Suppose f is smooth. Then $F(W) = f(X_1)$ is an L^p-smooth functional for all $p \ge 2$.

Proof. For simplicity of notation, we suppose $b \equiv 0$; only trivial changes are needed to include the case where b is nonzero. We also make the observation that if F_n is a sequence of L^p-smooth functionals such that F_n and $DF_n(H)$ converge to F and $DF(H)$ in L^p, respectively, when h satisfies (2.1), then F is L^p-smooth.

Suppose for now that σ is bounded. Suppose for each t that Y_t is an L^p-smooth functional on $C[0,1]$ with the L^p norms of Y_t and $D_W Y_t(H)$ uniformly bounded in t for each h satisfying (2.1). Suppose also that Y_t and $D_W Y_t(H)$ are continuous a.s. for each h satisfying (2.1). For each k, it is easy to see that

$$\sum_{j=1}^{k} \sigma(Y_{j/k})[W_{(j+1)/k} - W_{j/k}]$$

is Fréchet differentiable. As $k \to \infty$, this sum and its Fréchet derivative converge in L^p by the Burkholder-Davis-Gundy inequalities, so the functional $\int_0^t \sigma(Y_s)\,dW_s$ is L^p-smooth for each $t \le 1$.

Let $X_0(t) \equiv x_0$ and define by induction,

$$X_{i+1}(t) = x_0 + \int_0^t \sigma(X_i(t))\,dW_t.$$

We see that

$$DX_{i+1}(t) = I + \int_0^t \sigma'(X_i(s))DX_i(s)\,dW_s,$$

where we use the notation of (I.10.4). We saw in Corollary I.3.2 that $X_i(t)$ converges in L^p to the solution to

$$dX_t = \sigma(X_t)\,dW_t, \qquad X_0 = x_0.$$

Since σ' is smooth, $\sigma'(X_i(s)) \to \sigma'(X_s)$ in L^p for all $p \geq 2$, so it is easy to see that $DX_i(t)$ converges in L^p to the solution to

$$dDX_t = \sigma'(X_t)DX_t\,dW_t, \qquad DX_0 = I.$$

By induction, each $X_i(t)$ is L^p-smooth, and we therefore conclude that X_t is L^p-smooth when σ is bounded.

Finally, if σ is not bounded, let σ^M be smooth bounded approximations to σ. If X_t^M is the solution to (2.3) with σ replaced by σ^M, we see as above that X^M converges to X and DX^M converges to DX, so X_t is L^p-smooth. It follows by the chain rule that $F(W) = f(X_t)$ is L^p-smooth. $\qquad\square$

Note that the pair (X_t, DX_t) solves an SDE of the form (2.3), namely,

$$dX_t = \sigma(X_t)\,dW_t + b(X_t)\,dt,$$
$$dDX_t = \sigma'(X_t)DX_t\,dW_t + b'(X_t)DX_t\,dt,$$

with $(X_0, DX_0) = (x_0, I)$. So if f is smooth on $\mathbb{R}^d \times \mathbb{R}^{d \times d}$, Theorem 2.3 shows that $f(X_t, DX_t)$ is an L^p-smooth functional. Higher derivatives may be handled similarly.

3. A criterion for smooth densities

In this section we define the Malliavin covariance matrix Γ_t and give a criterion in terms of it for X_1 to have a smooth density. We need first some preliminary results.

(3.1) Proposition. *Suppose* $X : \Omega \to \mathbb{R}^d$. *Suppose for each k and all $j_1, \ldots, j_k \in \{1, \ldots, d\}$ there exists C_k such that*

$$\left| \mathbb{E}\left[\partial_{j_1 \cdots j_k} g(X) \right] \right| \leq C_k \|g\|_\infty$$

whenever $g \in C^k$. *Then there exists f smooth such that*

$$\mathbb{P}(X \in A) = \int_A f(x)\,dx$$

for all Borel sets A.

Proof. Let $\mu(dx) = \mathbb{P}(X \in dx)$ and let $\hat{\mu}(u) = \int e^{iu \cdot x} \mu(dx)$. If we let $u = (u_1, \cdots, u_d)$ and $g(x) = e^{iu \cdot x}$, then

$$\partial_{j_1 \cdots j_k} g = i^k u_{j_1} \cdots u_{j_k} e^{iu \cdot x},$$

and

$$|\mathbb{E}\,\partial_{j_1\cdots j_k}g(X)| = |u_{j_1}\cdots u_{j_k}|\,|\mathbb{E}\,e^{iu\cdot X}| \tag{3.1}$$

$$= |u_{j_1}\cdots u_{j_k}|\left|\int e^{iu\cdot x}\,\mu(dx)\right| = |u_{j_1}\cdots u_{j_k}|\,|\widehat{\mu}(u)|.$$

Since for each k, the left-hand side is bounded by $C_k\|g\|_\infty = C_k < \infty$, we conclude there exists c_1 such that

$$|\widehat{\mu}(u)| \le c_1/|u|^{d+1}.$$

Clearly $|\widehat{\mu}(u)|$ is bounded by 1, so $\widehat{\mu}$ is in $L^1(\mathbb{R}^d)$. By the Fourier inversion formula, μ has a bounded and continuous density. Let us denote it by f.

We now have

$$\widehat{\mu}(u) = \int e^{iu\cdot x}\mu(dx) = \int e^{iu\cdot x}f(x)\,dx = \widehat{f}(u).$$

Using (3.1) and the argument above, if $n > 0$, there exists c_2 such that

$$|u_{j_1}\cdots u_{j_n}\widehat{f}(u)| \le c_2/|u|^{d+1}.$$

Since $\widehat{f}$ is bounded, $i^n u_{j_1}\cdots u_{j_n}\widehat{f}(u)$ is in L^1. Its inverse Fourier transform is then continuous and bounded. However, the inverse Fourier transform of $i^n u_{j_1}\cdots u_{j_n}\widehat{f}(u)$ is $\partial_{j_1\cdots j_n}f$. This holds for each n and for each sequence $j_1,\ldots,j_n$. $\qquad\square$

If X_t is the solution to an SDE such as (2.3), then $Y_t = DX_t$ solves a linear SDE. If we were in one dimension, then Y_t would be given by an exponential; hence its reciprocal would be given by an exponential, and therefore its reciprocal should solve a linear SDE. There is a higher-dimensional generalization of this.

It is more convenient at this stage to use the Stratonovich integral; see Section I.9. Recall that if

$$dX_t = \sigma(X_t)\,dW_t + b(X_t)\,dt,$$

that is,

$$dX_t^i = \sum_j \sigma_{ij}(X_t)\,dW_t^j + b_i(X_t)\,dt,$$

then

$$d\langle\sigma_{ij}(X), W^j\rangle_t = \sum_{k,m}\partial_k\sigma_{ij}(X_t)\sigma_{km}(X_t)\,d\langle W^m, W^j\rangle_t$$

$$= \sum_k \partial_k\sigma_{ij}(X_t)\sigma_{kj}(X_t)\,dt.$$

So

$$dX_t^i = \sum_j \sigma_{ij}(X_t)\circ dW_t^j + \widetilde{b}_i(X_t)\,dt,$$

where

$$\widetilde{b}_i(x) = b_i(x) - \frac{1}{2}\sum_k \sigma_{kj}(x)\partial_k\sigma_{ij}(x).$$

We abbreviate this by

$$dX_t = \sigma(X_t)\circ dW_t + \widetilde{b}(X_t)\,dt.$$

The procedure to convert an SDE written in terms of Stratonovich integrals into an SDE written in terms of Itô integrals is similar.

(3.2) Proposition. *Suppose $Y_t : \Omega \to \mathbb{R}^{d\times d}$ solves*

$$dY_t = \sigma'(X_t)Y_t\circ dW_t + b'(X_t)Y_t\,dt, \qquad Y_0 = I. \tag{3.2}$$

If $Z_t : \Omega \to \mathbb{R}^{d\times d}$ is the solution to

$$dZ_t = -Z_t\sigma'(X_t)\circ dW_t - Z_t b'(X_t)\,dt, \qquad Z_0 = I, \tag{3.3}$$

then $Z_tY_t = I$ for all t.

We first explain what the notation means. Y_t solves

$$dY_{ij}(t) = \sum_{k,n=1}^{d} \partial_n\sigma_{ik}(X_t)Y_{nj}\circ dW_t^k + \sum_{n=1}^{d}\partial_n b_i(X_t)Y_{nj}\,dt, \tag{3.4}$$

whereas Z_t solves

$$dZ_{\ell i}(t) = -\sum_{m,k=1}^{d} Z_{\ell m}(t)\partial_i\sigma_{mk}(X_t)\circ dW_t^k \tag{3.5}$$

$$-\sum_{m=1}^{d} Z_{\ell m}\partial_i b_m(X_t)\,dt.$$

Proof. The proof is an application of the product formula (I.9.2). To see what is going on, let us look at the one-dimensional case. $Z_0Y_0 = I$ and

$$\begin{aligned}
d(Z_tY_t) &= Z_t\circ dY_t + Y_t\circ dZ_t \\
&= Z_t\sigma'(X_t)Y_t\circ dW_t + Z_t b'(X_t)Y_t\,dt \\
&\quad + Y_t(-Z_t)\sigma'(X_t)\circ dW_t + Y_t(-Z_t)b'(X_t)\,dt \\
&= 0.
\end{aligned}$$

Hence $Z_tY_t = Z_0Y_0 = I$.

The higher-dimensional version is the same idea exactly, but with more complicated notation. $\qquad\square$

As a consequence of Proposition I.7.4, observe that

$$\sup_{s\leq t}|Y_s| \in L^p \quad\text{and}\quad \sup_{s\leq t}|Z_s| \in L^p \tag{3.6}$$

for all p. One final preliminary proposition is necessary.

(3.3) Proposition. *Suppose*

$$dX_t = \sigma(X_t) \circ dW_t.$$

If $k_s : \Omega \times [0, 1] \to \mathbb{R}^d$ is adapted and satisfies $\int_0^1 |k_s|^2 \, ds < \infty$ and

$$R_t = Y_t \int_0^t Z_s \sigma(X_s) k_s \, ds, \tag{3.7}$$

then

$$R_t = \int_0^t \sigma'(X_s) R_s \circ dW_s + \int_0^t \sigma(X_s) k_s \, ds.$$

Proof. We again do the one-dimensional case for simplicity; no additional ideas are needed for the higher-dimensional case.

Recall

$$dY_t = \sigma'(X_t) Y_t \circ dW_t.$$

By the definition of R_t and Proposition 3.2, we have

$$\int_0^t Z_s \sigma(X_s) k_s \, ds = R_t Y_t^{-1} = R_t Z_t.$$

By the product formula (I.9.2) and (3.7),

$$\begin{aligned}
dR_t &= Y_t Z_t \sigma(X_t) k_t \, dt + \left(\int_0^t Z_s \sigma(X_s) k_s \, ds \right) \circ dY_t \\
&= \sigma(X_t) k_t \, dt + R_t Z_t \circ dY_t \\
&= \sigma(X_t) k_t \, dt + R_t Z_t \sigma'(X_t) Y_t \circ dW_t \\
&= \sigma(X_t) k_t \, dt + R_t \sigma'(X_t) \circ dW_t,
\end{aligned}$$

as required. $\qquad\square$

We need to calculate $D_W X_1(H)$ when $H_t = \int_0^t h_s \, ds$ and X_t satisfies $dX_t = \sigma(X_t) \circ dW_t$:

$$\begin{aligned}
X_1(W + \varepsilon H) &= x_0 + \int_0^1 \sigma(X_t(W + \varepsilon H)) \circ d(W_t + \varepsilon H_t) \\
&= x_0 + \int_0^1 \sigma(X_t(W + \varepsilon H)) \circ dW_t \\
&\qquad + \varepsilon \int_0^1 \sigma(X_t(W + \varepsilon H)) h_t \, dt
\end{aligned}$$

and

$$X_1(W) = x_0 + \int_0^1 \sigma(X_t(W)) \circ dW_t.$$

Taking the difference, dividing by ε, and letting $\varepsilon \to 0$, we obtain similarly as in Section I.10 that

$$D_W X_1(H) = \int_0^1 \sigma'(X_t) D_W X_t(H) \circ dW_t + \int_0^1 \sigma(X_t) h_t \, dt. \qquad (3.8)$$

Let us now define the *Malliavin covariance matrix* to be the random $d \times d$ matrix

$$\Gamma_t = \int_0^t Z_s \sigma(X_s) \sigma(X_s)^T Z_s^T \, ds. \qquad (3.9)$$

(3.4) Proposition. *Suppose X_t solves $dX_t = \sigma(X_t) \circ dW_t$, where σ is smooth. Suppose $\Gamma_1^{-1} \in L^p$ for all p. There exists c_1 such that if $f \in C^\infty$ is bounded, then*

$$|\mathbb{E} \, \partial_i f(X_1)| \le c_1 \|f\|_\infty, \qquad i = 1, \dots, d.$$

Proof. Let $f \in C^\infty$ be bounded. Define F on $C[0,1]$ by

$$F(W) = f(X_1)G,$$

where G is an L^p-smooth functional to be defined later. Applying the chain rule and product rule,

$$D_W F(H) = \sum_{n=1}^d \partial_n f(X_1) D_W X_1^n(H) G + f(X_1) D_W G(H).$$

Applying Theorem 2.1, we have for $H_t = \int_0^t h_s \, ds$

$$\mathbb{E}\left[\sum_{n=1}^d \partial_n f(X_1) D_W X_1^n(H) G \right] = \mathbb{E}\left[f(X_1) G \int_0^1 h_s \, dW_s \right] \qquad (3.10)$$
$$- \mathbb{E}\left[f(X_1) D_W G(H) \right].$$

Let e_i be the unit vector in the x_i direction and let us take

$$h_k(s) = (Z_s \sigma(X_s))^T e_k, \qquad H_k(t) = \int_0^t h_k(s) \, ds. \qquad (3.11)$$

Since $Z_t \in L^p$ for all p and σ is bounded, h_k satisfies $\int_0^1 |h_k(s)|^2 \, ds < \infty$. Define V to be the $d \times d$ matrix-valued random variable given by

$$V_{ik} = D_W X_1^i(H_k).$$

From (3.8) and Proposition 3.3,

$$D_W X_1(H_k) = Y_1 \int_0^1 Z_s \sigma(X_s) h_k(s) \, ds. \qquad (3.12)$$

Combining with (3.11) and the definition of Γ, we have $V = Y_1 \Gamma_1$, hence

$$V^{-1} = \Gamma_1^{-1} Z_1.$$

We apply (3.10) with H replaced by H_j and G replaced by $G_j = (V^{-1})_{ji}$. We then sum over $j = 1, \ldots, d$ to obtain

$$\mathbb{E}\left[\partial_i f(X_1)\right] = \mathbb{E}\left[\sum_{n=1}^{d} \partial_n f(X_1)\delta_{ni}\right] \tag{3.13}$$

$$= \mathbb{E}\left[\sum_{j,n=1}^{d} \partial_n f(X_1) V_{nj} G_j\right]$$

$$= \mathbb{E}\left[f(X_1)\sum_{j=1}^{d} G_j \int_0^1 h_j(s)\,dW_s\right]$$

$$- \mathbb{E}\left[f(X_1)\sum_{j=1}^{d} D_W G_j(H_j)\right].$$

We now need to check integrability on the right-hand side of (3.13). By our hypothesis, $\Gamma_1^{-1} \in L^p$ for all p. This and (3.6) show $V^{-1} \in L^p$ for all p. We have by the Burkholder-Davis-Gundy inequalities and Hölder's inequality that

$$\mathbb{E}\left[\int_0^1 h_j(s)\,dW_s\right]^p \le c_2 \mathbb{E}\left(\int_0^1 |h_j(s)|^2\,ds\right)^{p/2} \le c_2 \mathbb{E}\int_0^1 |h_j(s)|^p\,ds$$

if $p \ge 2$. Since σ is bounded, $\mathbb{E}\,|h_j(s)|^p$ is bounded by a constant independent of $s \le 1$. Therefore for each j,

$$\left|\mathbb{E}\left[f(X_1)G_j \int_0^1 h_j(s)\,dW_s\right]\right| \tag{3.14}$$

$$\le \|f\|_\infty (\mathbb{E}\,G_j^2)^{1/2}\left(\mathbb{E}\left(\int_0^1 h_j(s)\,dW_s\right)^2\right)^{1/2}$$

$$\le c_3\|f\|_\infty.$$

Since $I = VV^{-1}$, then $0 = D_W I(H_j) = V(D_W V^{-1}(H_j)) + (D_W V(H_j))V^{-1}$, or

$$D_W V^{-1}(H_j) = -V^{-1}(D_W V(H_j))V^{-1}. \tag{3.15}$$

We have already seen that V^{-1} is in L^p for all p. By our choice of h_j and the fact that σ is C^∞ with bounded derivatives, $D_W(D_W X_1(H_j))(H_j)$ is in L^p for all p, and we conclude that $D_W V(H_j) \in L^p$ for all p. Hence $D_W V^{-1}(H_j) \in L^p$ for all p. So

$$|\mathbb{E}\left[f(X_1)D_W G_j(H_j)\right]| \le \|f\|_\infty \mathbb{E}\,|D_W G_j(H_j)| \tag{3.16}$$

$$\le c_4\|f\|_\infty, \qquad j = 1, \ldots, d.$$

Combining (3.13), (3.14), and (3.16), we have

$$|\mathbb{E}\,\partial_i f(X_1)| \leq c_5\|f\|_\infty. \qquad\qquad \square$$

Let us now look at the higher-order partial derivatives. We continue the notation of the proof of Proposition 3.4.

(3.5) Proposition. *Suppose X_t solves $dX_t = \sigma(X_t)\circ dW_t$, where σ is smooth. Suppose $\Gamma_1^{-1} \in L^p$ for all p. For each n there exists C_n such that if $f \in C^\infty$ and $j_1,\ldots,j_n \in \{1,\ldots,d\}$, then*

$$|\mathbb{E}\,\partial_{j_1\cdots j_n} f(X_1)| \leq C_n\|f\|_\infty.$$

Proof. We will show how to do the case $n = 2$. The higher derivatives are done similarly.

If we apply (3.13) with f replaced by $\partial_k f$, we have

$$\mathbb{E}\,[\partial_{ik} f(X_1)] = \mathbb{E}\left[\partial_k f(X_1) \sum_{j=1}^{d} G_j \int_0^1 h_s\, dW_s\right] \qquad (3.17)$$

$$- \mathbb{E}\left[\partial_k f(X_1) \sum_{j=1}^{d} D_W G_j(H)\right].$$

Writing

$$K = \sum_{j=1}^{d}\left[G_j \int_0^1 h_s\, dW_s - D_W G_j(H)\right],$$

the right-hand side is

$$\mathbb{E}\,[\partial_k f(X_1)K].$$

Let us set $F = f(X_1)LK$, where L is an L^p-smooth functional that will be chosen in a moment. By Theorem 2.1,

$$\mathbb{E}\left[\sum_{n=1}^{\infty}\partial_n f(X_1)^T D_W X_1^n(H)LK\right] = \mathbb{E}\left[f(X_1)LK \int_0^1 h_s\, dW_s\right]$$

$$- \mathbb{E}\,[f(X_1)D_W(LK)(H)].$$

Let us replace H by H_j and let $L_j = (V^{-1})_{jk}$ for $j = 1,\ldots,d$, and sum. As in the proof of Proposition 3.4, we obtain

$$\mathbb{E}\,[\partial_k f(X_1)K]$$

$$= \mathbb{E}\left[f(X_1)\sum_{j=1}^{d} L_j K \int_0^1 h_j(s)\, dW_s\right] - \mathbb{E}\left[f(X_1)\sum_{j=1}^{d} K D_W L_j(H_j)\right]$$

$$- \mathbb{E}\left[f(X_1)\sum_{j=1}^{d} L_j D_W K(H_j)\right].$$

We have already seen that $K \in L^p$ for all p in the proof of Proposition 3.4. We also saw there that L_j, $D_W L_j(H_j)$, and $\int_0^1 h_j(s)\, dW_s$ are all in L^p for all p. Suppose we show that $D_W K(H_j)$ is in L^p for all p. We then have that

$$|\mathbb{E}\left[\partial_k f(X_1) K\right]| \le c_1 \|f\|_\infty,$$

and substituting this in (3.17),

$$|\mathbb{E}\left[\partial_{ik} f(X_1)\right]| \le c_1 \|f\|_\infty.$$

So it remains to show that $D_W K(H_j) \in L^p$ for all p. Now $D_W V(H_j)$ is in L^p for all p by the proof of Proposition 3.4. Since the σ are C^∞, then $D_W(D_W V(H_j))(H_j)$ is also in L^p for all p. By (3.15), this implies that $D_W(D_W V^{-1}(H_j))(H_j)$ is also in L^p for all p. Finally, if we write $M = \int_0^1 h_j(s)\, dW_s$, then

$$M(W + \varepsilon H_j) - M(W) = \int_0^1 \left[h_j(W + \varepsilon H_j) - h_j(W)\right] dW_s$$

$$+ \varepsilon \int_0^1 h_s(W + \varepsilon H_j) \cdot h_j \, ds.$$

Dividing by ε and letting $\varepsilon \to 0$, it follows that

$$D_W M(H_j) = \int_0^1 D_W h_j(H_j)\, dW_s + \int_0^1 |h_j(s)|^2 \, ds.$$

Since $D_W h_j(H_j) \in L^p$ for all p by the definition of h_j in (3.11), $D_W M(H_j) \in L^p$ for all p. Putting these facts together with Hölder's inequality, we have $D_W K(H_j) \in L^p$ for all p, and the proof is complete. $\qquad\square$

(3.6) Theorem. *Suppose X_t solves $dX_t = \sigma(X_t) \circ dW_t$, $X_0 = x_0$, where σ is smooth. If $\Gamma_1^{-1} \in L^p$ for all p, then X_1 has a smooth density.*

Proof. We combine Proposition 3.1 and Proposition 3.5. $\qquad\square$

4. Vector fields

In Section 5 we will prove Hörmander's theorem, which gives sufficient conditions for the distribution of X_1 to have a C^∞ density. This section has some preliminaries.

In PDE terms, the question of when $\mathbb{P}^{x_0}(X_1 \in dx) = f(x)\, dx$ with $f \in C^\infty$ is equivalent to the question of when the fundamental solution $p(t, x, y)$ to $\partial_t u = \mathcal{L}u$ is C^∞ in the variable y (see Section II.7). This property is closely related to the PDE notion of hypoellipticity.

If the coefficients of $\mathcal{L}$ are smooth and $\mathcal{L}$ is uniformly elliptic, this will always be the case. The interesting cases that require Hörmander's theorem are when $\mathcal{L}$ can be degenerate, i.e., not strictly elliptic at each point.

Let us first look at the uniformly elliptic case.

(4.1) Proposition. *Suppose the coefficients of $\mathcal{L}$ are bounded and C^∞, $\mathcal{L}$ is uniformly elliptic, and $x_0 \in \mathbb{R}^d$. Then the $\mathbb{P}^{x_0}$ distribution of X_1 has a C^∞ density.*

Proof. We show that the Malliavin covariance matrix (3.9) has an inverse that is in L^p for all p. The result then will follow by Theorem 3.6.

Recall $\Gamma_t = \int_0^t Z_s \sigma(X_s) \sigma^T(X_s) Z_s^T \, ds$. Γ_t is symmetric from its definition, and to prove Γ_1^{-1} is in L^p, it suffices to show that if λ is the smallest eigenvalue of Γ_1, then $\lambda^{-1} \in L^p$. Let v be a unit vector. Since $\sigma\sigma^T = a$ is uniformly elliptic,

$$v^T \Gamma_t v \geq c_1 \int_0^t |v^T Z_s|^2 \, ds.$$

Let $S = \inf\{t > 0 : |Z_t - I| \geq 1/2\}$. For $s \leq S$ the coefficients of the SDE defining Z_s are bounded, and so by Proposition I.8.1, if $p \geq 2$ and $\varepsilon \leq 1$,

$$\mathbb{P}(S < \varepsilon) = \mathbb{P}(\sup_{s \leq \varepsilon} |Z_s - I| \geq 1/2) \leq c_2 \varepsilon^p.$$

If $t \leq S$, then

$$|v^T Z_t| \geq |v^T| - |v^T(Z_t - I)| \geq \frac{1}{2}$$

since $|Z_t - I| \leq 1/2$. So

$$v^T \Gamma_1 v \geq c_1 \int_0^{S \wedge 1} |v^T Z_s|^2 \, ds \geq c_1(S \wedge 1)/4.$$

This is true for any unit vector v, so

$$\lambda = \inf_{v \in \partial B(0,1)} v^T \Gamma_1 v \geq c_1(S \wedge 1)/4.$$

Then

$$\mathbb{P}(\lambda^{-1} > y) = \mathbb{P}(\lambda < y^{-1}) \leq \mathbb{P}(S \wedge 1 \leq 4y^{-1}/c_1).$$

If $y > 4/c_1$, this is

$$\mathbb{P}(S < 4y^{-1}/c_1) \leq c_2\left(\frac{4y^{-1}}{c_1}\right)^p = \frac{c_3}{y^p}.$$

This shows $\lambda^{-1} \in L^{p-2}$. Since $p \geq 2$ is arbitrary, $\lambda^{-1} \in L^p$ for all p. $\qquad\square$

To state Hörmander's theorem, we need to express $\mathcal{L}$ in *Hörmander form*. Let $V(x) = \sum_{j=1}^d v_j(x) e_j$ be a smooth vector field. Here $v_j(x)$ are smooth bounded functions of x and $e_j = (0, \ldots, 0, 1, 0, \ldots, 0)$ is the unit vector in the x_j direction. We will need to follow the standard differential geometry convention that identifies vector fields with differential operators. So we also consider V as a first-order linear differential operator on C^∞ functions defined by

$$Vf(x) = \sum_{j=1}^{d} v_j(x)\partial_j f(x).$$

We then calculate

$$V^2 f(x) = \sum_{j=1}^{d} v_j(x)\partial_j \Big(\sum_{k=1}^{d} v_k \partial_k f \Big)(x) \tag{4.1}$$

$$= \sum_{j,k=1}^{d} v_j(x)v_k(x)\partial_{jk} f(x) + \sum_{k=1}^{d} \Big(\sum_{j=1}^{d} v_j(x)\partial_j v_k(x) \Big)\partial_k f(x),$$

which is a second-order differential operator.

We now suppose we have $V_1, \ldots, V_d$ and V_0 with $V_i = \sum_{j=1}^{d} v_{ij}(x)\partial_j$ and we set

$$\mathcal{L}f(x) = \frac{1}{2} \sum_{i=1}^{d} V_i^2 f(x) + V_0 f(x).$$

Using (4.1) with v_j replaced by v_{ij},

$$\mathcal{L}f(x) = \frac{1}{2} \sum_{i=1}^{d} \sum_{j,k=1}^{d} v_{ij}(x)v_{ik}(x)\partial_{jk} f(x) \tag{4.2}$$

$$+ \frac{1}{2} \sum_{i=1}^{d} \sum_{j,k=1}^{d} v_{ij}(x)\partial_j v_{ik}(x)\partial_k f(x) + \sum_{j=1}^{d} v_{0j}(x)\partial_j f(x)$$

$$= \frac{1}{2} \sum_{j,k=1}^{d} \Big(\sum_{i=1}^{d} v_{ij}(x)v_{ik}(x) \Big)\partial_{jk}(x)$$

$$+ \sum_{k=1}^{d} \Big(\frac{1}{2} \sum_{i,j=1}^{d} v_{ij}(x)\partial_j v_{ik}(x) + v_{0k}(x) \Big)\partial_k f(x).$$

Although this appears cumbersome, it ties in neatly with SDEs in Stratonovich form.

(4.2) Proposition. *Suppose $v_{ij}(x) = v_{ji}(x)$ for all i, j, x. Suppose X_t solves*

$$dX_t = v(X_t) \circ dW_t + v_0(X_t)\, dt, \qquad X_0 = x_0. \tag{4.3}$$

Then the operator $\mathcal{L}$ associated to X_t is

$$\frac{1}{2} \sum_{i=1}^{d} V_i^2 + V_0.$$

Equation (4.3) means

$$dX_t^i = \sum_{j=1}^{d} v_{ij}(X_t) \circ dW_t^j + v_{0i}(X_t)\, dt, \qquad i = 1, \ldots, d. \tag{4.4}$$

Proof. If we write (4.4) in Itô form, we have

$$dX_t^i = \sum_{j=1}^{d} v_{ij}(X_t)\, dW_t^j + \frac{1}{2} \sum_{j=1}^{d} \langle v_{ij}(X), W^j \rangle_t + v_{0i}(X_t)\, dt.$$

By Itô's formula,

$$v_{ij}(X_t) = \sum_{k=1}^{d} \partial_k v_{ij}(X_t) dX_t^k + \text{ bounded variation term}$$

$$= \sum_{k=1}^{d} \partial_k v_{ij}(X_t) \sum_{m=1}^{d} v_{km}(X_t) \circ dW_t^m + \text{ bounded variation term}$$

$$= \sum_{k=1}^{d} \partial_k v_{ij}(X_t) \sum_{m=1}^{d} v_{km}(X_t)\, dW_t^m + \text{ bounded variation term.}$$

So

$$\langle v_{ij}(X), W^j \rangle_t = \sum_{k=1}^{d} \partial_k v_{ij}(X_t) v_{kj}(X_t)\, dt.$$

Using Proposition I.2.1, the operator associated to X_t is

$$\mathcal{L}f(x) = \frac{1}{2} \sum_{i,k=1}^{d} \sum_{j=1}^{d} v_{ij}(x) v_{kj}(x) \partial_{ik} f(x) + \frac{1}{2} \sum_{j=1}^{d} \sum_{i,k=1}^{d} v_{kj}(x) \partial_k v_{ij}(x) \partial_i f(x)$$

$$+ \sum_{j=1}^{d} v_{0j}(x) \partial_j f(x)$$

$$= \frac{1}{2} \sum_{i,j,k} v_{ji}(x) v_{ki}(x) \partial_{jk} f(x)$$

$$+ \sum_{k} \Big(\frac{1}{2} \sum_{i,j} v_{ij}(x) \partial_j v_{ki}(x) + v_{0k}(x) \Big) \partial_k f(x).$$

Comparing with (4.2) completes the proof. $\qquad \square$

If V and W are two vector fields, define the *Lie bracket* of V and W by

$$[V, W] = VW - WV. \tag{4.5}$$

A calculation shows that if $V = \sum_j v_j \partial_j$ and $W = \sum_k w_k \partial_k$, then

$$[V, W] = \left(\sum_{j,k} v_j w_k \partial_{jk} + \sum_{j,k} v_j (\partial_j w_k) \partial_k \right)$$

$$- \left(\sum_{j,k} v_j w_k \partial_{jk} + \sum_{j,k} w_k (\partial_k v_j) \partial_j \right)$$

$$= \sum_{j,k} (v_j (\partial_j w_k) \partial_k - w_k (\partial_k v_j) \partial_j),$$

another first-order operator.

Let

$$\mathcal{S} = \mathcal{S}(V_1, \cdots, V_d) = \{V_i, [V_i, V_j], [V_i, [V_j, V_k]], \ldots, \tag{4.6}$$

$$i = 1, \ldots, d, j = 1, \ldots, d, \ldots\}.$$

So $\mathcal{S}$ is the smallest collection of vector fields containing $V_1, \ldots, V_n$ and closed under the operation $[\cdot, \cdot]$. We define $\mathcal{S}_1 = \{V_1, \ldots, V_d\}$, $\mathcal{S}_{i+1} = \mathcal{S}_i \cup \{[V_i, W] : i = 1, \ldots, d, W \in \mathcal{S}_i\}$. So $\mathcal{S}_i$ is the set of vector fields generated by the Lie brackets of $V_1, \ldots, V_d$ of length at most i, and $\mathcal{S} = \cup_{i=1}^{\infty} \mathcal{S}_i$.

As we mentioned above, the elements of $\mathcal{S}$ can also be considered as vectors in the linear algebra sense: if $W = \sum_{k=1}^{d} w_k \partial_k$, let $W(x) = \sum_{k=1}^{d} w_k(x) e_k$. For each x, $\mathcal{S}(x) = \{W(x) : W \in \mathcal{S}\}$ is a collection of vectors. Define

$$\mathcal{S}_i(x) = \{W(x) : W \in \mathcal{S}_i\}. \tag{4.7}$$

For each x, $\mathcal{S}(x)$ is a collection of vectors that may or may not span $\mathbb{R}^d$. Hörmander's condition is that $\mathcal{S}(x)$ spans $\mathbb{R}^d$ at each point x. Since the tangent space to $\mathbb{R}^d$ at a point x is again $\mathbb{R}^d$, Hörmander's condition is sometimes phrased by saying "$\mathcal{S}$ spans the tangent space at each point."

The meaning of Hörmander's condition is the following. If a particle moves according to a vector field V for a short time, then moves according to W for the same length of time, then $-V$ and then $-W$, its net motion is a short move along $\pm[V, W]$. If V and W do not commute, there will be a net motion different from 0. A particle diffusing according to V^2 can be considered as a particle moving for a short period of time along $\pm V$, and a particle diffusing according to $V^2 + W^2$ will thus also have some diffusion in the direction $\pm[V, W]$. So Hörmander's condition says that a particle starting at x diffusing under the operator $\mathcal{L}$ will diffuse in all directions (at least in an infinitesimal sense), and so the paths of X_t do not initially lie in some $(d-1)$-dimensional surface contained in $\mathbb{R}^d$. If the paths did lie in some $(d-1)$-dimensional surface, X_t could not have a density.

5. Hörmander's theorem

In this section we prove Hörmander's theorem, Theorem 5.6 below, which says that X_1 will have a C^{∞} density if $\mathcal{S}(x_0)$ spans $\mathbb{R}^d$. For simplicity we will consider only the case where the vector field V_0 is 0.

We first need two lemmas.

(5.1) Lemma. *There exist c_1 and c_2 such that if M_t is a continuous martingale, T a bounded stopping time, and $\varepsilon > 0$, then*

$$\mathbb{P}(\sup_{t \le T} |M_t| < \delta, \langle M \rangle_t > \varepsilon) \le c_1 e^{-c_2 \varepsilon / \delta^2}.$$

Proof. M_t is a time change of a Brownian motion W_t (cf. [PTA, Theorem I.5.11]). So the desired probability is bounded by

$$\mathbb{P}(\sup_{t \le U} |W_t| < \delta, \langle W \rangle_U > \varepsilon),$$

where U is a stopping time. Since $\langle W \rangle_U = U$, the probability above is in turn bounded by

$$\mathbb{P}(\sup_{t \le \varepsilon} |W_t| < \delta) = \mathbb{P}(\sup_{t \le \varepsilon/\delta^2} |W_t| < 1) \le c_1 e^{-c_2 \varepsilon / \delta^2}$$

by scaling and [PTA, Section II.4].   □

(5.2) Lemma. *Suppose T is a stopping time bounded by 1, W_t is a d-dimensional Brownian motion, and there exists c_1 such that $|C_s|$ and $|D_s| \le c_1$ for $s \le T$, where C_s is $\mathbb{R}^d$-valued, D_s is real-valued, and C_s and D_s are adapted. Let*

$$G_t = G_0 + \int_0^t C_s \cdot dW_s + \int_0^t D_s \, ds.$$

There exist c_2 and c_3 depending only on c_1 such that

$$\mathbb{P}\left(\int_0^T G_t^2 \le \varepsilon^{20}, \int_0^T |C_t|^2 \, dt \ge \varepsilon \right) \le c_2 e^{-c_3/\varepsilon}$$

for all ε sufficiently small.

Proof. Let

$$F_1 = \left\{ \sup_{s \le T} |G_s| > \varepsilon^4/4 \right\}.$$

Our first step is to prove

$$\mathbb{P}\left(\int_0^T |C_s|^2 \, ds \ge \varepsilon, F_1^c \right) \le c_4 e^{-c_5/\varepsilon}. \tag{5.1}$$

Let $a_i = i\varepsilon^5 \wedge T$ and $M_i(t) = \int_{a_i}^{t \wedge T} C_s \cdot dW_s$. If $\int_0^T |C_s|^2 \, ds \ge \varepsilon$, there must exist at least one i less than or equal to ε^{-5} such that

$$\int_{a_i}^{a_{i+1}} |C_s|^2 \ge \varepsilon^6.$$

By Lemma 5.1 applied to $M_i(t)$,

$$\mathbb{P}\left(\sup_{a_i \leq t \leq a_{i+1}} |M_i(t)| < \varepsilon^4, \int_{a_i}^{a_{i+1}} |C_s|^2 \, ds \geq \varepsilon^6\right)$$

$$\leq c_6 \exp(-c_7 \varepsilon^6 / (\varepsilon^4)^2) = c_6 \exp(-c_7 \varepsilon^{-2}).$$

So

$$\mathbb{P}\left(\exists i \leq \varepsilon^{-5} : \sup_{a_i \leq t \leq a_{i+1}} |M_i(t)| < \varepsilon^4, \int_{a_i}^{a_{i+1}} |C_s|^2 \, ds \geq \varepsilon^6\right) \tag{5.2}$$

$$\leq \varepsilon^5 c_6 \exp(-c_7 \varepsilon^{-2}) \leq c_8 \exp(-c_9/\varepsilon)$$

if ε is sufficiently small.

Suppose for some i we have $\sup_{t \in [a_i, a_{i+1}]} |M_i(t)| > \varepsilon^4$. Since

$$\int_{a_i}^{a_{i+1}} |D_s| \, ds \leq c_1 \varepsilon^5 < \varepsilon^4/2$$

if ε is sufficiently small,

$$\sup_{t \in [a_i, a_{i+1}]} |G_t - G_{a_i}| \geq \sup_{t \in [a_i, a_{i+1}]} |M_i(t)| - \varepsilon^4/2 > \varepsilon^4/2.$$

Hence for some $t \in [a_i, a_{i+1}]$, we have $|G_t| \geq \varepsilon^4/4$. This proves (5.1).

Let

$$F_2 = \left\{ \sup_{s,t \leq T, |t-s| \leq \varepsilon^{10}} |G_t - G_s| \leq \varepsilon^4/4 \right\}.$$

By Proposition I.8.1,

$$\mathbb{P}(F_2^c) \leq c_{10} \exp(-c_{11}/\varepsilon).$$

Since $|C_s| \leq c_1$, on the event where $\int_0^T |C_s|^2 \, ds \geq \varepsilon$, we have $T \geq \varepsilon/c_1^2$. We deduce that if $\omega \in F_2$ and $\sup_{s \leq T} |G_s| > \varepsilon^4/4$, then there is an interval of length at least ε^{10} contained in $[0, T(\omega)]$ on which $|G_s| \geq \varepsilon^4/8$, which implies that

$$\int_0^T G_s^2 \, ds \geq \varepsilon^{18}/64 \geq \varepsilon^{20}$$

for ε sufficiently small. $\qquad\qquad\qquad\qquad\qquad\qquad\qquad\qquad\square$

We now proceed to the proof of Hörmander's theorem. We break the proof into several steps. Recall the definition of $\mathcal{S}_m(x)$ from (4.7). We suppose X_t solves

$$dX_t = v(X_t) \circ dW_t, \qquad X_0 = x_0. \tag{5.3}$$

We define Z_t by (3.3) with $\sigma_{ij} = v_{ij}$ and $b \equiv 0$.

(5.3) Proposition. *Suppose $\mathcal{S}_m(x_0)$ spans $\mathbb{R}^d$. There exists a stopping time T bounded by 1 with the following properties.*

(a)   If $s \leq T$, then $|Z_s - I| \leq 1/2$;

(b) $T^{-1} \in L^p$ for all p;

(c)   for each $v \in \partial B(0,1)$, there exists $U \in \mathcal{S}_m$ and $r > 0$ such that

$$\sup_{u\in B(v,r)\cap\partial B(0,1)} \mathbb{P}^{x_0}\left(\int_0^T (Z_s U(X_s)\cdot u)^2\, ds < \varepsilon\right) = o(\varepsilon^p) \qquad (5.4)$$

as $\varepsilon \to 0$ for all $p \geq 2$. Moreover, r can be chosen independently of v.

Recall that we are considering vector fields both as collections of vectors and as first-order differential operators. Z_s is a $d\times d$ matrix, and U can be considered as a $d\times 1$ matrix, so $Z_t U(X_s)$ is a $d\times 1$ matrix, and hence the dot product with $u\in\partial B(0,1)$ makes sense. The notation $f(\varepsilon) = o(\varepsilon^p)$ means $f(\varepsilon)/\varepsilon^p \to 0$ as $\varepsilon \to 0$.

Proof. Since $\mathcal{S}_m(x_0)$ spans $\mathbb{R}^d$, setting

$$\delta = \inf_{v\in\partial B(0,1)}\left(\sup_{U\in\mathcal{S}_m}(U(x_0)\cdot v)^2\right),$$

we see that $\delta > 0$. Let $M > 2$ and let

$$T = \inf\{t > 0 : |X_t - x_0| > 1/M \text{ or } |Z_t - I| > 1/M\}\wedge 1. \qquad (5.5)$$

(a) follows immediately. For $\varepsilon > 0$,

$$\mathbb{P}(T \leq \varepsilon) \leq \mathbb{P}(\sup_{s\leq\varepsilon}|X_s - x_0| > 1/M) + \mathbb{P}(\sup_{s\leq\varepsilon}|Z_s - I| > 1/M),$$

which is $o(\varepsilon^p)$ as $\varepsilon \to 0$ for all p by Proposition I.8.1. Then

$$\mathbb{E}\,T^{-p} = p\int_0^\infty \lambda^{p-1}\mathbb{P}(T^{-1} > \lambda)\,d\lambda$$

$$= p\int_0^\infty \lambda^{p-1}\mathbb{P}(T < 1/\lambda)\,d\lambda < \infty$$

for all p, which proves (b).

The continuity of X_t and Z_t and of the vector fields in $\mathcal{S}_m$ allows us to conclude that if M is large enough,

$$\sup_{U\in\mathcal{S}_m}\sup_{s\leq T}|Z_s U(X_s) - U(x_0)| \leq \sqrt{\delta/8}.$$

If $v \in \partial B(0,1)$, there exists $U \in \mathcal{S}_m$ and $r > 0$ such that if $N = B(v,r)\cap \partial B(0,1)$,

$$\inf_{u\in N}\left(U(x_0)\cdot u\right)^2 \geq \delta/2,$$

and so

$$\inf_{s\leq T, u\in N}(Z_s U(X_s)\cdot u)^2 \geq (\sqrt{\delta/2} - \sqrt{\delta/8})^2 = \delta/8.$$

Thus using (b),

$$\sup_{u\in N}\mathbb{P}\left(\int_0^T (Z_s U(X_s)\cdot u)^2\, ds \leq \varepsilon\right) \leq \mathbb{P}(\delta T/8 < \varepsilon) = o(\varepsilon^p)$$

for all p, which is (c). $\qquad\square$

Next we have the proposition where the definition of Lie brackets is used.

(5.4) Proposition. *Let T be the stopping time defined in Proposition 5.3. Then for each $v \in \partial B(0,1)$, there exist $r > 0$ (not depending on v) and an integer i with $1 \leq i \leq d$ with*

$$\sup_{u \in B(v,r) \cap \partial B(0,1)} \mathbb{P}\Big(\int_0^T (Z_s V_i(X_s) \cdot u)^2 < \varepsilon \Big) = o(\varepsilon^p) \tag{5.6}$$

as $\varepsilon \to 0$ for all $p \geq 2$.

Proof. Let $v \in \partial B(0,1)$ and choose $U \in \mathcal{S}_m$ and $N = B(v,r) \cap \partial B(0,1)$ such that the conclusion of Proposition 5.3 holds. If $m = 1$, then $U = V_{i_0}$ for some i_0. Otherwise $U = [V_{i_0}, \widetilde{U}]$ for some $\widetilde{U} \in \mathcal{S}_{m-1}$ and some i_0.

If $\widetilde{U} = \sum_{j=1}^d a_j \partial_j$ and $V_i = \sum_{j=1}^d v_{ij} \partial_j$, then

$$[V_i, \widetilde{U}] = \sum_{k=1}^d \Big(\sum_{j=1}^d (v_{ij} \partial_j a_k - a_j \partial_j v_{ik}) \Big) \partial_k.$$

$\widetilde{U}_i(X_s) = a_i(X_s)$, and by Itô's formula,

$$d(\widetilde{U}_i(X_s)) = \sum_j \partial_j a_i(X_s) \circ dX_s^j = \sum_{j,k} \partial_j a_i(X_s) v_{jk}(X_s) \circ dW_s^k.$$

We have by (3.5),
$$dZ_s = -Z_s v' \circ dW_s,$$
that is,
$$dZ_{\ell i}(s) = - \sum_{m,k} Z_{\ell m}(s) \partial_i v_{mk}(X_s) \circ dW_s^k.$$

By the product formula,

$$d((Z_s \widetilde{U}(X_s))_\ell) = d\Big(\sum_i Z_{\ell i}(s) \widetilde{U}_i(X_s) \Big)$$

$$= \sum_i Z_{\ell i}(s) \circ d\widetilde{U}_i(X_s) + \sum_i \widetilde{U}_i(X_s) \circ dZ_{\ell i}(s)$$

$$= \sum_i Z_{\ell i}(s) \sum_{j,k} \partial_j a_i(X_s) v_{jk}(X_s) \circ dW_s^k$$

$$\quad - \sum_i a_i(X_s) \sum_{m,k} Z_{\ell m}(s) \partial_i v_{mk}(X_s) \circ dW_s^k$$

$$= \sum_k \sum_{i,j} \Big\{ Z_{\ell i}(s) \partial_j a_i(X_s) v_{jk}(X_s)$$

$$\quad - a_j(X_s) Z_{\ell i}(s) \partial_j v_{ik}(X_s) \Big\} \circ dW_s^k$$

$$= \sum_k \sum_i Z_{\ell i}(s) [V_k, \widetilde{U}]_i(X_s) \circ dW_s^k.$$

In order to apply Lemma 5.2, we need to see what this looks like in terms of Itô integrals. By (3.5) and the fact that $[V_k, \widetilde{U}]$ is smooth, the product formula tells us that $d(Z_{\ell i}(s)[V_k, \widetilde{U}]_i(X_s))$ can be written in the form $\sum_j A_{likj}(s)\, dW_s^j + B_{lik}(s)\, ds$, where $|A_{likj}(s)|$ is bounded by $c_1|Z_s|$ for all l, i, k, and j. The definition of Stratonovich integral then implies

$$d((Z_s\widetilde{U}(X_s))_\ell) = \sum_{i,k} Z_{\ell i}(s)[V_k, \widetilde{U}]_i(X_s)\, dW_s^k + \frac{1}{2}\sum_{i,k} A_{likk}(s).$$

We thus have for any u

$$d((Z_s\widetilde{U}(X_s))\cdot u) = \sum_k Z_s[V_k, \widetilde{U}](X_s)\cdot u\, dW_s^k + D_s\, ds,$$

where $|D_s| \leq c_2|Z_s|$ and c_2 is independent of u. Let $C_k(s) = Z_s[V_k, \widetilde{U}](X_s)\cdot u$ and $G_s = Z_s\widetilde{U}(X_s)\cdot u$. Recall $|Z_s - I| \leq 1/2$ if $s \leq T$. By Proposition 5.3(a), $|C_k(s)|$ and $|D_s|$ are bounded by c_3, a constant not depending on u, if $s \leq T$. Then

$$\mathbb{P}\left(\int_0^T G_s^2\, ds < \varepsilon^{20}\right) \tag{5.7}$$

$$\leq \mathbb{P}\left(\int_0^T G_s^2\, ds < \varepsilon^{20}, \int_0^T |C_s|^2\, ds \geq \varepsilon\right) + \mathbb{P}\left(\int_0^T |C_s|^2\, ds < \varepsilon\right).$$

By our choice of T and r and the definition of U,

$$\mathbb{P}\left(\int_0^T |C_s|^2\, ds < \varepsilon\right) \leq \mathbb{P}\left(\int_0^T (Z_s[V_{i_0}, \widetilde{U}]\cdot u)^2 < \varepsilon\right) = o(\varepsilon^p)$$

for all p. By Lemma 5.2, the first term on the right-hand side of (5.7) is $o(\varepsilon^p)$ for all p. Therefore

$$\mathbb{P}\left(\int_0^T G_s^2\, ds < \varepsilon\right) = o(\varepsilon^{p/20})$$

for all p, hence $o(\varepsilon^p)$ for all p.

We therefore have (5.4) with the condition $U \in \mathcal{S}_m$ replaced by the condition $U \in \mathcal{S}_{m-1}$. Repeating m times, we have (5.4) with the condition $U \in \mathcal{S}_m$ replaced by $U \in \mathcal{S}_1$, which gives the conclusion. $\qquad\square$

(5.5) Proposition. *Let X_t and T be as in Propositions 5.3 and 5.4 and suppose (5.6) holds. Then $\Gamma_1^{-1} \in L^p$ for all p.*

Proof. Recall the definition of Γ_1 from (3.9). Since Y_s and $\sigma(X_s)$ are in L^p for all p, to show $\Gamma_1^{-1} \in L^p$ for all p, it suffices to show $\lambda^{-1} \in L^p$ for all p, where λ is the smallest eigenvalue of Γ_1. Now

$$\lambda^2 = \inf_{v\in\partial B(0,1)} \|\Gamma_1 v\|^2.$$

So we need to show

$$\mathbb{P}\left(\inf_{v\in\partial B(0,1)}\int_0^1\sum_{i=1}^d (Z_s V_i(X_s)\cdot v)^2\,ds < \varepsilon\right) = o(\varepsilon^p)$$

for all p. This probability is bounded by

$$\mathbb{P}\left(\inf_{v\in\partial B(0,1)}\int_0^T\sum_{i=1}^d Z_s V_i(X_s)\cdot v)^2\,ds < \varepsilon\right).$$

We consider only $\varepsilon < 1$. The map $v \to \int_0^T\sum_{i=1}^d (Z_s V_i(X_s)\cdot v)^2\,ds$ is a quadratic form, say $v \to v^T Q v$, where Q depends on ω. For $s \leq T$, we have $|Z_s - I| \leq 1/2$ and $|X_s - x_0| \leq 1/2$, so by the definition of T in (5.5), the entries of Q are bounded, say by R. We can cover $\partial B(0,1)$ by finitely many balls with centers v_j and radii $r\wedge(\varepsilon/dR)$, and no more than $c_1\varepsilon^{-d}$ such balls will be needed. Note that

$$|(v-v')^T Q(v-v')| \leq d^2 R^2 |v-v'|^2.$$

So if $v^T Q v < \varepsilon$ for some $v \in \partial B(0,1)$, choose one of the points v_j with $|v - v_j| < \varepsilon/dR$, and then $v_j^T Q v_j < 2\varepsilon + 2\varepsilon^2 < 4\varepsilon$ for some j. Therefore

$$\mathbb{P}\left(\inf_{v\in\partial B(0,1)}\int_0^T\sum_{i=1}^d (Z_s V_i(X_s)\cdot v)^2\,ds < \varepsilon\right)$$

$$\leq c_1\varepsilon^{-d}\sup_j \mathbb{P}\left(\int_0^T\sum_{i=1}^d (Z_s V_i(X_s)\cdot v_j)^2\,ds < 4\varepsilon\right).$$

This will be $o(\varepsilon^p)$ for all p. $\qquad\square$

We can now state and prove Hörmander's theorem.

(5.6) Theorem. *Suppose X_t solves (4.3) with the V_j smooth and $V_0 = 0$. Suppose $\mathcal{S}(x_0)$ spans $\mathbb{R}^d$. Then there exists a bounded C^∞ function f such that*

$$\mathbb{P}^{x_0}(X_1 \in A) = \int_A f(y)\,dy.$$

Proof. Since $\mathcal{S}(x_0)$ spans $\mathbb{R}^d$, then $\mathcal{S}_m(x_0)$ spans $\mathbb{R}^d$ for some integer m. By Propositions 5.3 through 5.5, $\Gamma_1^{-1} \in L^p$ for all p. By Proposition 3.4, the theorem follows. $\qquad\square$

6. An alternative approach

In this section we give an alternative approach to the theorem that if the inverse of the Malliavin covariance matrix is in L^p for all p, then X_t has a smooth density. This approach uses the Ornstein-Uhlenbeck operator.

(6.1) Lemma. *If $\mathcal{L}f(x) = f''(x) - xf'(x)$, $m(dx) = (2\pi)^{-1/2}e^{-x^2/2}\,dx$, and f and g are smooth, then*

$$\int g(\mathcal{L}f)\,m(dx) = -\int f'g'\,m(dx) = \int f(\mathcal{L}g)\,m(dx). \tag{6.1}$$

Proof. Since f and g are smooth, $\mathcal{L}f$ is bounded by a constant times $1+|x|$, whereas $e^{-x^2/2}f'$ tends to 0 rapidly as $|x| \to \infty$. We have

$$\int g(\mathcal{L}f)\,m(dx) = \frac{1}{\sqrt{2\pi}}\int g(x)(e^{-x^2/2}f'(x))'\,dx,$$

and the first equality follows by integration by parts. The second equality follows from the first by interchanging the roles of f and g. $\qquad\square$

$(1/2)\mathcal{L}f(x)$ is the operator corresponding to the Ornstein-Uhlenbeck operator. The above lemma says that $\mathcal{L}$ is self-adjoint with respect to $m(dx)$.

If we define $\mathcal{L}f(x) = \Delta f(x) - x\cdot\nabla f(x)$, f and g are smooth, and $m(dx) = (2\pi)^{-d/2}e^{-|x|^2/2}\,dx$ is a measure on $\mathbb{R}^d$, the same proof shows

$$\int g(\mathcal{L}f)\,m(dx) = -\int \nabla f \cdot \nabla g\,m(dx) = \int f(\mathcal{L}g)\,m(dx). \tag{6.2}$$

We also observe that

$$\mathcal{L}(fg)(x) = g(x)\mathcal{L}f(x) + 2\nabla f(x) \cdot \nabla g(x) + f(x)\mathcal{L}g(x). \tag{6.3}$$

If $f = (f_1,\ldots,f_d) : \mathbb{R}^d \to \mathbb{R}^d$ is smooth and $u \in C^2(\mathbb{R}^d)$,

$$\nabla(u \circ f)(x) = \sum_{j=1}^{d}(\partial_j u \circ f)(x)\nabla f_j(x) \qquad \text{and} \tag{6.4}$$

$$\mathcal{L}(u \circ f)(x) = \sum_{i,j=1}^{d}(\partial_{ij}u \circ f)(x)\nabla f_i(x) \cdot \nabla f_j(x)$$

$$+ \sum_{i=1}^{d}(\partial_i u \circ f)(x)\mathcal{L}f_i(x).$$

We now define a Hilbert space $\mathcal{H}$ that is contained in $C[0,1]$, where here $C[0,1]$ is the set of continuous functions with domain $[0,1]$ and values in $\mathbb{R}^d$. Let

$$\mathcal{H} = \Big\{ f \in C[0,1] : f \text{ is absolutely continuous,}$$

$$\int_0^1 |f'(t)|^2 \, dt < \infty, f(0) = x_0 \Big\}.$$

Observe that $f' : [0,1] \to \mathbb{R}^d$. Define

$$\langle f, g \rangle = \langle f, g \rangle_{\mathcal{H}} = \int_0^1 f'(t) \cdot g'(t) \, dt. \tag{6.5}$$

If $f \in \mathcal{H}$, we can extend the definition of $\langle f, w \rangle$ to Brownian paths W_t. Note that if $w \in \mathcal{H}$, then $\langle f, w \rangle = \int_0^1 f'(t) \cdot dw(t)$. So if W_t is a Brownian motion, define

$$\langle f, W.(\omega) \rangle = \int_0^1 f'(t) \cdot dW_t. \tag{6.6}$$

This is defined only up to almost sure equivalence.

Clearly $\mathcal{H}$ is dense in $\{ f \in C[0,1] : f(0) = x_0 \}$ under the supremum norm. If $F : C[0,1] \to \mathbb{R}$ is a smooth functional, let us define $D_w F(h)$ as in Section 1 and define

$$\mathcal{L}F(w) = \sum_{i=1}^{\infty} D_w(D_w F(h_i))(h_i) - \sum_{i=1}^{\infty}(D_w F(h_i))\langle w, h_i \rangle, \tag{6.7}$$

where $\{h_i\}$ is an orthonormal basis for $\mathcal{H}$. This definition applies to $w \in \mathcal{H}$, but by the above remark, we can extend it to Brownian motion paths W_t. It turns out that definition of $\mathcal{L}F$ is independent of the choice of orthonormal basis (see Ikeda and Watanabe [1]), but we do not need this fact.

(6.2) Proposition. *If F and G are smooth functionals, then*

$$\mathbb{E}\big[((\mathcal{L}F)(W.))G(W.)\big] = -\mathbb{E}\sum_{i=1}^{\infty} D_W F(h_i) D_W G(h_i)$$

$$= \mathbb{E}\big[((\mathcal{L}G)(W.))F(W.)\big].$$

Proof. First, suppose

$$F(w) = f(\langle w, h_1 \rangle, \ldots, \langle w, h_n \rangle) \tag{6.8}$$

and

$$G(w) = g(\langle w, h_1 \rangle, \ldots, \langle w, h_n \rangle),$$

where f and g are smooth. The map $w \to \langle w, h_i \rangle$ is a linear functional on $C[0,1]$. Let us calculate $D_w(\langle w, h_i \rangle)(h)$. We have

$$\int h_i' \cdot (w + \varepsilon h)' - \int h_i' \cdot w' = \varepsilon \int h_i' \cdot h' \, dt = \varepsilon \langle h, h_i \rangle.$$

So $D_w(\langle w, h_i \rangle)(h) = \langle h, h_i \rangle$.

By the chain rule and the orthogonality of the h_i,

$$D_w F(h_i) = \sum_{j=1}^{n} \partial_j f(\langle w, h_1 \rangle, \ldots, \langle w, h_n \rangle) D_w(\langle w, h_j \rangle)(h_i)$$

$$= \sum_{j=1}^{n} \partial_j f(\langle w, h_1 \rangle, \ldots, \langle w, h_n \rangle)\langle h_i, h_j \rangle$$

$$= \partial_i f(\langle w, h_1 \rangle, \ldots, \langle w, h_n \rangle).$$

Similarly,

$$D_w(D_w F(h_i))(h_i)) = \partial_{ii} f(\langle w, h_1 \rangle, \ldots, \langle w, h_n \rangle),$$

and so

$$\mathcal{L}F(w) = \sum_{i=1}^{n} \partial_{ii} f(\langle w, h_1 \rangle, \ldots, \langle w, h_n \rangle) \tag{6.9}$$

$$- \sum_{i=1}^{n} \partial_i f(\langle w, h_1 \rangle, \ldots, \langle w, h_n \rangle).$$

The quantities $\langle W, h_i \rangle = \int h_i' \cdot dW$ are stochastic integrals with respect to Brownian motion with integrands that are deterministic functions, so are mean 0 Gaussian random variables with variance $\int_0^1 |h_i'|^2 \, dt = \langle h_i, h_i \rangle = 1$. Similarly, the covariance of $\langle W, h_i \rangle$ and $\langle W, h_j \rangle$ is given by

$$\mathrm{Cov}\left(\int_0^1 h_i' \cdot dW, \int_0^1 h_j' \cdot dW \right) = \int_0^1 h_i' \cdot h_j' \, dt = \langle h_i, h_j \rangle = 0$$

if $i \neq j$. Therefore $\langle W, h_i \rangle$ are i.i.d. mean 0 variance 1 normal random variables. Let $m_n(dx) = (2\pi)^{-n/2} e^{-|x|^2/2} \, dx$ be a probability measure on $\mathbb{R}^n$. Then for $H : \mathbb{R}^n \to \mathbb{R}$,

$$\mathbb{E}\left[H(\langle W, h_1 \rangle, \ldots, \langle W, h_n \rangle) \right] = \int H(x_1, \ldots, x_n) \, m_n(dx). \tag{6.10}$$

Combining (6.9) and (6.10), we have

$$\mathbb{E}\left[((\mathcal{L}F)(W))G(W) \right]$$

$$= \int \left(\sum_{i=1}^{n} \partial_{ii} f(x_1, \ldots x_n) \right.$$

$$\left. - \sum_{i=1}^{n} \partial_i f(x_1, \ldots, x_n) x_i \right) g(x_1, \ldots, x_n) \, m_n(dx)$$

$$= \int (\Delta f - x \cdot \nabla f)(x_1, \ldots, x_n) g(x_1, \ldots, x_n) \, m_n(dx).$$

By (6.2), this is equal to

$$-\int \nabla f(x_1,\ldots,x_n)\cdot \nabla g(x_1,\ldots,x_n)\, m_n(dx)$$

$$= -\int \sum_{i=1}^{n} \partial_i f(x_1,\ldots,x_n)\partial_i g(x_1,\ldots,x_n)\, m_n(dx)$$

$$= -\mathbb{E}\left[\sum_{i=1}^{n} \partial_i f(\langle W,h_1\rangle,\ldots,\langle W,h_n\rangle)\partial_i g(\langle W,h_1\rangle,\ldots,\langle W,h_n\rangle)\right]$$

$$= -\mathbb{E}\left[\sum_{i=1}^{n} D_W F(h_i)D_W G(h_i)\right],$$

as required. The second equality follows by reversing the roles of F and G. Finally, smooth functionals of

$$(W_{t_1},\ldots,W_{t_m}) = \left(\int 1_{[0,t_1]}\, dW,\ldots,\int_0^1 1_{[0,t_m]}\, dW\right)$$

can be obtained as limits of functionals of the form (6.8), and by a limit procedure, we have the proposition for all smooth functionals. $\square$

If F and G are L^p-smooth functionals on $C[0,1]$ for all p, then the conclusion of Proposition 6.2 applies to F and G by a straightforward limit argument.

If $F, G : C[0,1] \to \mathbb{R}^d$ are L^p-smooth functionals, let

$$\langle\langle F,G\rangle\rangle = \sum_{k=1}^{\infty} F(h_k)G(h_k). \tag{6.11}$$

(6.3) Proposition. *If $u \in C^2(\mathbb{R}^d)$ and $F = (F^{(1)},\ldots,F^{(d)}) : C[0,1] \to \mathbb{R}^d$ are L^p-smooth functionals, then*

$$D_W(u \circ F)(h) = \sum_{i=1}^{d} \partial_i u \circ F(W)D_W F^{(i)}(h),$$

$$\mathcal{L}(u \circ F)(W) = \sum_{i,j=1}^{d} \partial_{ij} u \circ F(W)\langle\langle D_W F^{(i)}, D_W F^{(j)}\rangle\rangle$$

$$+ \sum_{i=1}^{d} \partial_i u \circ F(W)\mathcal{L}F^{(i)}(W), \qquad and$$

$$\mathcal{L}(F^{(i)}F^{(j)}) = (\mathcal{L}F^{(i)}(W))F^{(j)}(W) + 2\langle\langle D_W F^{(i)}, D_W F^{(j)}\rangle\rangle$$
$$+ F^{(i)}(W)(\mathcal{L}F^{(j)}(W)).$$

Proof. The proof follows for smooth F from (6.3) and (6.4) by summing over h_i. Taking a limit gives the result for L^p-smooth functionals. $\square$

(6.4) Theorem. *Suppose F is L^p-smooth and let*

$$\Lambda_{ij} = \langle\langle D_W F^{(i)}, D_W F^{(j)} \rangle\rangle.$$

If $\Lambda^{-1} \in L^p$ for all p, then F has a C^∞ density: there exists $f \in C^\infty$ such that

$$\mathbb{P}(F \in A) = \int_A f(x)\,dx, \qquad A \subseteq \mathbb{R}^d.$$

Proof. As in (3.15), $D_W(\Lambda^{-1}) = -\Lambda^{-1}(D_W\Lambda)\Lambda^{-1}$. Similarly, since $0 = \mathcal{L}(I) = \mathcal{L}(\Lambda\Lambda^{-1})$,

$$\mathcal{L}(\Lambda^{-1}) = -\Lambda^{-1}(\mathcal{L}\Lambda)\Lambda^{-1} + 2\langle\langle \Lambda^{-1}D_W\Lambda, \Lambda^{-1}(D_W\Lambda)\Lambda^{-1} \rangle\rangle.$$

If Q is L^p-smooth for all p, then from Proposition 6.3 and the definition of Λ, we have

$$\mathbb{E}\left[\Lambda^{-1}\langle\langle D_W F, D_W(u \circ F)\rangle\rangle Q\right] \tag{6.12}$$
$$= \mathbb{E}\left[\Lambda^{-1}\langle\langle D_W F, D_W F\rangle\rangle(\nabla u \circ F)Q\right]$$
$$= \mathbb{E}\left[\Lambda^{-1}\Lambda(\nabla u \circ F)Q\right]$$
$$= \mathbb{E}\left[(\nabla u \circ F)Q\right].$$

We also have

$$\mathcal{L}(F(u \circ F)) = F(\mathcal{L}(u \circ F)) + (\mathcal{L}F)(u \circ F) + 2\langle\langle D_W F, D_W(u \circ F)\rangle\rangle. \tag{6.13}$$

So from (6.13) and Proposition 6.2,

$$\mathbb{E}\left[\Lambda^{-1}\langle\langle D_W F, D_W(u \circ F)\rangle\rangle Q\right] \tag{6.14}$$
$$= \frac{1}{2}\mathbb{E}\left[\Lambda^{-1}\left\{\mathcal{L}(F(u \circ F)) - F\mathcal{L}(u \circ F) - (\mathcal{L}F)(u \circ F)\right\}Q\right]$$
$$= \frac{1}{2}\mathbb{E}\left[(F)(u \circ F)\mathcal{L}(\Lambda^{-1}Q) - (u \circ F)\mathcal{L}(\Lambda^{-1}FQ)\right.$$
$$\left. - (u \circ F)\Lambda^{-1}Q(\mathcal{L}F)\right]$$
$$= \mathbb{E}\left[(u \circ F)R(Q))\right],$$

where

$$R(Q) = \frac{1}{2}\{F\mathcal{L}(\Lambda^{-1}Q) - \mathcal{L}(\Lambda^{-1}FQ) - \Lambda^{-1}Q(\mathcal{L}F)\}. \tag{6.15}$$

Recall that e_i is the unit vector in the x_i direction. So (6.12) and (6.14) imply

$$\mathbb{E}\left[(\partial_i u \circ F)Q\right] = \mathbb{E}\left[(u \circ F)\{e_i \cdot R(Q)\}\right].$$

From our assumptions, we conclude $R(Q)$ is in L^p for all p, so taking $Q = 1$,

$$|\mathbb{E}\left[\partial_i u \circ F\right]| \leq c_1 \|u\|_\infty.$$

If we now take $Q = e_j \cdot R(1)$ and repeat,

$$\mathbb{E}\left[\partial_{ji}u \circ F\right] = \mathbb{E}\left[(\partial_i u \circ F)\{e_j \cdot R(1)\}\right]$$
$$= \mathbb{E}\left[(u \circ F)e_i \cdot (R(e_j \cdot R(1)\}\right].$$

Again our assumptions imply $R(e_j \cdot R(1)) \in L^p$ for all p; hence

$$\left|\mathbb{E}\left[\partial_{ji}u \circ F\right]\right| \leq c_2\|u\|_\infty.$$

We continue by induction to obtain our result for the higher-order partials. We then apply Proposition 3.1. $\qquad\square$

We want to apply Theorem 6.4 to obtain Theorem 3.6. We saw in Section 2 that if $F^{(i)} = X_1^i$, then each $F^{(i)}$ is L^p-smooth. It remains to calculate

$$\langle\langle D_W F^{(i)}, D_W F^{(j)}\rangle\rangle.$$

(6.5) Proposition. *We have*

$$\langle\langle D_W F^{(i)}, D_W F^{(j)}\rangle\rangle = (Y_1 \Gamma_1 Y_1^T)_{ij}.$$

Proof. Recall from (3.12) that

$$D_W X_1(h) = Y_1 \int_0^1 Z_s \sigma(X_s) h'(s)\, ds.$$

This means

$$D_W X_1^i(h_k) = \sum_{a,b,c=1}^d Y_{ia}(1) \int_0^1 Z_{ab}(s)\sigma_{bc}(X_s)(h_k')_c(s)\, ds.$$

Now if A^i is $\mathbb{R}^d$-valued and defined by

$$A_c^i(t) = \int_0^t \sum_{a,b=1}^d Y_{ia}(1)Z_{ab}(s)\sigma_{bc}(X_s)\, ds,$$

then

$$\int_0^1 \sum_{c=1}^d (A_c^i)'(h_k')_c\, ds = \langle A^i, h_k\rangle.$$

So

$$D_W X_1^i(h_k)D_W X_1^j(h_k) = \langle A^i, h_k\rangle\langle A^j, h_k\rangle.$$

Since $\{h_k\}$ is an orthonormal basis, we obtain

$$\sum_{k=1}^\infty \langle A^i, h_k\rangle\langle A^j, h_k\rangle = \langle A^i, A^j\rangle.$$

We then have

$$\langle A^i, A^j \rangle = \int_0^1 (A^i)' \cdot (A^j)' \, ds = \int_0^1 \sum_{c=1}^d (A_c^i)'(A_c^j)' \, ds$$

$$= \sum_{c=1}^d \int_0^1 [Y(1)Z(s)\sigma(X_s)]_{ic}[Y(1)Z(s)\sigma(X_s)]_{jc} \, ds$$

$$= \int_0^1 \{[Y_1 Z_s \sigma(X_s)][Y_1 Z(s)\sigma(X_s)]^T\}_{ij} \, ds.$$

Therefore

$$\sum_{k=1}^\infty D_W F^{(i)}(h_k) D_W F^{(j)}(h_k) = \left\{ Y_1 \left[\int_0^1 Z_s \sigma(X_s)\sigma(X_s)^T Z_s^T \, ds \right] Y_1^T \right\}_{ij}.$$

We thus obtain

$$\Lambda_{ij} = \langle\langle D_W F^{(i)}, D_W F^{(j)} \rangle\rangle = (Y_1 \Gamma_1 Y_1^T)_{ij}, \qquad (6.16)$$

as required. $\qquad\qquad\square$

(6.6) Corollary. *If $\Gamma_1^{-1} \in L^p$ for all p, then F has a C^∞ density.*

Proof. Since $Y_1^{-1} = Z_1$ and both Y_1 and Z_1 are in L^p for all p, we have from (6.16) that Λ^{-1} is in L^p for all p if and only if Γ_1^{-1} is in L^p for all p. Now apply Theorem 6.4. $\qquad\qquad\square$

7. Notes

Two approaches evolved from Malliavin's seminal work (see, e.g., Malliavin [1]): the Girsanov approach pioneered by Bismut [1] and the Ornstein-Uhlenbeck operator approach developed by Stroock [1]. We have only looked at one application of the Malliavin calculus. For much more, see the book by Nualart [1] and the references therein.

For Sections 1 through 3 we followed Bichteler and Fonken [1] and Norris [1]. Sections 4 and 5 are based on Norris [1]. Section 6 is derived from Ikeda and Watanabe [1].

BIBLIOGRAPHY

The reference PTA refers to Bass [1].

D.G. Aronson
[1]   Bounds on the fundamental solution of a parabolic equation. *Bull. Amer. Math. Soc.* **73** (1967) 890–896.

M.T. Barlow
[1]   One-dimensional stochastic differential equations with no strong solution. *J. London Math. Soc.* **26** (1982) 335–345.

M.T. Barlow and R.F. Bass
[1]   Transition densities for Brownian motion on the Sierpinski carpet. *Probab. Th. rel. Fields* **91** (1992) 307–330.

R.F. Bass
[1]   *Probabilistic Techniques in Analysis.* Springer, New York, 1995.

R.F. Bass and K. Burdzy
[1]   A critical case for Brownian slow points. *Probab. Th. rel. Fields* **105** (1996) 85–108.

R.F. Bass and P. Hsu
[1]   Some potential theory for reflecting Brownian motion in Hölder and Lipschitz domains. *Ann. Probab.* **19** (1991) 486–508.

R.F. Bass and E. Pardoux
[1]   Uniqueness for diffusions with piecewise constant coefficients. *Probab. Th. rel. Fields* **76** (1987) 557–572.

K. Bichteler and D. Fonken
[1]   A simple version of the Malliavin calculus in dimension N. In *Seminar on Stochastic Processes, 1982*, 97–110. Birkhäuser, Boston, 1983.

P. Billingsley
[1]   *Convergence of Probability Measures.* Wiley, New York, 1968.

J.M. Bismut
[1]   Martingales, the Malliavin calculus, and hypoellipticity under general Hörmander conditions. *Zeit. f. Wahrsch.* **56** (1981) 469–505.

R.M. Blumenthal and R.K. Getoor
[1]   *Markov Processes and Potential Theory.* Academic Press, New York, 1968.

L. Breiman, *Probability.* Addison-Wesley, Reading, MA, 1968.

L. Caffarelli
[1]   *Métodos de continuação em equações eliticas não-lineares.* Inst. Mat. Pure e Apl. Rio de Janeiro, 1986.

M.C. Cerutti, L. Escauriaza, and E.B. Fabes
[1]   Uniqueness for some diffusions with discontinuous coefficients. *Ann. Probab.* **19** (1991) 525–537.

Z.Q. Chen, R.J. Williams, and Z. Zhao
[1]   On the existence of positive solutions of semilinear elliptic equations with Dirichlet boundary conditions. *Math. Ann.* **298** (1994) 543–556.

K.L. Chung and Z. Zhao
[1]   *From Brownian Motion to Schrödinger's Equation.* Springer, New York, 1995.

E.B. Davies
[1]   *Heat Kernels and Spectral Theory.* Cambridge Univ. Press, Cambridge, 1989.

R. Durrett
[1]   *Brownian Motion and Martingales in Analysis.* Wadsworth, Belmont, CA, 1984.

E.B. Dynkin
[1]   *Markov Processes, vol. 1, 2.* Springer, New York, 1965.
[2]   Superprocesses and parabolic nonlinear differential equations. *Ann. Probab.* **20** (1992) 942–962.

L.C. Evans

[1] Some estimates for nondivergence structure second order elliptic equations. *Trans. Amer. Math. Soc.* **287** (1985) 701–712.

E.B. Fabes and D.W. Stroock

[1] The L^p integrability of Green's functions and fundamental solutions for elliptic and parabolic operators. *Duke Math. J.* **51** (1984) 997–1016.

[2] A new proof of Moser's parabolic Harnack inequality via the old ideas of Nash. *Arch. Mech. Rat. Anal.* **96** (1986) 327–338.

A. Friedman

[1] *Partial Differential Equations of Parabolic Type.* Prentice-Hall, Englewood Cliffs, NJ, 1964.

M. Fukushima, Y. Oshima, and M. Takeda

[1] *Dirichlet Forms and Symmetric Markov Processes.* Berlin, de Gruyter, 1994.

T. Funaki

[1] Probabilistic construction of the solutions of some higher order parabolic differential equations. *Proc. Japan Acad. Ser. A Math. Sci.* **55** (1979) 176–179.

P. Gao

[1] The martingale problem for a differential operator with piecewise continuous coefficients. In *Seminar on Stochastic Processes, 1992.* Birkhäuser, Boston, 1993.

J.B. Garnett

[1] *Bounded Analytic Functions.* Academic Press, New York, 1981.

D. Gilbarg and J. Serrin

[1] On isolated singularities of solutions of second order elliptic differential operators. *J. Analyse Math.* **4** (1955/56) 309–340.

D. Gilbarg and N.S. Trudinger

[1] *Elliptic Partial Differential Equations of Second Order, 2nd ed.* Springer, New York, 1983.

U. Haussmann and E. Pardoux

[1] Time reversal of diffusions. *Ann. Probab.* **14** (1986) 1188–1205.

N. Ikeda and S. Watanabe

[1] *Stochastic Differential Equations and Diffusion Processes.* North Holland/Kodansha, Tokyo, 1981.

K. Itô and H.P. McKean

[1] *Diffusion Processes and Their Sample Paths.* Springer, New York, 1965.

F. Knight

[1] On invertibility of martingale time changes. In *Seminar on Stochastic Processes, 1987.* Birkhäuser, Boston, 1988.

M.G. Krein and M.A. Rutman
[1] Linear operators leaving invariant a cone in a Banach space. *Amer. Math. Soc. Sel. Translations, Series 1,* **10** (1962) 199–325.

N.V. Krylov
[1] An inequality in the theory of stochastic processes. *Th. Probab. Applic.* **16** (1971) 438–448.
[2] Certain bounds on the density of distributions of stochastic integrals. *Izv. Akad. Nauk* **38** (1974) 228–248.
[3] Once more about the connection between elliptic operators and Itô's stochastic equations. *Statistics and Control of Stochastic Processes,* Moscow, 1984.
[4] On one-point weak uniqueness for elliptic equations. *Comm PDE* **17** (1992) 1759–1784.

N.V. Krylov and M.V. Safonov
[1] An estimate of the probability that a diffusion process hits a set of positive measure. *Soviet Math. Dokl.* **20** (1979) 253–255.
[2] A certain property of solutions of parabolic equations with measurable coefficients. *Math. USSR Izv.* **16** (1981) 151–164.

Y. Kwon
[1] Reflected Brownian motion in Lipschitz domains with oblique reflection. *Stoch. Proc. Applic.* **51** (1994) 191–205.

J.-F. LeGall
[1] The Brownian snake and solutions of $\Delta u = u^2$ in a domain. *Probab. Th. rel. Fields* **102** (1995) 395–432.

T. Leviatan
[1] Perturbations of Markov processes. *J. Funct. Anal.* **10** (1972) 309–325.

P.-L. Lions and A.-S. Sznitman
[1] Stochastic differential equations with reflecting boundary conditions. *Comm. Pure Appl. Math.* **37** (1984) 511–537.

W. Littman, G. Stampacchia, and H.F. Weinberger
[1] Regular points for elliptic equations with discontinuous coefficients. *Ann. Scuola Norm Sup. Pisa* **17** (1963) 43–77.

P. Malliavin
[1] Stochastic calculus of variation and hypoelliptic operators. In *Proc. of the International Symp. on SDEs, Kyoto 1976.* Tokyo, 1978.

V.G. Maz'ja

[1]   *Sobolev Spaces.* Springer, New York, 1985.

H.P. McKean

[1]   Elementary solutions for certain parabolic partial differential equations. *Trans. Amer. Math. Soc.* **82** (1956) 519–548.

J. Moser

[1]   On Harnack's inequality for elliptic differential equations. *Comm. Pure Appl. Math.* **14** (1961) 577–591.

[2]   On pointwise estimates for partial differential equations. *Comm. Pure Appl. Math.* **24** (1971) 727–740.

N. Nadirashvili

[1]   Nonuniqueness in the martingale problem and Dirichlet problem for uniformly elliptic operators. Preprint.

J. Nash

[1]   Continuity of solutions of parabolic and elliptic equations. *Amer. Math. J.* **80** (1958) 931–954.

L. Nirenberg

[1]   On elliptic partial differential equations. *Ann. Scuola Norm. Sup. Pisa* **13** (1959) 1–48.

J. Norris

[1]   Simplified Malliavin calculus. In *Séminaire de Probabilités XX*, 101–130. Springer, New York, 1986.

D. Nualart

[1]   *The Malliavin Calculus and Related Topics.* Springer, New York, 1995.

R. Pinsky

[1]   *Positive Harmonic Functions and Diffusion.* Cambridge Univ. Press, Cambridge, 1995.

S.C. Port and C.J. Stone

[1]   *Brownian Motion and Classical Potential Theory.* Academic Press, New York, 1978.

P. Protter

[1]   *Stochastic Integration and Differential Equations.* Springer, New York, 1990.

PTA

R.F. Bass, *Probabilistic Techniques in Analysis.* Springer, New York, 1995.

C. Pucci
[1]   Limitazioni per soluzioni di equazioni ellitiche. *Ann. Mat. Pura Appl.* **74** (1966) 15–30.

D. Revuz and M. Yor
[1]   *Continuous Martingales and Brownian Motion.* Springer, New York, 1991.

F. Riesz and B. Sz.-Nagy
[1]   *Functional Analysis.* Ungar, New York, 1955.

L.C.G. Rogers and D. Williams
[1]   *Diffusions, Markov Processes, and Martingales, vol. 2: Itô calculus.* Wiley, New York, 1987.

M.V. Safonov
[1]   On weak uniqueness for some elliptic equations. *Comm. PDE* **19** (1994) 943–957.

E.M. Stein
[1]   *Singular Integrals and Differentiability Properties of Functions.* Princeton Univ. Press, Princeton, 1970.

D.W. Stroock
[1]   The Malliavin calculus, functional analytic approach. *J. Funct. Anal.* **44** (1981) 212–257.
[2]   Diffusion semigroups corresponding to uniformly elliptic divergence form operators. In *Séminaire de Probabilités XXII*, 316–347. Springer, New York, 1988.

D.W. Stroock and S.R.S. Varadhan
[1]   Diffusion processes with boundary conditions. *Comm. Pure Appl. Math.* **24** (1971) 147–225.
[2]   *Multidimensional Diffusion Processes.* Springer, New York, 1979.

B. Tsirelson
[1]   An example of a stochastic differential equation having no strong solution. *Th. Probab. Appl.* **20** (1975) 427–430.

T. Yamada and S. Watanabe
[1]   On the uniqueness of solutions of stochastic differential equations. *J. Math. Kyoto* **11** (1971) 155–167.

INDEX

GPSR Compliance
The European Union's (EU) General Product Safety Regulation (GPSR) is a set
of rules that requires consumer products to be safe and our obligations to
ensure this.

If you have any concerns about our products, you can contact us on

ProductSafety@springernature.com

In case Publisher is established outside the EU, the EU authorized
representative is:

Springer Nature Customer Service Center GmbH
Europaplatz 3
69115 Heidelberg, Germany

www.ingramcontent.com/pod-product-compliance
Ingram Content Group UK Ltd.
Pitfield, Milton Keynes, MK11 3LW, UK
UKHW020917130726
13719UKWH00012B/169